Springer Proceedings in Physics 17

Springer Proceedings in Physics

Managing Editor: H. K. V. Lotsch

Volume 1 *Fluctuations and Sensitivity in Nonequilibrium Systems*
Editors: W. Horsthemke and D. K. Kondepudi

Volume 2 *EXAFS and Near Edge Structure III*
Editors: K. O. Hodgson, B. Hedman, and J. E. Penner-Hahn

Volume 3 *Nonlinear Phenomena in Physics* Editor: F. Claro

Volume 4 *Time-Resolved Vibrational Spectroscopy*
Editors: A. Laubereau and M. Stockburger

Volume 5 *Physics of Finely Divided Matter* Editors: N. Boccara and M. Daoud

Volume 6 *Aerogels* Editor: J. Fricke

Volume 7 *Nonlinear Optics: Materials and Devices* Editors: C. Flytzanis and J. L. Oudar

Volume 8 *Optical Bistability III* Editors: H. M. Gibbs, P. Mandel, N. Peyghambarian, and S. D. Smith

Volume 9 *Ion Formation from Organic Solids (IFOS III)* Editor: A. Benninghoven

Volume 10 *Atomic Transport and Defects in Metals by Neutron Scattering*
Editors: C. Janot, W. Petry, D. Richter, and T. Springer

Volume 11 *Biophysical Effects of Steady Magnetic Fields*
Editors: G. Maret, J. Kiepenheuer, and N. Boccara

Volume 12 *Quantum Optics IV* Editors: J. D. Harvey and D. F. Walls

Volume 13 *The Physics and Fabrication of Microstructures and Microdevices*
Editors: M. J. Kelly and C. Weisbuch

Volume 14 *Magnetic Properties of Low-Dimensional Systems*
Editors: L. M. Falicov and J. L. Morán-López

Volume 15 *Gas Flow and Chemical Lasers* Editor: S. Rosenwaks

Volume 16 *Photons and Continuum States of Atoms and Molecules*
Editors: N. K. Rahman, C. Guidotti, and M. Allegrini

Volume 17 *Quantum Aspects of Molecular Motions in Solids*
Editors: A. Heidemann, A. Magerl, M. Prager, D. Richter, and T. Springer

Volume 18 *Electro-optic and Photorefractive Materials* Editor: P. Günter

Springer Proceedings in Physics is a new series dedicated to the publication of conference proceedings. Each volume is produced on the basis of camera-ready manuscripts prepared by conference contributors. In this way, publication can be achieved very soon after the conference and costs are kept low; the quality of visual presentation is, nevertheless, very high. We believe that such a series is preferable to the method of publishing conference proceedings in journals, where the typesetting requires time and considerable expense, and results in a longer publication period. Springer Proceedings in Physics can be considered as a journal in every other way: it should be cited in publications of research papers as *Springer Proc. Phys.*, followed by the respective volume number, page number and year.

Quantum Aspects of Molecular Motions in Solids

Proceedings of an ILL-IFF Workshop
Grenoble, France, September 24–26, 1986

Editors: A. Heidemann, A. Magerl,
M. Prager, D. Richter, and T. Springer

With 117 Figures

Springer-Verlag Berlin Heidelberg New York
London Paris Tokyo

Dr. Anton Heidemann
Dr. Andreas Magerl
Dr. Dieter Richter
Institut Laue-Langevin, Avenue des Martyrs, Centre de Tri 156X,
F-38042 Grenoble Cedex, France

Dr. Michael Prager
Professor Tasso Springer
Institut für Festkörperforschung, Kernforschungsanlage Jülich GmbH, Postfach 1913,
D-5170 Jülich, Fed. Rep. of Germany

ISBN-13: 978-3-642-71916-5 e-ISBN-13: 978-3-642-71914-1
DOI: 10.1007/978-3-642-71914-1

Softcover reprint of the hardcover 1st edition 1987

Offset printing: Weihert-Druck GmbH, D-6100 Darmstadt

2153/3150-543210

Preface

The Institut Max von Laue-Paul Langevin (ILL) in Grenoble regularly organizes workshops that deal with various applications of neutrons in physics, chemistry, biology and materials science. The workshop "Quantum Aspects of Molecular Motions in Solids" was jointly organized by the Institut Laue-Langevin and the Institut für Festkörperforschung at the Kernforschungsanlage Jülich and took place in September 1986 in Grenoble.

Tunneling phenomena in molecular crystals were first observed with macroscopic methods like specific heat experiments and later also with NMR. Finally, the development of high resolution neutron scattering techniques like neutron backscattering led to direct spectroscopic observation of the tunnel split ground state. This breakthrough was achieved in 1975 at Jülich. Since then the large variety of high-resolution techniques available in combination with high neutron flux have turned ILL into the leading laboratory in the field of tunneling spectroscopy. Since 1980 regular meetings of scientists involved in this topic have been organized every two years (Jülich, Braunschweig, Nottingham) and have led to an intense exchange of ideas and experimental results. The present workshop is the fourth of this series and the first with published proceedings presenting the state of the art in this field. The eight review articles introduce scientists not involved in the subject to the actual discussion. Sessions on translational tunneling of light interstitials in metals as well as on tunneling phenomena in amorphous substances provide bridges to adjacent fields.

The first session of the workshop dealt with rotational tunneling in relation to the molecular potential and phonon interactions. Important progress has been achieved in the precise determination of rotational potentials by combining spectroscopic measurements of librational levels and tunnel splitting of deuterated and protonated molecules of the same species. By varying external parameters like the pressure one can obtain further information. The transition from quantum rotation at low temperatures to thermally activated classical jump reorientation at high temperatures provides one of the most fundamental and thus most interesting features in rotational tunneling. Experimental data have been available for a decade but the theoretical description is still a matter of controversy. A very general treatment of the coupling of a quantum rotor to the phonons by second-order perturbation theory provides a consistent description of the general observations. Often more insight is obtained by using simplified models with coupling Hamiltonians of special symmetry. Another approach to the problem consists in solving numerically the time-dependent Schrödinger equation.

In the session on coupled systems, new experimental results were presented which establish important rotor-rotor coupling effects in many systems. In the simplest case a coupling of just two rotors was found, while in the most complex case a three-dimensional network of coupled rotors has been considered. Experimental progress is stimulating the development of adequate theoretical models.

In the third session on rotational tunneling the use of tunneling molecules as a probe to study their environment was emphasized. Site symmetries in solid phases and on surfaces, phase transitions, the structure of rare gas matrices used for matrix isolation spectroscopy, the chemical environment, intra- and intermolecular interactions were investigated. The sensitivity of the tunnel splitting to small changes of the rotational potential makes this probe especially useful.

The session on translational tunneling concentrated on the motion of light interstitials in metals. The transport phenomena of the hydrogen isotopes, including the positive muon, constitute a fascinating area of research dealing with motional properties in the whole range between classical hopping and electron-like band propagation. New experimental results revealed hydrogen tunneling states associated with oxygen or nitrogen in niobium. In muon spin rotation experiments, muon diffusion coefficients exhibiting a power-law increase with decreasing temperature were measured. Recent theoretical considerations suggest that these transport phenomena are dominated by the interaction with conduction electrons. This limits the existence of hydrogen tunneling states in niobium and causes the weakly negative temperature power laws observed in muon diffusion.

Finally, in the session on disordered systems the low-temperature anomalies of amorphous substances were discussed and the two-state tunneling model was presented. However, a microscopic understanding of the tunneling entities has still to be achieved.

The editors hope that this survey of quantum motions in molecular crystals, metals and glasses may stimulate further interest in this field. In particular, organic chemistry could establish the tunneling entities as a new highly sensitive probe to investigate local environments. Further perspectives are also linked to new developments in instrumentation. Very-high-resolution spectroscopy at large energy transfers employing, e.g., the neutron spin echo technique or advanced time-of-flight methods at spallation sources, may allow the study of tunneling splitting of excited states in more detail. In addition, a variety of new spectroscopic NMR methods have been developed which greatly extend the range of detectable tunneling frequencies and which provide information complementary to that obtained from inelastic neutron scattering.

The editors would like to thank the speakers for promptly preparing their manuscripts; they hope that readers will excuse the small "imperfections" in the manuscripts, which were accepted in order to speed publication. We express our gratitude to the KFA-Jülich and to the Institut Laue-Langevin for their technical and financial help and to the ILL staff who helped to organize the workshop, particularly to Mrs. Katja Mayer-Jenkins for finalizing the manuscript.

Grenoble, Jülich
November 1986

A. Heidemann A. Magerl B. Maier
M. Prager D. Richter T. Springer

Contents

Part I	**Rotational Tunneling I – Potentials and Phonon Interaction**

The Temperature Dependence of Rotational Tunneling – Simulation of a Quantum System at Finite Temperatures
By A. Hüller and L. Baetz (With 1 Figure) 2

The Rotational Potential of Solid Nitromethane
By S.F. Trevino (With 9 Figures) 9

The Orientation of the Rotational Potential in the Crystal Frame: $(NH_4)_2PtCl_6$
By A. Hoser, W. Prandl, and G. Heger (With 2 Figures) 19

Inelastic Neutron Scattering Study of Methyl Tunnelling in α-Crystallised Toluene. By D. Cavagnat, A. Magerl, C. Vettier, and S. Clough (With 3 Figures) 24

Tunnelling Spectroscopy and Dynamics of CH_3 in a Homologous Series of Ketones
By A.J. Horsewill, R.M. Green, and A.M. Alsanoosi (With 2 Figures) 28

NH_4^+ Motions in NH_4MF_3 Cubic Perovskites
By R. Navarro, E. Palacios, J. Bartolomé, R. Burriel, and D. González (With 2 Figures) 33

Spectroscopic Raman Study of Trifluoroperovskites of Ammonium Under High Pressure
By E. Palacios, J. Bartolomé, F. Agulló-Rueda, J.M. Calleja, M. Cardona, K. Syassen, and K. Strößner (With 2 Figures) 38

Part II	**Rotational Tunneling II – Coupled Systems**

Rotational Tunneling of Coupled Groups Studied by Inelastic Neutron Scattering. By A. Heidemann (With 7 Figures) 44

Study of the Dipolar Proton Spin System in Lithium Acetate with NMR Dispersion Spectroscopy
By E. Crits and L. Van Gerven (With 5 Figures) 54

Tunnelling of Ammine Ligands in Nickel Hexammine Iodide
By G.J. Kearley, H. Blank, and J.K. Cockcroft (With 3 Figures) ... 58

The Free Rotor, NH_3, in Metal Hexammine Ions
By H. Blank, J.K. Cockcroft, and G.J. Kearley (With 2 Figures) ... 62

Tunnel Spectroscopy by Dipolar-Driven NMR. By K.J. Abed, G.J. Barker, S. Clough, and A.J. Horsewill (With 3 Figures) 65

Zeeman-Tunnel Resonance Line Shapes in NMRRR Investigations of Rotational Tunneling. By G. Vandemaele, A. Buekenhoudt, and L. Van Gerven (With 4 Figures) 72

Coupled Tunneling of Methyl Groups in Solids as Studied by Proton NMR Method. By S. Takeda and H. Chihara (With 4 Figures) 76

Tunneling of CH_3 Groups in Molecular Crystals of the $X(CH_3)_n$ Type. By J. Peternelj, I. Jencic, B. Cvikl, and M.M. Pintar (With 2 Figures) 81

Part III **Rotational Tunneling III – Surfaces, Isotope Effect, Matrix Isolation**

Hindered Rotation and Tunnelling of Adsorbed Species
By S.J.M. Collinson, G.M. Hunt, M.V. Smalley, M.L. Statham, and R.K. Thomas (With 6 Figures) 86

Rotational Tunneling of Matrix-Isolated Tetramethyl Compounds
By W. Langel, M. Prager, O. Schrems, and E. Knözinger (With 5 Figures) 93

Rotational Excitations in $(CH_4)_{1-x}Kr_x$ Solid Mixtures
By S. Grondey, M. Prager, W. Press, and A. Heidemann (With 5 Figures) 98

Quantum Rotation of Ammonia Molecules in Disordered $Ca(NH_3)_6$
By F. Leclercq, P. Damay, A.J. Dianoux, and W. Press (With 3 Figures) 103

The Methyl Rotational Potentials of Various Isotopic Methyl Iodides and Their Coupling to Phonons
By M. Prager, J. Stanislawski, and W. Häusler (With 3 Figures) ... 107

Partially Deuterated Methyl Groups in Molecular Solids: The System $4\text{-}CH_2D\text{-}C_5H_4N$ with an Original Sixfold Symmetric Barrier
By H.B. Brom, H. den Adel, and B. Alefeld (With 2 Figures) 113

Calorimetric and Dielectric Study of Partly Deuterated Ammonium Hexachlorostannate
By J. Gilchrist, J. Odin, and J. Peyrard (With 2 Figures) 118

Tunneling to Reorientation Transition Effects in High Field ^{2}H-NMR Spectra of ND_4^+ Ions. By L.P. Ingman, E. Koivula, Z.T. Lalowicz, M. Punkkinen, and E.E. Ylinen (With 1 Figure) 122

Part IV **Translational Tunneling**

Influence of Phonons and Electrons on Low-Temperature Translational Tunneling. By H. Grabert (With 3 Figures) 130

Quantum Diffusion of Light Interstitials in Metals
By D. Richter (With 6 Figures) 140

The Influence of Electrons on the Tunneling State of a Hydrogen Atom in a Metal. By H. Wipf, D. Steinbinder, K. Neumaier, P. Gutsmiedl, A. Magerl, and A.J. Dianoux (With 2 Figures) 153

H and D Tunneling in Nb: Dependence on Defect Concentration and Trapping Center. By P. Gutsmiedl, M. Schiekhofer, K. Neumaier, and H. Wipf (With 3 Figures) 158

Sound Velocity Measurements on Hydrogen-Doped Niobium in the Superconducting and Normal States
By W. Morr, G. Weiss, and H. Wipf (With 1 Figure) 163

Quantitative Theory of the Tunnel Level Splitting for Hydrogen Trapped at Oxygen in Niobium. By H. Teichler 167

Debye-Waller Peculiarity in Phonon-Assisted Tunneling
By M. Wagner (With 2 Figures) 172

Quantum Diffusion Above $\Theta_{Debye}/2$. By A. Klamt (With 2 Figures) 177

Hydrogen Diffusion by Tunneling in Metals
By I. Svare (With 2 Figures) 182

Optical Studies of Proton Tunneling and Relaxation in Benzoic Acid
By M. Pierre, H.P. Trommsdorff, and R.M. Hochstrasser (With 3 Figures) 186

Part V **Disordered Systems**

How Universal Is the Spectral Density of Tunneling States in Glasses?
By S. Hunklinger (With 5 Figures) 194

Very Low Frequency Elastic Loss of PMMA Below 5 K
By M. v. Schickfus, H. Tietje, and E. Gmelin (With 3 Figures) 204

Ultrasonic Attenuation Study of Two-Level Systems in Neutron-Irradiated Quartz upon Heat Treatment
By A. Vanelstraete and C. Laermans (With 1 Figure) 208

The One-Dimensional Quantum Rotator H_2S^- in Several Alkali Halides. By D. Schoemaker and E. Goovaerts (With 2 Figures) 212

Relaxation Times of $k \approx 0$ Rotons in Parahydrogen Crystals at Low Concentrations of Orthohydrogen Impurities
By E. Goovaerts, X.Y. Chen, C. Sierens, A. Bouwen, and D. Schoemaker (With 3 Figures) 217

Index of Contributors 221

Part I

Rotational Tunneling I – Potentials and Phonon Interaction

The Temperature Dependence of Rotational Tunneling – Simulation of a Quantum System at Finite Temperatures

A. Hüller and L. Baetz

Institut für Theoretische Physik der Universität Erlangen-Nürnberg, Glückstr. 6, D-8520 Erlangen, Fed. Rep. of Germany

Rotational tunneling is a manifestation of the symmetry properties of the molecular wave function under proton or deuteron exchange in the solid state. Symmetry effects of this kind are well known in the gas phase e.g. of molecular hydrogen where an antisymmetric spatial wave function combines with a symmetric spin function and vice versa, in order to yield a total wave function which is antisymmetric under the exchange of the two protons.

In the solid state the nuclei are usually well localized at their lattice sites, around which they perform small oscillations. There is only little overlap between the single particle states at neighbouring sites and consequently spatial wave functions of different symmetry w.r.t. the interchange of identical nuclei do not differ in energy by an appreciable amount. This is different for solids containing small molecules, polyatomic ions or sidegroups as CH_4, SiH_4, NH_3, $NH_4{}^+$ and $-CH_3$ or their deuterated or partly deuterated homologues. In all these examples the overlap of wave functions with just two protons exchanged is again extremely small and thus the symmetry requirements under a pairwise proton exchange can safely be neglected. On the other hand, states which differ from each other by such a permutation of the protons (or deuterons) which corresponds to a whole body rotation of the molecule may have considerable overlap. Whole body rotations which transform a YX_4 tetrahedron onto itself or rotations of a YX_3 group around its threefold symmetry axis correspond to even permutations of identical particles. Therefore the total wave function must be symmetric under these rotations both for Bose and Fermi particles, i.e. for YH_4, YD_4, YH_3 and YD_3. The energy splitting of states with different symmetry of the spatial part is maximal in the gas phase (a few $B = \hbar^2/2\Theta$ where Θ is the moment of inertia and $B \approx 650\mu eV$ for CH_3) where the molecules rotate freely. In the solid state the splitting decreases exponentially with increasing rotational potential and for the range of potential strengths encountered in real systems one finds tunnel splittings which range from the level spacings of free rotors over many orders of magnitude down to unobservably small energy differences [1]. Neutrons provide a means to measure tunneling frequencies above 0.3μeV, with NMR-level crossing experiments one has observed splittings which are several orders of magnitude smaller.

As long as we do not consider interactions which depend on the spin of the protons or deuterons involved, the spatial part of the Hamiltonian is by itself symmetric under permutations of the protons. Then it cannot induce transitions between states with different symmetry of the space part under the exchange of the protons or deuterons. In particular in a crystal without magnetic ions, phonons cannot cause transitions between these states. Even if magnetic ions are present, the interaction of the nuclear magnetic moments with the fluctuating field gradient is small and transitions are rare. The intra- and intermolecular dipole-dipole interactions are even weaker. This explains why rotational tunneling in the μeV-regime (~0.01K) is observable for temperatures up to 50K.

Transitions between the tunneling states are observed in neutron scattering. The spin-dependent part of the interaction between the neutron and the nuclei (protons and deuterons) of the target can change the permutation symmetry of a state. For simplicity we will restrict the discussion to methyl groups. Let φ_i be the rotation angle of one of the three protons around the rotation axis of the methyl group. The permutation symmetry corresponds to a $2\pi/3$ periodicity in the collective coordinate $\alpha = (\varphi_1+\varphi_2+\varphi_3)/3$. Because of the $2\pi/3$-periodicity of the rotational potential and the 2π-periodicity of the wave function, the latter can be classified according to Bloch's theorem:

$$\langle\alpha|n,K\rangle = \langle\alpha|\Psi_{n,K}\rangle\, e^{iK\alpha} \tag{1}$$

where $\langle\alpha|\Psi_{n,K}\rangle$ is itself $2\pi/3$ periodic, $K = 0,\pm1$ is the wavenumber which distinguishes states of different permutation symmetry, and n is the band index. States with K=0 are called A-states, those with K=$\pm$1 are called E-states. K is conserved in a crystal without magnetic interactions, but the interaction with a neutron can cause a transition. The spin-dependent part of the interaction of a methyl group with a neutron which is scattered with a momentum transfer $\underset{\sim}{Q}$ may be written:

$$2A_{sd}(I(I+1))^{-1/2}\, \underset{\sim}{S}\cdot \sum_{K=0,\pm1} \underset{\sim}{I}_{-K}G_K \tag{2}$$

where A_{sd} is the spin-dependent scattering length, $\underset{\sim}{S}$ the neutron spin. $\underset{\sim}{I}_K$ and G_K are operators of a given permutation symmetry formed according to

$$G_K = \sum_{\gamma=1}^{3} G_\gamma \exp(iK\gamma\, 2\pi/3) \tag{3}$$

and in the same way for $\underset{\sim}{I}_K$ where $\underset{\sim}{I}_\gamma$ is the proton spin and $G_\gamma = \exp(i\underset{\sim}{Q}\cdot\underset{\sim}{R}_\gamma)$ with $\underset{\sim}{R}_\gamma$ the proton position. For small momentum transfers $G_1 \approx 3iQR\exp(i\alpha)$. The operator which multiplies the wave function by $\exp(i\alpha)$ will be denoted by X. The scattering function for a methyl group which was initially in a state with the symmetry label K is obtained as the Fourier transform with respect to time of the self-correlation function of X:

$$\sum_n p_n \langle Kn|U(0,t)X^+U(t,0)X|Kn\rangle \tag{4}$$

where p_n is the probability of finding the state $|Kn\rangle$ and $U(t,0)$ the time development operator of the system under study.

The temperature dependence of rotational tunneling has been studied experimentally and theoretically. At low temperatures the width of the tunneling lines is limited by instrumental resolution. The transitions are found at the frequency which corresponds to the energy difference between the states of different symmetry within librational ground state multiplet $|n=0,K=+1\rangle$ and $|n=0,K=0\rangle$. With increasing temperatures the lines usually broaden and shift to lower frequencies until the energy gain and loss parts merge into a single quasielastic peak, characteristic for the high temperature classical hopping regime [1]. There are, however, exceptions from the common rule as e.g. in $(NH_4)_2SnCl_6$ [2] where the line broadening remains relatively small for temperatures below 40K and CH_3I [3] where the initial shift is to higher frequencies.

We shall not review the many attempts to explain the temperature dependence of rotational tunneling by intuitive argumentation, but rather restrict ourselves to the papers where the Hamiltonian of a tunneling system, coupled to a phonon bath, is treated in some approximation. One way of doing this starts from the assumption that the tunneling system couples separately to the individual phonon modes. The Hamiltonian to be diagonalized then reads:

$$H = -\frac{\hbar^2}{2\Theta}\frac{\partial^2}{\partial\alpha^2} - V_0\cos 3\alpha - \frac{\hbar^2}{2m}\frac{\partial^2}{\partial x^2} + \frac{1}{2}m\omega^2_{\underset{\sim}{q}j}x^2 + g_1 x\sin 3\alpha + g_2 x\cos 3\alpha\,. \tag{5}$$

Here the first line describes the tunneling system, a methyl group with moment of inertia Θ in a potential of strength V_0, the second line describes one of the lattice oscillators with coordinate x, mass m, and frequency $\omega_{\underset{\sim}{q}j}$. The third line represents the interaction terms which both conserve the threefold symmetry of the rotational potential. For small x the first term corresponds to a displacement of the potential - it is therefore called the shaking term, the second one modulates the potential strength and is called the breathing term. The eigenstates $|n,K,n_{\underset{\sim}{q}j}\rangle$ of H are characterized by the band index n, the symmetry label K and the occupation number $n_{\underset{\sim}{q}j}$ of the lattice mode under consideration. Tunneling in the librational ground state occurs between two states of different symmetry K with n=0 and equal $n_{\underset{\sim}{q}j}$. The tunneling frequency ω_t depends on $n_{\underset{\sim}{q}j}$:

$$\omega_t = \omega_0 - \lambda_{\underset{\sim}{q}j}\, n_{\underset{\sim}{q}j}. \tag{6}$$

ω_0 is the tunneling frequency for $n_{\underset{\sim}{q}j}=0$. The reduction factors $\lambda_{\underset{\sim}{q}j}$ have been calculated from the overlap of harmonic oscillator test functions in the potential pockets [4,5]. The lowering of the tunneling frequency can be understood as a destructive interference in the overlap integral for test functions with an increasing number of nodes. When the lattice modes are interacting independently with the methyl group (6) has to be summed over the polarization vectors $\underset{\sim}{q}$ and band indices j of the modes involved. A thermal average yields the mean value of the tunneling frequency as a function of temperature.

$$\overline{\omega_t(T)} = \omega_0 - \sum_{\underset{\sim}{q}j} \lambda_{\underset{\sim}{q}j}\, \overline{n_{\underset{\sim}{q}j}}\ . \tag{7}$$

Recently HÄUSLER [6] has diagonalized the Hamiltonian (5) exactly. For $g_1 \neq 0, g_2 = 0$ his results confirm the earlier approximate calculations [4,5]. For the other case $(g_1=0, g_2 \neq 0)$ the $\lambda_{\underset{\sim}{q},j}$ turn out to be negative, leading to an increase of the tunneling frequency with temperature, a result which has been used to explain the experiment on CH_3I [3].

The number of modes contributing to the frequency shift grows with the number of atoms N in the crystal. The reduction factors must be proportional to 1/N in order to ensure a shift which does not depend on N for $N \to \infty$. The width of the tunneling line calculated from $\overline{[\omega_t(T) - \overline{\omega_t(T)}]^2}$ then goes to zero in the thermodynamic limit. This is different when a only few resonant modes contribute appreciably to the shift, as e.g. the phonons which match in energy with the librational transition from n=0 to 1 of the methyl group.

HEWSON [7] treats (4) in perturbation theory. The equation of motion for (4) is expressed in terms of higher order correlation functions which contain one or more phonon operators. He obtains a hierarchy of equations which is decoupled by factorization. This way he obtains results which are correct up to second order in the coupling between the methyl group and the phonons. As phonons do not induce transitions between states of different symmetry each symmetry species can be treated separately. Shift and width of the tunneling lines are found from the difference and the sum of imaginary and real parts of the single species self-energy. Indeed the width is due to the phonons in resonance with the librational excitation of the methyl group. The general agreement between Hewson's theory and experiments seems to be very good. We do, however, not possess enough information about the phonon density of states or the coupling constants of the systems under study to establish the parameters of the model a priori. They can be chosen such that agreement with experiment is obtained.

Recently WHITTALL and GEHRING [8] have applied the path integral formalism to the problem of rotational tunneling. At low temperatures they reproduce the results of Hewson. They claim that their theory correctly describes the transition to the classical high temperature regime, characterized by incoherent hopping across the barrier. In order to perform the path integrals, the methyl group is approximated by a harmonic oscillator. Oscillators of different frequency are used for the different symmetries to account for the different level spacing, but apart from that the symmetry species are treated equally. Again, the frequency shift is traced back to the negative tunnel splitting in the first excited state and the broadening is due to the resonant phonons.

We have recently started an investigation of rotational tunneling at finite temperature using computer simulation. To this end a methyl group has been coupled to a heat bath by means of an intermediate classical rotor. The coupling between the methyl group and the classical rotor is reversible, the influence of the bath is modelled in terms of a Langevin equation with stochastic force R(t) and friction. There is no direct coupling between the methyl group and the bath. The wave function of the methyl group and the angle X of the classical rotor fulfil the coupled equations:

$$i\hbar \frac{d}{dt} |\psi(t)\rangle = H(t) |\psi(t)\rangle \qquad (8)$$

$$I\ddot{X} + I\gamma\dot{X} + \frac{d}{dX}U(X) + \langle\psi(t)|\frac{\partial V(\alpha - X)}{\partial X}|\psi(t)\rangle = R(t) \tag{9}$$

where $I \approx 10\Theta$ is the moment of inertia of the classical particle, U(X) a potential, γ the friction constant, and

$$H(t) = -B\frac{\partial^2}{\partial\alpha^2} + V(\alpha - X(t)). \tag{10}$$

The time-dependent Schrödinger equation contains only the potential shaking term, the breathing term has been omitted. The potential V is required to respect the threefold symmetry of the problem. For simplicity we use $V(\alpha-X(t)) = -V_0 \cos(3\alpha-3X(t))$. It is a plausible but somewhat adhoc assumption to use the quantum mechanical expectation value in (9) to compute the torque exerted by the quantum rotor on the classical rotor. The assumption is justified by the conservation of angular momentum which results for the angular momentum exchange between the quantum rotor and classical rotor. In the computer simulation one tries to find a good approximation of δ-correlated noise with $\overline{R(t)R(0)} = G\,\delta(t)$. When G, γ, and T fulfil $G = 2I\gamma kT$, as suggested by the fluctuation-dissipation theorem for a single classical rotor, coupled to a heat bath at a temperature T, one finds that the kinetic energy of the classical particle exceeds kT/2 by roughly 10%. This is an effect of the reversible coupling to the quantum particle which is expected to influence the relation between G, γ, and T.

Equations (8) and (9) are integrated numerically to obtain the methyl group wave function as a function of time. To evaluate (4) the system is started in an A-state at t=0, it is transformed immediately into an E-state, developed to the time t and then retransformed with the result $|\tilde{\psi}(t)\rangle = X^+U(t,0)X|K=0,n\rangle$. Also calculated is the time development of the A-state itself $|\psi(t)\rangle = U(t,0)|K=0,n\rangle$. Then the correlation function $\langle\psi(t)|\tilde{\psi}(t)\rangle$ required in (4) is formed. The influence of the fluctuating potential is best exhibited by the phase C(t) and $\tilde{C}(t)$ of the wave functions:

$$\langle K=0,n|\psi(t)\rangle = A(t)\exp(iC(t)); \tag{11}$$

C(t) and $\tilde{C}(t)$ are plotted in Fig. 1 for n=0 and for one realization of the stochastic process. To this end the zero of energy has been chosen at the level of the unperturbed groundstate of A symmetry. Thus at zero temperature one would find C(t) = 0 and $\tilde{C}(t) = -\omega_t t$. The increasing separation between C(t) and $\tilde{C}(t)$ is still visible at

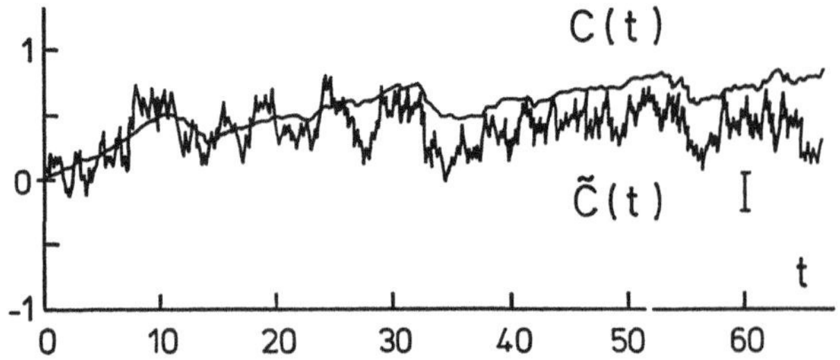

Fig. 1: The time development of the phases at temperature T = 1B. Time is measured in units of ħ/B. The potential strength is 30B which corresponds to a tunneling frequency $\omega_t = 0.0053B/\hbar$. At $t = 60\hbar/B$ we have indicated the phase difference $\omega_t t$ which would occur at T = 0

finite temperatures, but there are strong fluctuations around the zero temperature behaviour. It is surprising that $\tilde{C}(t)$, the phase of the E-state, fluctuates more strongly than C(t). This cannot be traced back to the small differences in the level spacings of the A and E-states. It is due to the action of the operators X and X^+ which are responsible for the symmetry changes of $\tilde{\Psi}(t)$.

A Fourier transform of (4) yields the scattering function. One finds broadened tunneling lines, but no frequency shift with temperature. The missing shift should not come as a surprise when one bears in mind that a shift results from an interference effect for excited lattice modes. In our model where the lattice modes are represented by a classical particle, there are no interference effects of this kind. The temperature dependence of the broadening will be studied in detail in the future.

The simulation exhibits the transition from tunneling to classical jump diffusion. Phase slips in C(t) or $\tilde{C}(t)$ which occur more frequently with increasing temperatures, in effect, produce rotations of the probability density of the methyl group.

1. W. Press: Single-Particle Rotations in Molecular Crystals, Springer Tracts in Modern Physics, Vol.92, (Springer, Berlin, Heidelberg 1981)
2. M. Prager, W. Press, B. Alefeld, A. Hüller: J. Chem. Phys. 67, 5126 (1977)
3. M. Prager, J. Stanislawski, W. Häusler: to be published
4. S. Clough, J.R. Hill: Ph.D. Thesis (J.R. Hill), Nottingham University
5. A. Hüller: Z. Phys. B 36, 215 (1980)
6. W. Häusler: to be published
7. A.C. Hewson: J. Phys. C 15, 3841 (1982) and 15, 3855 (1982)
8. M. Whittall, G. Gehring: preprint 1986

The Rotational Potential of Solid Nitromethane

S.F. Trevino

Energetics and Weapons Division, Army Research Development and Engineering Center, Dover, NJ 07801, USA and
Reactor Radiation Division, National Bureau of Standards, Gaithersburg, MD 20899, USA

The properties of the rotational potential of the methyl group in solid nitromethane have been studied, principally by neutron scattering. There are Raman scattering studies which corroborate several assignments of the energy levels obtained from the neutron measurements. The crystal structure has been determined by a combination of single crystal X-Ray measurements and neutron powder diffraction. The pressure dependence of the rotational energy levels exhibited a behavior different from that of most other systems previously studied and proved essential in the understanding of the potential in nitromethane.

1. INTRODUCTION

Interest in the study of the rotation of the methyl group of nitromethane in the solid state was first generated when it was made known to us that in the gas phase the barrier to rotation was very small (6 cal/mole)[1]. In such a system the inter-molecular interactions in the solid would dominate and their effect on the rotational potential would not be masked by a large intra-molecular contribution. In addition, the molecule (CH_3NO_2) is a simple one in which only one dimensional rotation exists. A literature search at that time revealed that there was no information on its crystal structure, knowledge of which can be essential in the interpretation of many subsequent measurements. It was incumbent on us to attempt the structure determination and this is reported in the next section. Of the various experimental tools available in the study of rotating molecular groups, quasi-elastic neutron scattering has been very useful in the past and the results of these measurements are reported in the second section. Of course, the most revealing information concerning the properties of the rotational potential is obtained by measuring the position of the various energy levels it determines. This information may be most readily obtained by inelastic neutron scattering if instruments are available of adequate resolution. The results of these measurements are reported in the third section along with their surprising pressure dependence. Finally in the last section we discuss the form of the rotational potential which is consistent with all the spectroscopic information and the possible source of various components of this potential.

2. THE CRYSTAL STRUCTURE

A sample of deuterated nitromethane was prepared by freezing the liquid which was contained in a cylindrical sample holder mounted in a cryostat. It was found that this procedure did not produce a sample suitable for powder neutron diffraction, in that severe preferred orientation obtained. The melting point of nitromethane is -28.5 C. It therefore was possible to grind the solid with a mortar and pestle which were placed on solid carbon dioxide. The sample was then loaded into the sample holder which was also in the carbon dioxide and rapidly transferred to a cold cryostat. This procedure produced a sample with no preferred orientation and a diffraction pattern was obtained suitable for study by the RIETVELD method [2] as modified by HEWAT [3] and PRINCE and FINGER [4]. It was, however, not possible to index the pattern or determine the space group due to overlap of lines at large angles. It was necessary to grow a single crystal *in situ* on the goniometer of a single crystal X-Ray diffractometer. The single crystal was quite readily grown by the BRIDGEMANN method on a $P2_1$ diffractometer which provided a stream of cold (-45 C) gas over the goniometer. The space group was determined to be $P2_12_12_1$ with one molecule in the asymmetric unit. The significance of this last discovery is that all methyl groups experience the same environment. There is but one potential to be found. A total of 289 observable reflections were used in the refinement. The nitrogen, carbon, and oxygen atoms were readily found, after which a difference electron map revealed the approximate positions of the hydrogen atoms. The neutron powder diffraction patterns were subsequently used to continue the refinement of the structure. In this case rigid body constraints were used to better model the curvilinear thermal vibrations. Figure 1a presents a view of the unit cell determined at 4.2 °K. The temperature dependence of the mean square amplitude of libration of the deuterated methyl group was determined from 4.2 K to 125 K and was found to be consistent with a simple harmonic oscillator with a level spacing of 5 meV. These data are shown in Fig. 1b. The significance of this finding will become apparent later. The last finding relevant to the present work is that one deuterium was found at each of three positions 120° apart, implying that when the methyl group performs thermally activated jumps they will be three-fold.

Details of this work can be found in Ref. [5].

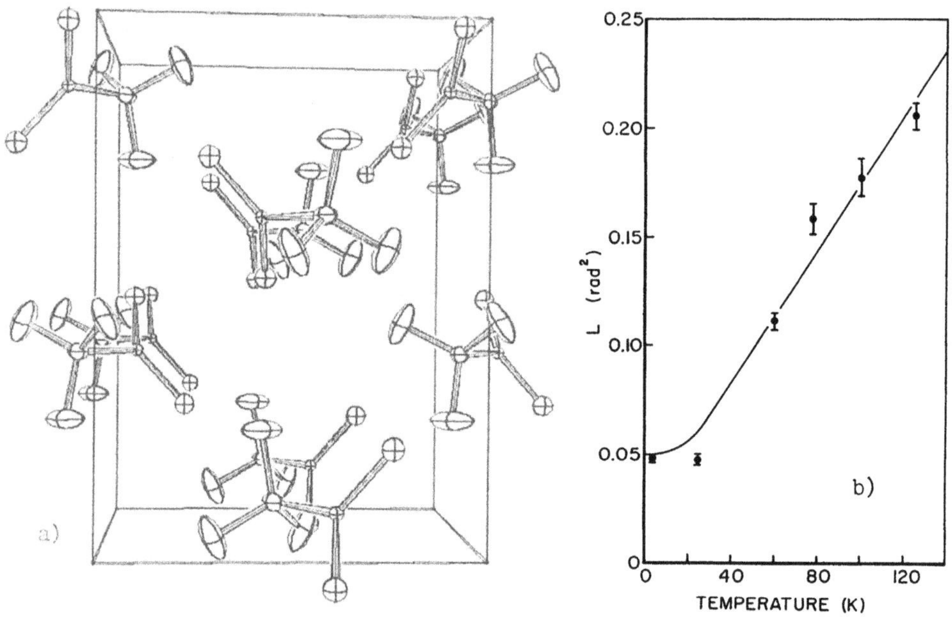

Figure 1. a) A view of the unit cell of CD_3NO_2 at 4.2 °K b) The mean square amplitude of libration of the methyl group in CD_3NO_2 as a function of temperature. The solid line is the amplitude of a simple harmonic oscillator with a level spacing of 5meV.

3. QUASIELASTIC NEUTRON SCATTERING

A simple model for thermally activated rotations about the nitrogen-carbon bond can be formulated [6]. In this model one assumes instantaneous jumps between the equilibrium positions followed by residence for a time τ at the equilibrium positions. From the crystal structure we know that the reorientations are by 120°. This model produces an incoherent neutron scattering law for a polycrystalline sample which has the form:

$$3S(Q,\omega) = \left\{\left(1 + \frac{2\sin Qr}{Qr}\right)\delta(\omega) + \frac{2\Gamma}{\pi(\Gamma^2+\omega^2)}\left(1-\frac{\sin Qr}{Qr}\right)\right\}e^{-2W}. \quad (1)$$

Where here Q is the momentum transfer, r is the distance between the equilibrium positions of the hydrogen atoms, W is the Debye-Waller factor, $\hbar\omega$ is the energy transfer in the scattering and $\Gamma=3/2\tau$. This model will be a good approximation to the scattering law if the residence time τ is long compared to the time it takes to perform the reorientational jump. Note that the relative intensities of the delta function and Lorentzian are prescribed by the model. In the measurements of the quasi-elastic spectra it was found that great care was needed in preparing the sample as with the powder diffraction measurements. If there exist substantial amounts of preferred orientation, the measured line shapes were found to be irreproducible from one sample preparation to the next and were therefore unreliable. The method described above for the preparation of the sample for the diffraction studies was employed here also.

The quasielastic scattering was measured at six temperatures between 50° and 150° K and for five values of Q between 1.0 and 3.8Å^{-1} for all but the two lowest temperatures. In fitting the data with the model, Eq. (1) was folded with the measured resolution (.44 meV) and corrected for the analyzer efficiency [viz $S(Q,\omega)_{meas}=S(Q,\omega)k_f^3\cot\theta_A$, where k_f is the wave vector of the scattered neutron and θ_A is the Bragg angle of the analyzer crystal]. The parameters varied in the fitting were an offset energy from the nominal elastic position, an overall intensity factor, the residence time and the background. At each temperature the residence time τ was determined from the data at a Q of 2.88 Å^{-1} where the Lorentzian intensity is a maximum. This value was used and held fixed in the fitting of the data for the other values of Q. Figure 2 presents the data taken at 78° K and the fit obtained as a solid line, the dashed line represents the contribution of the Lorentzian. The adequacy of the fit is more than satisfactory. As an instructive exercise an attempt to fit the data with a model which assumes a 60° reorientation was performed. This was done for the data at Q=3.81 Å^{-1} where the Lorentzian of a six-fold model would have maximum intensity. The data and the best fit obtained are shown in the upper right-hand corner of Fig. 2. It is readily apparent that the data are of sufficient quality to distinguish 3-fold from 6-fold reorientational motion.

In Fig 3. is presented the logarithm of Γ^{-1} vs 1000/T. A straight line is found to represent the data quite well, implying that the reorientational motion is by thermally activated process with an Arrhenius equation presented in the Figure. The activation energy is 10.2 meV. This is a rather low barrier, which encouraged us to attempt a measure of the ground state splitting due to quantum mechanical tunneling.

Reference to the details of this work can be found in Ref. [6].

4. SPECTROSCOPIC MEASUREMENTS

The first attempt [7] to measure the tunnel splitting of the ground state was made with the spectrometer BT 4 of the NBS research reactor with a resolution function of .053 meV. Figure 4a presents the data and a fit obtained with 3 gaussians of width equal to the resolution whose positions and intensities were allowed to vary. As is obvious from an inspection of this figure the resolution was not adequate to the job but a preliminary value of 45 μeV for the ground state tunnel splitting was obtained. Broadening at a temperature 4.2 K due to thermally activated reorientation would be much too narrow for a process with an activation energy of 10 meV and thus too small to explain the additional scattering evident in the figure. This evidence was sufficient to convince my colleagues at the Institute Laue-Langevin to perform the measurement with the backscattering spectrometer IN13 with a resolution of 7 μeV. The results are presented in Fig. 4b in which the tunnel-splitting line is clearly visible at 35 μeV [8]. The measurement of this property in a fully deuterated sample which was accomplished with the instrument IN10 with a resolution of 0.4 μeV, is also presented in this Figure. A value of 1.7 μeV was determined.

Other properties which can yield information on the nature of the potential describing the librational motions are the location of the excited states. The first measurements were performed with the instrument BT4 at the NBS reactor with a resolution of ~ 0.5 meV [6]. Figure 5a presents the data for both the

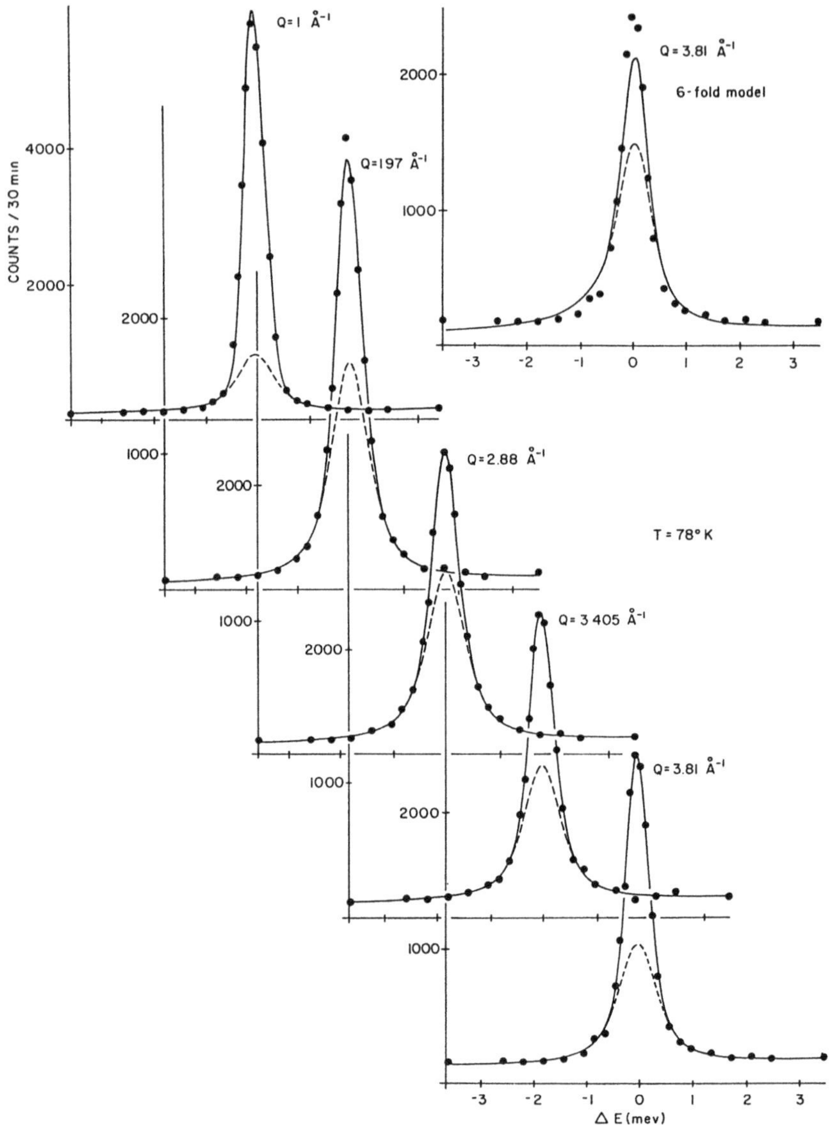

Figure 2. Quasielastic neutron scattering spectra for several values of momentum transfer of CH_3NO_2. The solid lines are the result of fits as discussed in the text with a three-fold jump model. The spectrum at the upper right represents the best fit obtained with a six-fold jump model. The dashed lines represent the Lorentzian contribution.

protonated and deuterated molecules. In this figure the librational levels are assigned to the most intense feature since they are expected to have the largest amplitude of vibration. In addition, the peak in the deuterated molecule (5.3 meV) is ~ 1/1.4 times the centroid of the peak (7.5 meV) in the protonated molecule (recall at this point that the temperature dependence of the mean square amplitude of libration of the deuterated group as determined by the diffraction study is consistent with a torsional level at 5 meV). The two dashed curves in the lower figure were thought to represent the two levels caused by tunnel splitting of the

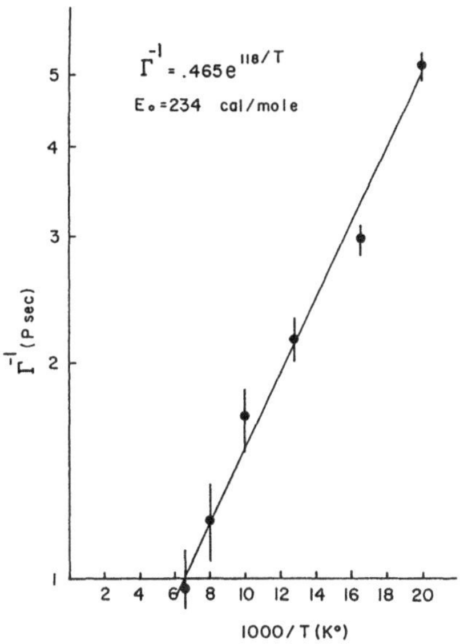

Figure 3. The logarithm of Γ^{-1} vs 1000/T as obtained from the quasielastic scattering data. The straight line is obtained by a least squares fit to the data which resulted in the Arrhenius relation presented in the figure.

a) b)

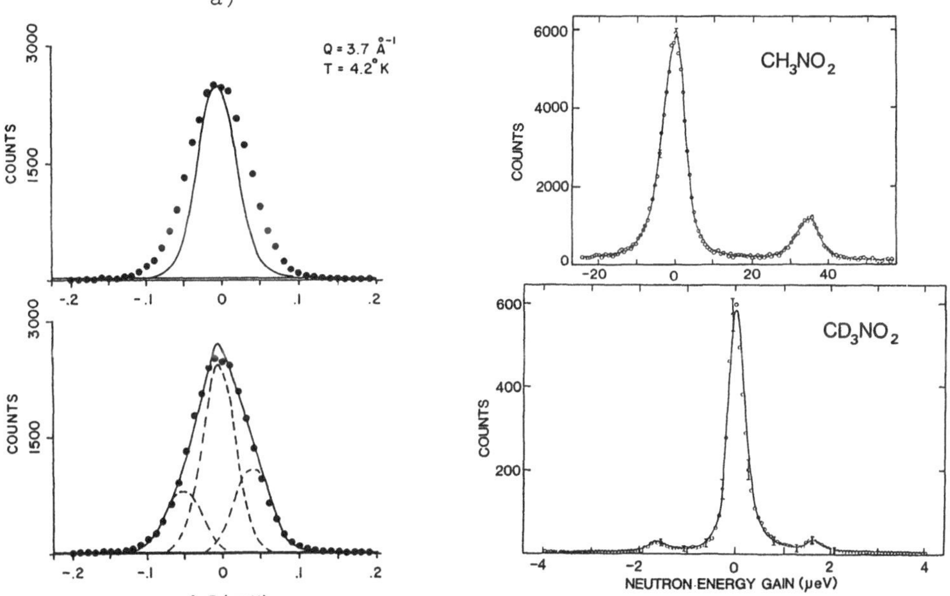

Figure 4. a) the first attempt to measure with inelastic neutron scattering the ground state tunnel splitting in CH_3NO_2 with an instrumental resolution of 53 μeV. The upper figure presents the data and the measured resolution (solid line) and the lower figure presents the fit with three gaussians. b) The successful measurement of the ground state tunnel splitting in both isotopic species of nitromethane

first excited state. As additional evidence for the correctness of the assignment of the first excited state of CH_3NO_2, data were obtained as a function of temperature and are presented in Fig. 5b. The feature which was thought to be due to the torsional level exhibited behavior characteristic of such highly anharmonic vibrations **viz** that they broaden with temperature until they are no longer discernible. Even at 4.2 K however, there remains a substantial intrinsic width to

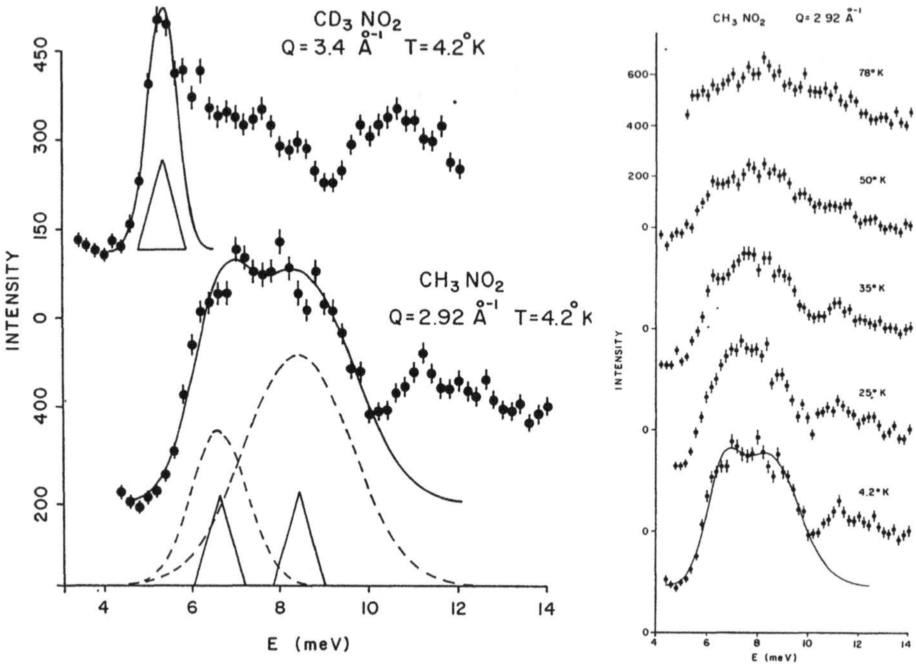

Figure 5. a) Inelastic neutron scattering measurement of the first excited torsional state in both isotopic species of nitromethane. The solid lines are fits to the data with gaussians and the dashed lines in the lower figure are the two gaussians as discussed in the text. b) The temperature dependence of the inelastic neutron scattering from CH_3NO_2

the peaks in the protonated sample, indicating either a finite life time to the states or dispersion of the torsional mode. We speculated that if the width was due to life time effects caused by interaction with the virtual phonon bath, the excited state being thought to be near the top of a small barrier of 10 meV, that increasing the barrier by applying pressure, would tend to decouple these modes by driving the torsional mode deeper into the potential well and thus change the width. Figure 6a presents the results for CH_3NO_2 at 4.2 K and various pressures up to 4.8 kbar. Of the two lines which were thought to be due to the tunnel split first excited state one, that corresponding to the largest peak moved down in energy, while the other remained in approximately the same position. We are forced to assign the peak with the largest intensity to the unresolved first excited state and the lower intensity feature to a density of phonon states whose pressure dependence should be less pronounced that the torsional state. However this assignment requires a wholly unexpected pressure dependence. The barrier was expected to increase, causing the first excited state to increase in energy, not decrease. A natural next measurement would be to determine the pressure dependence of the torsional level in CD_3NO_2 and these measurements are presented in Fig. 6b where again the level decreases in energy. We were led to speculate that if the first excited state misbehaved, the tunnel splitting of the ground state might increase instead of decreasing with pressure, as had been observed with all systems in which the pressure dependence of tunnel splitting had previously been studied. IN10 and IN13 were again used to perform the measurements which are presented in Fig. 7a for CH_3NO_2 and in Fig. 7b for CD_3NO_2. It is clear from inspection of the figures that the ground state tunnel splitting of both species has indeed increased with pressure. At this point there was total confusion.

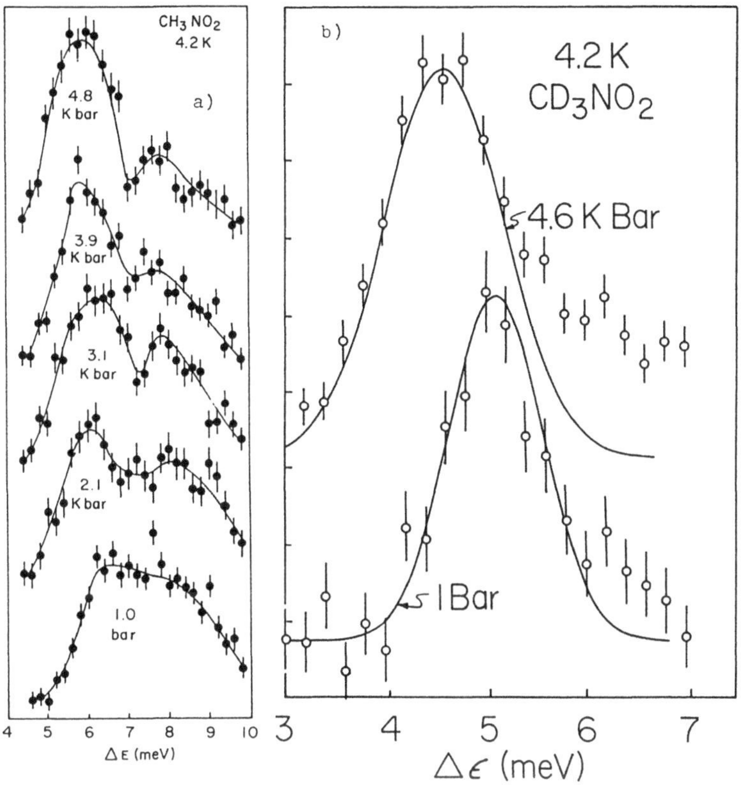

Figure 6. a) The pressure dependence of the first excited torsional state in CH_3NO_2 at 4.2 K as determined by inelastic neutron scattering b) the pressure dependence of the first excited state in CD_3NO_2 at 4.2 K measured with inelastic neutron scattering

5. THE POTENTIAL

During the course of this work, attempts to determine a potential which would reproduce the observations as they arrived were not successful. In particular the low activation energy determined by the quasielastic scattering data was too small for any potential which would yield the measured ground state tunnel splitting. The data which led to the resolution of the problem was the fact that the ground state tunnel splitting of the deuterated species increased almost exponentially with pressure, whereas that for the protonated species increased only linearly with pressure. Dominique Cavagnat reasoned that because the ground state of the deuterated molecule exists deeper in the potential well than that of the protonated molecule, the secret lay there. She added a small amount of six-fold potential to the larger three-fold term to obtain a potential of the form

$$V(\Theta)=V_3(1-\cos 3\Theta) + V_6[1-\cos(6\Theta+\delta)] \qquad (2)$$

in which Θ is the rotational angle of the methyl group about the C-N bond, V_3 and V_6 are the strengths of the three- and six-fold parts of the potential respectively, and δ is a phase angle between the three-fold and six-fold parts of the potential. The solution to the eigenvalue problem for such a potential is well known [9]. Using only the spectroscopic data, i.e. the ground state tunnel splitting and first excited state position of both isotopic molecules, a least square fit produced the following values for the parameters of the potential. V_3=25.5 meV, V_6=−15.5 meV and δ = 30 for the data at ambient pressure and V_3=22.5 meV, V_6=−17.2 meV and δ=25 for the data at 4.8 kbar. This potential is shown in Fig. 8 with the positions of the various levels for both isotopic species.

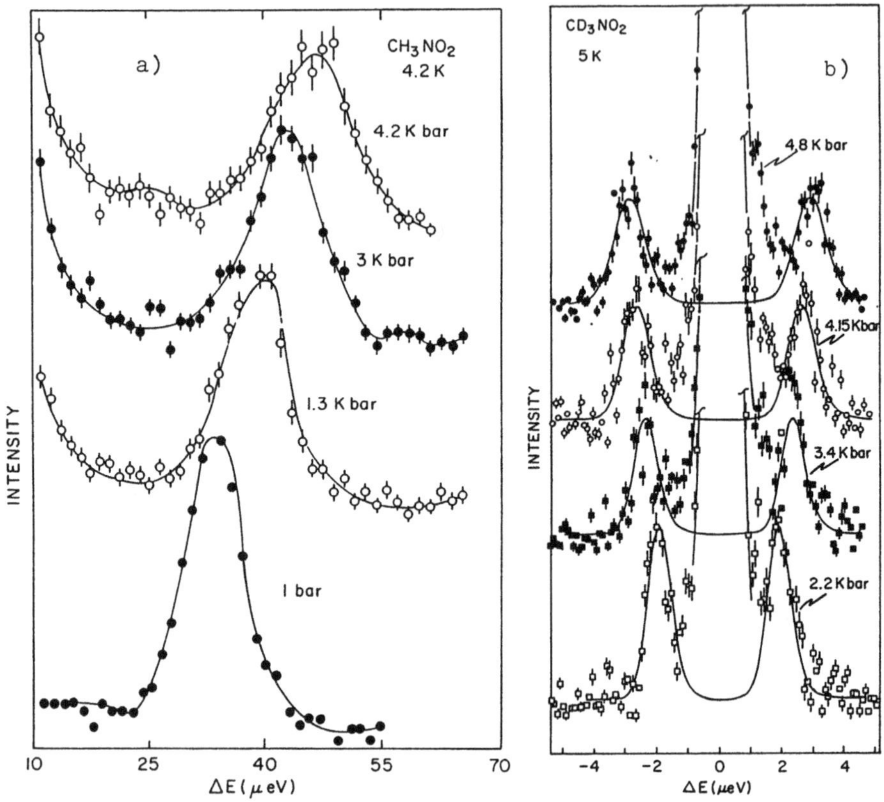

Figure 7. The pressure dependence of the ground state tunnel splitting in a) CH_3NO_2 and b) CD_3NO_2 measured with inelastic neutron scattering

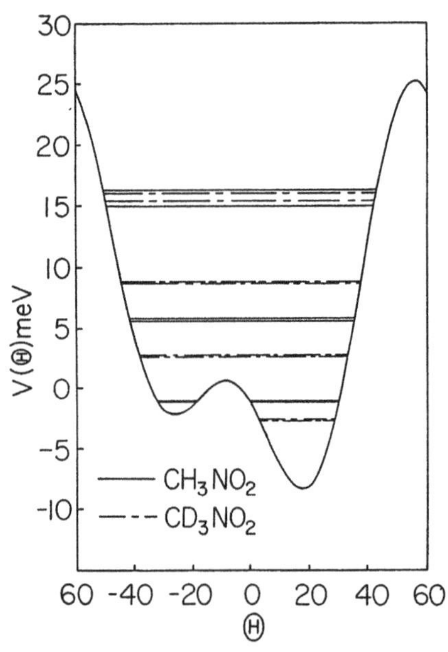

Figure 8. The rotational potential as a function of methyl group rotation with parameters corresponding to 1 bar (see text and table 1). Also presented are the torsional levels for CH_3NO_2 (solid lines) and CD_3NO_2 (dashed lines). The level splittings are only schematic for the lower levels.

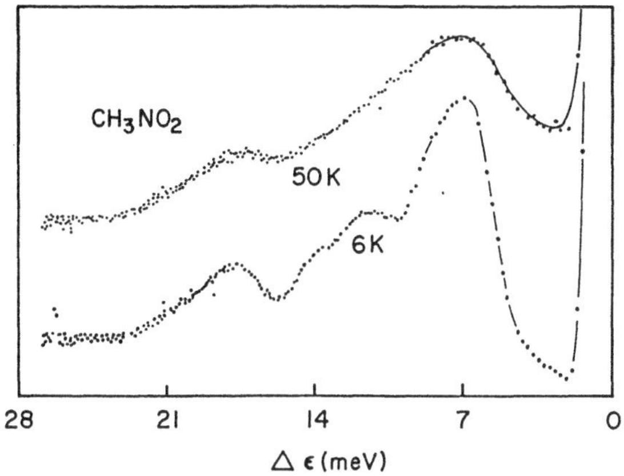

Figure 9. Inelastic neutron scattering spectra at two temperatures from CH_3NO_2. The first (7meV) and second (17.5meV) excited torsional levels are evident.

Inspection of this Fig. reveals the existence of a second bound excited state in CH_3NO_2 at approximately 17 meV. Very quickly a spectrum was obtained using the time-of-flight spectrometer IN4 at the ILL and it is shown in Fig. 9. Inspection of the figure reveals both the previously measured first excited state at 7 meV and the second excited state at 17.5 meV. Such decisive confirmation of the prediction requires acceptance of the essential correctness of Dominique's calculation [10]. Table 1 presents a comparison of the measured and calculated levels with this potential for two pressures.

Table 1. Measured and calculated levels of the two isotopic species of nitromethane at two pressures and 5 K. The values of the parameters are given in the text.

	Ground-state splitting(μeV)		First-excited state (meV)		Second-excited state (meV)		
	Meas.	Calc.	Meas.	Calc.	Meas.	Calc.	Pressure
CH_3NO_2	35	34	6.7	6.9	17.5	16.7	1 bar
CD_3NO_2	1.7	1.3	5.3	5.2	10.6	11.4	1 bar
CH_3NO_2	54	54	5.6	5.6	...	15.8	4.8 kbar
CD_3NO_2	2.8	2.5	4.4	4.1	...	10.7	4.8 kbar

The one piece of data which caused confusion in my mind was the activation energy 10meV which is smaller than even the second bound state. We now know that for systems with low barriers there is not a one-to-one correspondence between the barrier as determined by low temperature spectroscopic studies and an activation energy determined at elevated temperatures. As for the "anomalous" pressure dependence of the energy levels, it may be that the observed phenomenon is a result of the small value of the three-fold term,which does not obscure the effect of the six-fold term,and that other systems in which the overall barrier is small will also exhibit such behavior. All of the other systems whose pressure dependence had previously been studied had substantially larger barriers.

In order to determine the origin of the various terms in the potential,and in particular the origin of the six-fold term, we have used a simple Lennard-Jones potential to describe the interactions between the methyl hydrogens and the surrounding oxygen atoms. The rotational potential is obtained by calculating the total energy as a function of the orientation of one methyl group keeping the surrounding lattice fixed. It was found that with appropriate values of the Lennard-Jones potential, a rotational potential very similar to that of Fig. 9 could be obtained, and that the origin of the six-fold term is the repulsive H-O interaction. In fact, the asymmetric location of the six-fold term with respect to the three-fold term was well reproduced and is due to the asymmetric distribution of the oxygen atoms with respect to the equilibrium position of the methyl groups [10]. More recently CAVAGNAT and PESQUER [11] have performed similar calculations using potential parameters of Kitaigorodskii and have concluded that the three-fold term is due to H-H interactions and that the six-fold term is due to repulsive H-O interactions. The conclusion that the source of the six-fold term is due to H-O repulsive interactions is consistent with the observed pressure dependence of this term in that it increases vis-a-vis the three-fold term with pressure.

Finally a word concerning the identification of torsional modes with Raman spectroscopy. There are at least two published works in which a Raman spectrum of solid nitromethane is presented. In both of these works the torsional mode is observed as a very week band compared with the features due to other motions. CAVAGNAT et. al. [12] correctly assigned the first excited state with foreknowledge of the neutron scattering results but missed assigning the second excited torsional level. On the other hand KANESAKA et. al. [13] assigned a peak at 148 cm^{-1} (18 meV) to a torsional level but assumed that it was the first excited state instead of the second excited state. The lesson is that neutron scattering is the method of choice when studying torsional modes involving groups containing hydrogen.

6. CONCLUSION

It is clear that a complete understanding of even such a "simple" system as nitromethane requires a great deal of data. This potential is probably the most thoroughly characterized system and yet there is more to be done. We intend to perform at least two more types of measurements. In the first, the effects on the position and width of the several levels of diluting the protonated species in a deuterated sample will be determined. And secondly, a measurement of the Q dependence of the torsional levels and phonons will be measured on a single crystal of deuterated nitromethane. Considering the several surprises encountered in the past measurements, we do not dare to make any predictions on the results of these measurements.

REFERENCES

1. E. Tannenbaum, R. J. Myers, and W. D. Gwinn, J. Chem Phys. **25**, 42(1956)
2. H. M. Rietveld, J Appl. Cryst. **2**, 65(1969)
3. A. W. Hewat, J. Phys. C **6**, 2559(1973)
4. E. Prince and L. W. Finger, Acta Cryst. Sect B **29**, 179(1973)
5. S. F. Trevino, E. Prince and C. R. Hubbard, J. Chem. Phys. **73**, 2996(1980)
6. K. Sköld, J. Chem. Phys. **49**, 2443(1968); C. T. Chudley and R. J. Elliott, Proc. Phys. Soc. London **77**,, 353(1961)
7. S. F. Trevino, J. Chem. Phys. **71**, 1973(1979)
8. B. Alefeld, I. S. Anderson, A. Heidemann, A. Magerl and S. F. Trevino, J. Chem. Phys. **76**, 2758(1982)
9. Dudly R. Herschbach, J. Chem. Phys. **31**, 91(1959); R. F. Gloden, Rapport Euratom, EUR4349f and EUR4358f; James E. Woolrab, Rotational Spectra and Molecular Structure, ed. by Ernest M. Loebl (Academic, New York, 1967), Appendix 12
10. D. Cavagnat, A. Magerl, C. Vettier, I. S. Anderson and S. F. Trevino, Phys. Rev. Lett. **54**, 193(1985)
11. D. Cavagnat and M. Pesquer, J. Phys. Chem. **1986**, 90, 3289
12. D. Cavagnat, J. Lascombe, J. C. Lassegues, A. J. Horsewill A. Heidemann and J. B. Suck, J. Phys. (Paris) **45**, 67(1984)
13. Isao Kanesaka, Haruko Naka and Kiyoyasu Kawai, J. Chem. Phys. **70**, 5773(1979)

The Orientation of the Rotational Potential in the Crystal Frame: $(NH_4)_2PtCl_6$

A. Hoser[1], *W. Prandl*[1], *and G. Heger*[2]

[1]Institut für Kristallographie der Universität Tübingen, Charlottenstr. 33, D-7400 Tübingen, Fed. Rep. of Germany

[2]IAK/KFK P.O. Box 36 40, D-7500 Karlsruhe 1, Fed. Rep. of Germany and CEN Saclay, Lab. Leon Brillouin, F-91191 Gif-sur-Yvette, France

1. INTRODUCTION

Molecular librations of large amplitude in crystals may be described in many cases by a single-particle rotational potential $V(\omega)$ /1,2/. Rotational potentials have been derived from thermodynamic data, from librational energies, from the tunnel splitting and from Bragg scattering data, and from combinations of these methods. The $V(\omega)$ determined from an experiment can only be checked against a model potential calculated from any kind of atom-atom interaction if its absolute orientation is known: because $V(\omega)$ is due to the neighbouring atoms in their actual geometric arrangement. The phrase "absolute orientation" means in this context: the true orientation of $V(\omega)$ with respect to the crystal frame.

From all the experiments mentioned above - with the exception of the Bragg scattering method - only relative orientations of $V(\omega)$ can be obtained. A conceivable help in this situation might be the use of atom-atom potentials: one could calculate model rotational potentials and establish the absolute orientation by comparing $V_{MOD}(\omega)$ and $V_{EXP}(\omega)$. We will show that, with the present knowledge of the atom-atom potentials involved, this procedure does not work in the case of the (NH_4) rotational potential in $(NH_4)_2PtCl_6$.

2. ROTATIONAL POTENTIALS FROM TUNNEL SPLITTING DATA

Experiments using single crystals will be considered first. High-resolution inelastic neutron scattering from single crystals may be used to observe at the same time the tunnel splitting and the anisotropy of the squared non-diagonal matrix elements of the multipole operators $\mathcal{M}^{\ell}_{\varepsilon\varepsilon'}(\omega)$. The $\mathcal{M}^{\ell}_{\varepsilon\varepsilon'}(\omega)$ are symmetry-adapted Wigner functions, $\ell = 0,1,2, \ldots$, and ε, ε' are labels for unit representations of the site and molecular group of the molecule in question, respectively /3/. As long as the molecules move independently there is essentially no difference between inelastic coherent and incoherent scattering due to individual molecules because interference terms are absent. It is the technical aspect of the large incoherent cross-section of the proton which makes the investigations of molecules containing protons more attractive.

Unique information on the anisotropy of the scattering and the tunnel splitting of one molecule is sufficient, at least in principle, to extract $V(\omega)$ and its absolute orientation. This situation is given if all molecules in the crystal are translationally equivalent, i.e. with one orientationally disordered molecule per unit cell. We consider, as the next step, n (n > 1) molecules per unit cell which are transformed into one another by proper or improper rotations, glide planes or screw axes. These molecules are still equivalent, but not translationally. They will show the same tunnel splitting, and the observed anisotropy of the scattering is the sum of n anisotropic intensity distributions derived from one unique primary distribution by the proper or improper rotations referred to above. This case can be dealt with properly in calculations by comparing the observations with a superposition of n rotated images obtained from the model intensity of one $V(\omega)$.

It is, however, not possible to assign this $V(\omega)$ in a unique way to one of the n sites available: there are n ways to do so, and only one of them is right. Difficulties of this kind occur generally in solid state physics when microscopic tensors are analysed with local probes, e.g. the Mößbauer effect. We may carry on this analogy by remembering that in general, magnetic structures cannot be determined from Mößbauer absorption data alone.

If the experiment is performed with a powdered sample all angular anisotropies are averaged out of course, and the only experimental information used in the calculation of $V(\omega)$ is the unique pattern of levels split by tunneling.

The case of molecules occupying nonequivalent sites is a straightforward generalisation of the discussion given so far.

In passing, we emphasize that rotational potentials derived from an experiment can be compared with those calculated from atom-atom interactions only if the absolute orientation of $V_{EXP}(\omega)$ is known.

3. ROTATIONAL POTENTIALS DERIVED FROM BRAGG SCATTERING

The intensity J derives from the structure factor F of the whole unit cell /3,4,5/

$$J = |F|^2 \tag{1}$$

with

$$F = F_{RIG} + F_{ROT}. \tag{2}$$

F_{RIG} is the contribution of the rigid part of the structure, and F_{ROT} is given by

$$F_{ROT}(\underline{Q}) = \sum_{i=1}^{n} e^{-\beta^{(i)}Q^2} \cdot a_{ROT}^{(i)}(\underline{Q}) \cdot e^{i\underline{QR}^{(i)}} , \tag{3}$$

$$a_{ROT}^{(i)}(\underline{Q}) = 4\pi \sum_L i^L b_s j_L(Qr_s) P_{L\varepsilon}(\hat{\underline{Q}}^{(i)}) <\mathcal{M}^L_{\varepsilon\varepsilon'}> \Pi^s_{L\varepsilon'} . \tag{4}$$

Here b_s is the scattering length or the form factor of the atoms of the species s which are librating on the surface of radius r_s around the molecular center of mass located in $\underline{R}^{(i)}$. The index i counts symmetrically equivalent molecules. The $j_L(...)$, $P_{L\varepsilon}(\hat{\underline{Q}}^{(i)})$, $\Pi^s_{L\varepsilon}$, are spherical Bessel functions, site-symmetry adapted spherical harmonics with normalized scattering vectors $\hat{\underline{Q}}^{(i)}$ referred to the local site coordinate system i as the argument, and molecular structure constants, respectively /3/. The rotational potential enters via the thermal averages $<\mathcal{M}^L_{\varepsilon\varepsilon'}>$ which are in the Boltzmann approximation:

$$<\mathcal{M}^L_{\varepsilon\varepsilon'}> = Z^{-1} \int \mathcal{M}^L_{\varepsilon\varepsilon'}(\omega) e^{-\beta V(\omega)} d\tau \tag{5}$$

with

$$V(\omega) = \Sigma v^{\ell}_{\varepsilon\varepsilon'} \mathcal{M}^{\ell}_{\varepsilon\varepsilon'}(\omega). \tag{6}$$

The potential parameters $v^{\ell}_{\varepsilon\varepsilon'}$ have to be calculated from the experimental data. A quantum statistical form of $<\mathcal{M}^L_{\varepsilon\varepsilon'}>$ was given by BRÜCKEL et al. /5/.

It is clear from (1)-(6) that the absolute orientation of $V(\omega)$ enters into the (coherent elastic) Bragg scattering, and can therefore be extracted from experimental data. This argument applies also for cases with more than one disordered molecule (e.g.: $Ni(NH_3)_6(NO_3)_2$ /6/) and for plastic crystals /7/.

4. THE CASE IN QUESTION: $(NH_4)_2PtCl_6$ -

This compound is cubic with the space group Fm3m. One (primitive) unit cell contains two NH_4 molecules at the sites (1/4 1/4 1/4) and (1/4 1/4 3/4), both with site symmetry $\bar{4}3m$. Both sites and their environments which determine the local

$V(\omega)$ are transformed into one another by a 90° rotation and a shift of a/2 along $\underline{z}$.

SMITH /8/ has recently extracted atom-atom potentials from librational and tunneling frequencies of NH_4Cl, $(NH_4)_2SnCl_6$, and $(NH_4)_2PdCl_6$. He uses these potentials to predict the rotational potential of the isomorphic compounds $(NH_4)_2XCl_6$ with X = Pd, Re, Ir, Pt, Pb, and Te. His calculated tunnel splittings agree fairly well with the observations, in particular for X = Pd /9/, Ir, Pt /10/. The $V(\omega)$ suggested by SMITH are similar in most cases to the $V(\omega)$ found earlier in $(NH_4)_2SnCl_6$ /4,5/. For the compounds with X = Pd, Ir, Pt, however, the calculated potential coefficient v_{11}^3 has a positive sign which is different from all the other members in the series.

We calculate from SMITH's data the rotational potential for several axes of rotation (Fig. 1a) and the proton distribution at T = 300 K for the molecule occupying the site (1/4 1/4 1/4): Fig. 2a. In order to prove or disprove SMITH's conjecture we have studied $(NH_4)_2PtCl_6$ single crystals with neutron diffraction. The main numerical results are given in Table 1. It is obvious that only a negative v_{11}^3 agrees with the observations. The $V_{EXP}(\omega)$ and the proton density are given in Figs. 1b and 2b, respectively. We should mention that a Frenkel model using discrete sites for the protons and resulting in fairly large "thermal" amplitudes agrees fairly well with the distribution of Fig. 2b, however with a slightly poorer goodness-of-fit.

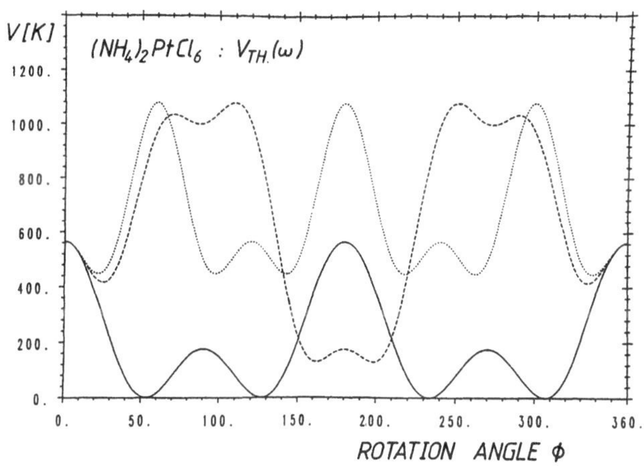

Fig. 1a

The rotational potential after SMITH /8/. The rotation axes are as designated in Fig. 1b.
For ϕ = 0 the protons are in $a\cdot(\frac{1}{4}\,\frac{1}{4}\,\frac{1}{4}) + \underline{r}_i$ with
$\underline{r}_1 = r_s(111)/\sqrt{3}$,
$\underline{r}_2 = r_s(\bar{1}\bar{1}1)/\sqrt{3}$,
$\underline{r}_3 = r_s(1\bar{1}\bar{1})/\sqrt{3}$,
$\underline{r}_4 = r_s(\bar{1}1\bar{1})/\sqrt{3}$;
a = 9.851 Å and
$r_s \equiv r_{N-H}$ = 1.038(3) Å

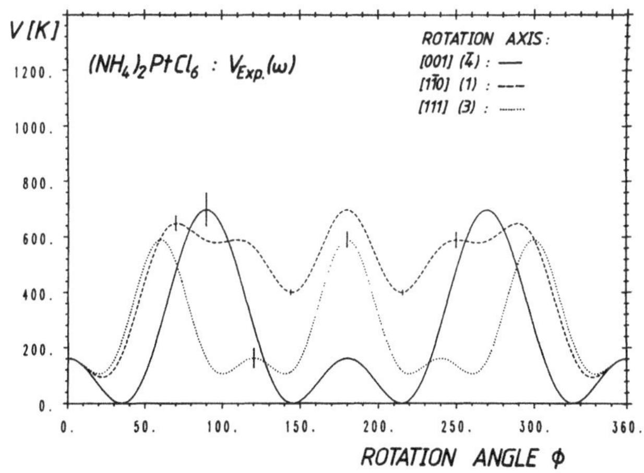

Fig. 1b

The observed rotational potential

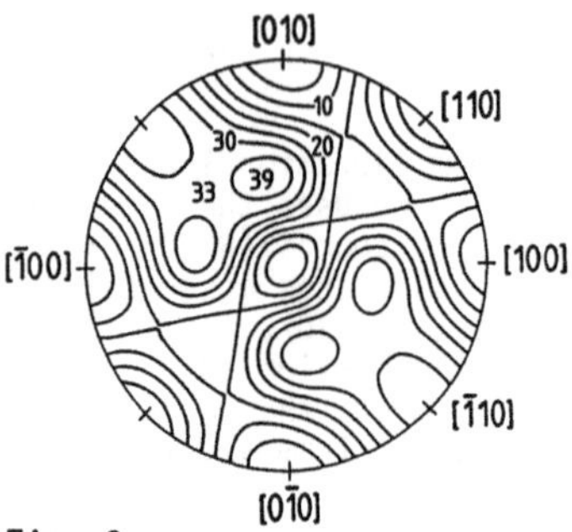

Fig. 2a

The proton distribution of the orientationally disordered NH_4 molecule at (1/4 1/4 1/4) calculated from SMITH's rotational potential

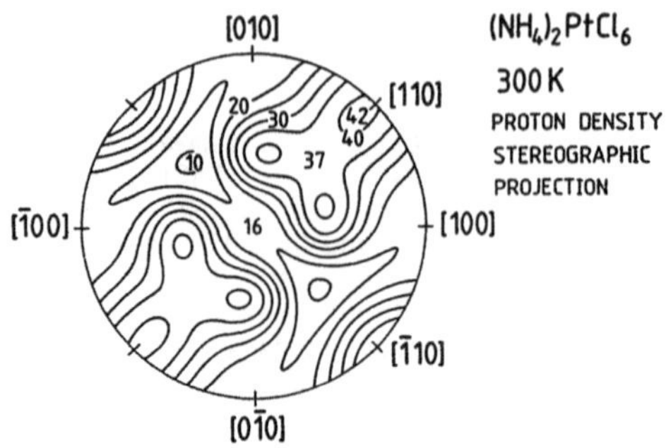

Fig. 2b

The observed proton density

Table 1 EXPERIMENTAL DATA, NUMERICAL RESULTS

			after SMITH /8/
instrument / reactor		P110 / ORPHEE	
wavelength	[Å]	0.832	
number of reflections/ independent reflections		1183/200	
sample temperature	[K]	300	
substance		$(NH_4)_2PtCl_6$	
a	[Å]	9.851	
Pt in (0 0 0): u	[$Å^2$]	0.0116(2)	
Cl in (x 0 0): x		0.23554(5)	
u_{11}	[$Å^2$]	0.0130(2)	
u_{22}	[$Å^2$]	0.0285(2)	
NH_4			
N in $(\frac{1}{4}\,\frac{1}{4}\,\frac{1}{4})$: u	[$Å^2$]	0.0273(2)	
r_{N-H}	[$Å^2$]	1.038(3)	
H: pot.par. v_{11}^{3}	[K]	- 268(18)	+ 196
v_{11}^{4}	[K]	- 261(38)	- 662
v_{11}^{6}	[K]	+ 343(37)	+ 512
$R = \Sigma\Delta F/\Sigma F_o$		0.026	
$R_W = \sqrt{\Sigma/w\Delta F)^2 / \Sigma(wF_o)^2}$		0.022	
goodness-of-fit = $\sqrt{\Sigma(w\Delta F)^2/(n-m)}$		1.81	

The general features of the $V_{EXP}(\omega)$ in Fig. 1b is indeed fairly similar to observations in $(NH_4)_2TeCl_6$ /11/, $(ND_4)_2SnCl_6$ /12/ and $(NH_4)_2SnCl_6$. It is, however, the softest rotational potential we have observed so far: this is in agreement with the wide tunnel splitting observed in $(NH_4)_2PtCl_6$ /10/. A comparison with Fig. 1a shows large differences which result essentially in a 90° rotation between the calculated and the observed easy orientation (Figs. 2a,b). A general feature of all the $V(\omega)$ observed in the $(NH_4)_2XCl_6$ series is the secondary maximum which occurs for the most symmetrical orientation of the NH_4 molecule at $\phi = 0$. It is the reason for the three-lobed density distribution in Fig. 2b. The density maxima correspond to the minima in $V(\omega)$ at $\phi = \pm 35°$ for a rotation by the $\bar{4}$ axis. The predominant motion at room temperature and above will therefore consist of 70° jumps: the hindering potential is only 160 K.

We draw two conclusions from our observations. First: the prediction of $V(\omega)$ from adjusted atom-atom potentials is, with the present knowledge, not yet possible. The rotational potential may be wrong even if it predicts the observed tunnel splitting. This argument applies in particular for weak rotational potentials. Second: the $V_{EXP}(\omega)$ may be used for a fairly detailed picture of the molecular motion. It reveals details beyond the a-priori assumptions on which the jump diffusion model is based.

REFERENCES

1. H.M. James, T.A. Keenan: J. Chem. Phys. 31, 12 (1959)
2. W. Press: Single-Particle Rotations in Molecular Crystals, Springer Tracts Mod. Phys. 92 (Springer, Berlin, Heidelberg, New York 1981)
3. W. Prandl: Acta Cryst. A37, 811 (1981)
4. K. Vogt, W. Prandl: J. Phys. C (Solid State Phys.) 16, 4753 (1983)
5. T. Brückel, W. Prandl, K. Vogt, C.M.E. Zeyen: J. Phys. C (Sol. State Phys.) 17, 4071 (1984)
6. A. Hoser, W. Joswig, W. Prandl, K. Vogt: Molec. Phys. 56, 853 (1986)
7. P. Gerlach, W. Prandl, K. Vogt: Molec. Phys. 52, 383 (1984)
8. D. Smith: J. Chem. Phys. 82, 5133 (1985)
9. M. Prager, W. Press, A. Heidemann, C. Vettier: J. Chem. Phys. 80, 2777 (1984)
10. M. Prager, A.M. Raaen, I. Svare: J. Phys. C (Sol. State Phys.) 16, L181 (1983)
11. G. Heger, A. Hoser, W. Prandl, to be published
12. D. Gregson, A. Hoser, W. Prandl, to be published

Inelastic Neutron Scattering Study of Methyl Tunnelling in α-Crystallised Toluene

D. Cavagnat[1], *A. Magerl*[2], *C. Vettier*[2], *and S. Clough*[3]

[1]Laboratoire de Spectroscopie Moléculaire et Cristalline, 351, Cours de la Libération, F-33405 Talence Cedex, France
[2]Institut Laue-Langevin, 156X, F-38042 Grenoble Cedex, France
[3]Department of Physics, University of Nottingham, Nottingham NG7 2RD, UK

INTRODUCTION

A considerable amount of information was collected over recent years about rotational dynamics of the methyl group in molecular crystals by using principally N.M.R. and neutron scattering techniques. This latter method is particularly powerful for the low rotational barrier cases by providing a complete level scheme in a large temperature and pressure range. Combining these results with precise structural data, a good determination of the potential energy function and of the responsible intermolecular forces may be reached, as illustrated recently with the nitromethane study /1-3/.

In the α-crystallised toluene also, the methyl group rotational barrier is relatively low and essentially from intermolecular origin /4/. Likewise, there is no phase transition over a wide temperature (from 1 K to the melting point 178 K) and pressure (up to 9 Kbar) range /5/. But it offers a particularly interesting case of a monomethylated compound containing methyl groups with two different molecular environments, as determined by structural study /6/.

In a previous study of this system, some difficulties were encountered in the analysis of the INS and Raman spectra of the rotational transitions due partly to the presence of two inequivalent rotators /7/. In order to remove these problems, a more thorough study of the INS tunnelling spectra is performed by investigating the pressure, temperature and isotopic effect /8/.

RESULT and DISCUSSION

Tunnelling study

As expected from the structural data /6/, two well-resolved tunnel peaks at ν_t = 28.5 and 26 µeV are observed in the INS spectrum of α-crystallised $C_6H_5CH_3$ at 5 K, corresponding to two different methyl rotators /8/ (fig.1).

a) Isotopic effect

The CD_3 tunnel frequency was previously measured at 1.1 µeV in $C_6D_5CD_3$ /7/.

When deuterating only the benzene ring, a shift toward the lower energy and a decrease of the CH_3 tunnel peaks splitting are observed. (ν_t = 24.7 and 25.8 µeV in $C_6D_5CH_3$) (fig.1). A zero point vibrational energy effect was proposed to explain these differences /8/. But recent isotopic dilution experiments show that this effect is not the main reason for the spectral changes. Indeed, the CH_3 tunnel spectrum of $C_6D_5CH_3$ is only broadened when diluting in $C_6D_5CD_3$ (fig.1). On the contrary, the CH_3 tunnel spectrum of $C_6H_5CH_3$ is strongly affected by the isotopic dilution and becomes very similar to that of diluted $C_6D_5CH_3$ (fig.1). The broadening can be due to some disorder in the crystalline methyl group environment induced by the isotopic dilution. But it appears that the CH_3 tunnel transitions are very dependent on the benzene ring deuteration of the surrounding molecules.

On the other hand, the width difference observed between the two peaks could indicate the presence of some coupling between the CH_3 rotators.

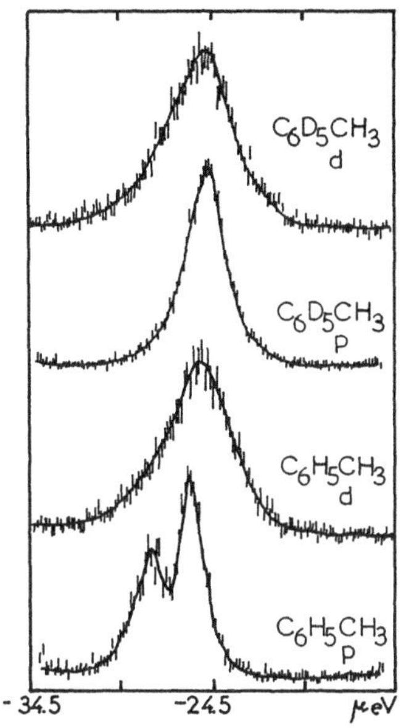

Fig.1 - INS tunnelling spectra of α-crystallised $C_6H_5CH_3$ and C D CH toluenes at 5 K (IN 10, Q = 4.0 Å^{-1}, P = 1 bar); p: pure compound; d: 5% diluted in $C_6D_5CD_3$

Fig.2 - The pressure dependence of the ground state splittings in α-crystallised $C_6H_5CH_3$ at 6 K (IN 13, Q = 4.0 Å^{-1})

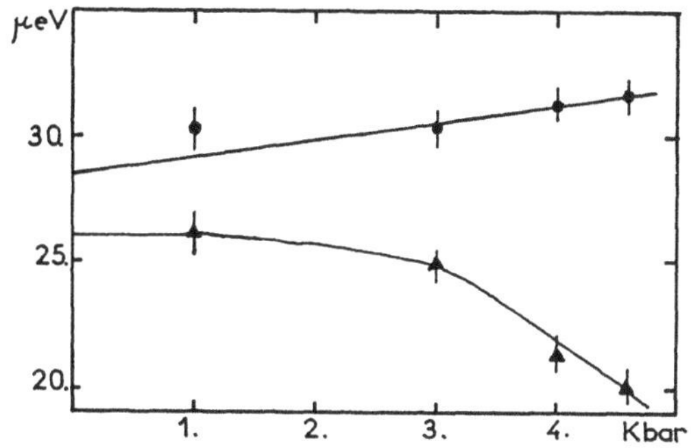

b) Pressure effect

When increasing pressure up to 4.6 Kbar, the tunnel peaks splitting of α-crystallised $C_6H_5CH_3$ increases up to 11.7 µeV : the 28.5 µeV peak shifts linearly up to 31.7 µeV, an anomalous pressure change very similar to that observed in nitromethane /2/, whereas the 26 µeV peak shifts quasi-exponentially down to 20 µeV, which is generally observed under pressure/9/,(fig.2). As no structural change occurs in the crystalline structure up to 9 Kbar /5/, this indicates that the pressure changes of the two methyl group molecular environments are different.

c) Temperature effects

When heating the pure samples, a broadening and a shift toward the elastic line of both tunnel peaks are observed. They can well be described by exponential forms :

$$\Gamma(T) = \Gamma_o \exp(-E_\Gamma/kT) \text{ for the broadening}$$

$$\Delta\hbar\omega(T) = \hbar\omega_o \exp(-E_s/kT) \text{ for the shift.}$$

Under 1 bar pressure, two different sets of E_s, E_Γ activation energy values are found for each tunnel peak ($E_s \leqq E_\Gamma$) : around 6.8 meV for the lower energy peak and around 10 meV for the higher energy peak (table) (fig.3). The E_Γ value is generally related by theory to the first torsional transition energy E_{01} /10,11/ and thus two torsional transitions may be expected around 6.8 and 10 meV. Indeed two lines around 5.8 and 10 meV can be assigned to torsional transitions in the Raman spectra of the CH_3 derivatives, but only the first one is observed in the INS spectra /7/.

Under 4.6 Kbar pressure, due to the exponential conditions, only the tunnel shift can be satisfactorily analysed. Two very similar E_s values are found for the two tunnel peaks around 6.5 meV /8/.

Potential energy function

A rather large amount of information is now obtained about the rotational levels scheme of the methyl group in α-crystallised toluene, allowing usually a full description of the rotational behaviour. However, the problem arises of combining correct tunnelling and torsional transitions from inequivalent CH_3 rotators.

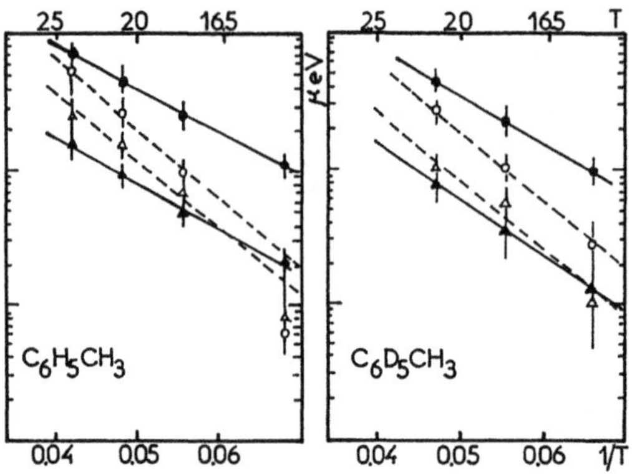

Fig.3
The temperature dependence of the tunnelling shift (o,●) and width (Δ,▲) in α-crystallised toluenes (P = 1 bar) (IN 10)
(●▲ 28.5 and 25.8 μeV)
(oΔ 26.0 and 24.7 μeV)

In consideration of the difficulties of the torsional spectra assignment /7/, a simple interpretation, implying a mainly V_3 potential for both methyl groups, may be considered by identifying only the 10 meV line with the first torsional transition (the 5.8 meV one being assigned to a lattice vibration).

However, this assignment does not account for the different temperature and pressure dependence of the two tunnelling transitions. Despite the present controversies /10-12/, all the theories relate the tunnelling temperature dependence to the first excited torsional levels. This suggests here to associate the higher tunnelling and torsional transitions (table) /13/. On the other hand, referring to previous analogous studies /2,9/, the tunnelling pressure dependence suggests that one potential is largely of V_3 type whereas the other one is a mixed V_3, V_6 potential. On these bases, potential parameters were derived for both inequivalent methyl groups reproducing correctly the experimentally measured transitions /8/. The potential V_I corresponding to the lower energy tunnelling excitation is a mixed V_3, V_6 potential ($|V_3/V_6| = 2$) and can be related to the methyl group in the more compact interatomic network (mean distance 4.42 Å /6/). The potential V_{II} is largely a V_3 potential ($|V_3/V_6| = 12$) with a lower height. It can correspond to the methyl group in the more distant molecular surroundings (mean distance 5.2 Å /6/).

TABLE : Experimental values of the methyl group rotational transitions(ν_t = tunnelling, ν_τ = torsion) and parameters characterizing the tunnelling temperature dependence in α-crystallised $C_6D_5CH_3$ and $C_6H_5CH_3$ for two pressure values. (* : only observed in Raman spectra /7/).

Pure isotopic derivatives	Pressure kbar.	measured transitions at 5 K		Tunnelling temperature dependence			
				Width		Shift	
		ν_t μeV	ν_τ meV	E_Γ meV	Γ_o μeV	E_s meV	$\hbar\omega_o$ μeV
$C_6D_5CH_3$	10^{-3}	24.7	5.8	7.6	51	6.5	159
		25.8	9.8*	9.8	159	9.6	506
$C_6H_5CH_3$	10^{-3}	26.0	5.9	6.8	41	6.2	146
		28.5	10.1*	10.0	274	9.6	535
	4.6	20.0	/	/	/	6.6	104
		31.7				6.2	118

Even so, supplementary results are necessary to confirm this potential model, in particular, the pressure dependence of the torsional transitions and of the CD_3 tunnelling. On the other hand, a theoretical calculation of these potentials, as done with nitromethane /3/, could get information about the intermolecular forces at the origin of these potentials and about the isotopic dilution effect. But this requires a very good low-temperature structure knowledge.

CONCLUSION

The α-crystallised toluene structure supplies a very interesting case of a monomethylated compound with two inequivalent methyl groups. This more complicated situation is well known to impose severe problems to interpret correctly the experimental data /13/.

However, toluene lends itself to various isotopic changes and to a large pressure and temperature-range study. Thus, a large amount of information, but still incomplete, has been obtained principally by INS. Though they are embedded in the same phonon spectrum, the two inequivalent CH_3 rotators are found to have different temperature and pressure dependence. The isotopic dilution experiments show that the INS tunnel peaks are greatly affected by the surrounding benzene rings isotopic change.

These results and the available structural data lead to the determination of potential parameters for both inequivalent methyl groups.

The authors are very much indebted to Dr A.Heidemann and to Dr.J.C.Lassègues for clarifying discussions. They are also grateful for the technical assistance received at the ILL Grenoble.

BIBLIOGRAPHY

1. S.F.Trevino and W.H.Rymes : J.Chem.Phys., 73, 3001 (1980)
2. D.Cavagnat,A.Magerl,C.Vettier,I.S.Anderson and S.Trevino, Phys.Rev.Letters, 54, 193 (1985)
3. D.Cavagnat and M.Pesquer, J.Phys.Chem., 90, 3289 (1986).
4. W.Müller-Warmuth, R.Schüler, M.Prager and A.Kollmar, J.Magn.Res., 34, 83 (1979)
5. L.Bosio,A.Defrain and G.Folcher, J.Chim.Phys.Chim.Phys.Biol., 73, 813 (1976)
6. M.Anderson,L.Bosio,J.Bruneaux-Poulle and R.Fourme, J.Chim.Phys.Chim.Phys.Biol., 74, 68 (1977)
7. D.Cavagnat,J.Lascombe,J.C.Lassègues,A.J.Horsewill,A.Heidemann and J.B.Suck, J.Phys., 45, 97 (1984)
8. D. Cavagnat,A.Magerl,C.Vettier and S.Clough, J.Phys.C., in press
9. S.Clough,A.Heidemann,M.Paley and C.Vettier, J.Phys., C12, L 781 (1979)
10. A.C.Hewson, J.Phys., C15, 3885 (1982)
11. J.C.Huggins, Thesis, Oxford (1983)
12. S.Clough, J.Phys., C14, 1009 (1981)
13. M.Prager,M.Müller-Warmuth, Z.Naturforsch., 39a, 1187 (1984)

Tunnelling Spectroscopy and Dynamics of CH_3 in a Homologous Series of Ketones

A.J. Horsewill, R.M. Green, and A.M. Alsanoosi

Department of Physics, University of Nottingham, Nottingham NG7 2RD, UK

The barrier heights and tunnel frequencies for CH_3 groups in a homologous series of ketones and some related organic solids have been measured by NMR and INS spectroscopy. The observed barrier heights for chemically and physically distinguishable methyl groups have been interpreted and correlated with the molecular and electronic structure.

1. Introduction

In the solid state at low temperature the reorientation of a methyl group about its C_3-axis is dominated by quantum mechanical tunnelling through a potential barrier. The barrier has both intra-molecular and inter-molecular contributions and reflects the details of the molecular and electronic environment of the particular methyl group under investigation. The methyl tunnelling frequency, ν_t, is an approximately exponential function of the barrier height and forms a particularly sensitive probe of the molecular environment. One aim of the investigations reported here has been to employ methyl tunnelling spectroscopy and dynamics in an experimental survey of some simple organic solids with a view to gaining insight into the physical and chemical principles which determine the barrier heights and shapes.

At low temperature the tunnelling motion is a coherent process and the tunnelling spectrum of CH_3 is simply determined by the eigenvalues of a time-independent Hamiltonian,

$$\hat{\mathcal{H}} = -\frac{\hbar^2}{2I}\frac{\partial^2}{\partial\phi^2} + \frac{V_3}{2}(1 - \cos 3\phi). \qquad (1)$$

As the temperature of the lattice is raised then the motion smoothly evolves into the incoherent classical regime. The eigenstates accessible to CH_3 are determined by the Hamiltonian (1), consequently it is to be expected that the dynamics of CH_3 at elevated temperatures will reflect the magnitude of the static potential barrier to which a particular CH_3 is exposed. In fact, as has been demonstrated experimentally, the reorientation rate is determined in a comparatively simple way from a Boltzmann weighted expectation value of the angular momentum and may be deduced from the eigenstates of (1). The clearest demonstration of this fact comes from the observed correlation between the tunnel frequency and the temperature, θ_{min}, at which the proton spin-lattice relaxation time T_1 is a minimum, [1]. A further aim of these experiments has been to investigate the validity of simple dynamical models of the type formulated by CLOUGH et al [2].

2. A Homologous Series of Ketones

The investigations reported here concern nuclear magnetic resonance (NMR) and inelastic neutron scattering (INS) experiments upon a homologous series of ketones for which the general molecular structure is given by the following:

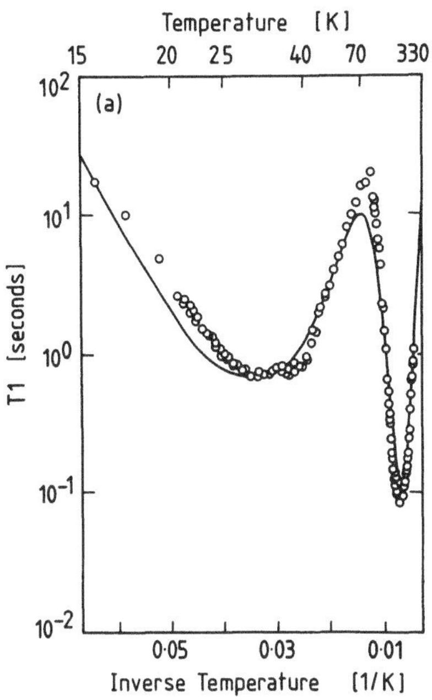

Figure 1: The temperature dependence of T_1 for 2-hexanone.

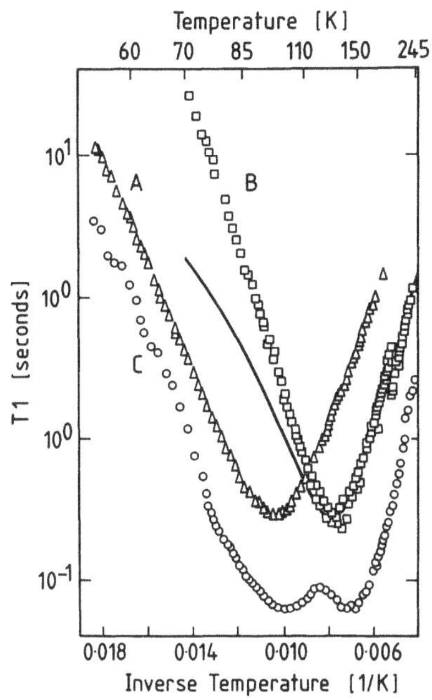

Figure 2: The temperature dependence of T_1 for (A) 3-pentanone, (B) 4-heptanone and (C) 3-hexanone.

$$CH_3 - (CH_2)_n - \overset{\overset{\large O}{\|}}{C} - (CH_2)_m - CH_3 \quad .$$

We first consider a series for which n = 0, m = 1,2,3 ..., one typical member of which is 2-hexanone. The temperature dependence of T_1 is displayed in Fig. 1. We observe two minima in T_1. According to the correlation referred to earlier [1] we conclude that there are at least two physically distinguishable methyl groups in the crystal, each of which experience quite different barrier heights. It has already been established in an earlier paper [3] that the low temperature minimum belongs to a CH_3 group with a comparatively low barrier ($V_3 \approx 450$ K) and is indeed the CH_3 group which is adjacent to the carbonyl group. The CH_3 group at the other end of the carbon chain in an alkyl chain-like environment experiences a high barrier ($V_3 \approx 1700$ K) and gives rise to the minimum in T_1 at 140 K.

The tunnel splitting of the low barrier CH_3 is accessible to measurement by the high-resolution INS spectrometer IN10 at ILL Grenoble. The IN10 spectrum for 2-hexanone displays four individual tunnel peaks, each belongs to a CH_3 with a slightly different barrier. The results and the corresponding 3-fold barrier heights, V_3, deduced from diagonalisation of (1) are displayed in the Table. All four tunnel peaks belong to a CH_3 adjacent to the carbonyl group, the splittings within the tunnel spectrum reflect the different inter-molecular contributions to the barrier at four molecular sites in the unit cell. The fifth tunnel splitting recorded in the table belongs to the strongly hindered methyl group and has been measured by NMR spectroscopy [5].

Figure 2 displays the results of NMR measurements upon (A) 3-pentanone, (B) 4-heptanone and (C) 3-hexanone. In 3-pentanone n = m = 1, each CH_3 group is separated from the carbonyl group by a single CH_2 and the minimum in T_1 occurs at 96 K. In 4-heptanone n = m = 2, the CH_3 groups are remote from the carbonyl group and the minimum in T_1 occurs at 131 K, a similar value to the CH_3 groups which

TABLE: Measured tunnel frequencies and barrier heights (* Predictions from Θ_{min})

Material		Θ_{min}	$h\nu_t$	V_3
2-hexanone	1)	29 K	5.8 ± 0.1 μeV	403 ± 2 K
	2)	29 K	5.0 μeV	417 K
	3)	29 K	2.3 μeV	499 K
	4)	29 K	1.5 μeV	546 K
	5)	145 K	$(6.3 \pm 0.2) \cdot 10^{-4}$ μeV	1708 ± 10 K
3-pentanone		96 K	0.015 μeV*	1170 K*
3-hexanone	1)	101 K	0.01 μeV*	1215 K*
	2)	142 K	0.001 μeV*	1630 K*
4-heptanone		131 K	0.00199 μeV	1502 K [4]
Acetyl acetone	1)	11 K	100 μeV*	160 K*
	2)	40 K	1 μeV*	590 K*
Tiglic acid	1)	23 K	10.8 μeV	344 K
	2)	52 K	0.47 μeV	687 K
Acetophenone		98 K	0.014 μeV*	1180 K*

occupy alkyl-chain-like environments in 2-hexanone. In 3-hexanone n = 1, m = 2 and we observe two minima in T_1, one at 101 K, the other at 142 K. By comparison with the above the high temperature T_1 minimum is assigned to the CH_3 group on the longest alkyl-chain fragment. The low temperature T_1 minimum belongs to a slightly less hindered CH_3 group which is one CH_2 removed from the carbonyl group. The predicted barrier heights are summarised in the table.

3. Discussion

We may write down an expression for the potential barrier to reorientation of CH_3 (assumed 3-fold for simplicity) as follows:

$$V_3 = V_{intra} + V_{inter} = (V_{attr} + V_{rep}) + V_{inter} \qquad (2)$$

It is with the term V_{intra} that we begin our discussion and attempt to correlate the qualitative trends in the intra-molecular contribution to the barrier observed above with identifiable features in the molecular and electronic structure of the molecule.

When the CH_3 group is in an alkyl-chain environment then the barrier is high ($V_3 \approx 1700$ K). In this case the methyl group is directly bonded to a sp^3 carbon atom. Similar results are obtained for the n-alkanes [6] and other related materials. The explanation of the rotational barrier in the similar molecule ethane has been the subject of theoretical studies for some years. Some of the more successful calculations [7] [8] demonstrate that the primary origin of the barrier lies in the exchange interaction between the C-H bond orbitals on CH_3 with the bond orbitals on the carbon atom to which the methyl group is bonded. Consequently it is in the Pauli exclusion principle that we find the existence of the barrier in this case. This is perhaps not surprising as we are concerned here with steric hindrance, indeed the barrier is dominated by the repulsive term, V_{rep}, and the potential energy minimum in these materials corresponds to the staggered conformation. So long as the wavefunctions are constrained to obey the Pauli principle then the calculated barriers represent the experimental data well.

Indeed it is significant that the results of the calculations are relatively insensitive to bond polarities and detailed molecular geometry.

When the CH_3 group is adjacent to a carbonyl group then the barrier to rotation is relatively low ($V_3 \approx 450$ K). In fact, as the splittings within the IN10 spectrum of 2-hexanone demonstrate, V_{inter} forms a substantial part of the total barrier. In this case the methyl group is bonded to a sp^2 carbon atom. Allen [9] has performed an "energy component" analysis of V_{intra} on acetaldehyde and pentene. His calculations demonstrate the importance of the term V_{attr} in the potential energy function. This is an attractive component to the barrier, is consequently negative in sign, and arises as a result of bonding type interactions between the C-H bond orbital and the sigma-pi double bond structure. The net result is a low barrier compared with alkyl CH_3 groups, as repulsive terms in the barrier are replaced partially by attractive terms of opposite sign. Consequently the familiar principles of chemical bonding can be seen to be applicable on a smaller scale to the origin of rotational barriers.

In 3-pentanone and 3-hexanone minima in T_1 are observed at lower temperature than for longer alkyl chain fragments. Here the CH_3 groups are one CH_2 removed from the carbonyl group, and the barrier is lower than expected for CH_3 bonded to a sp^3 carbon. When account is taken of the "staggered" nature of the carbon chain it is clear that the CH_3 group is physically close (~2.1 A) to the carbonyl group though not directly bonded to it. We propose that bonding interactions between CH_3 and the carbonyl group have the effect of reducing the barrier with the introduction of terms in V_{attr}.

The term V_{inter} represents the inter-molecular contribution to the barrier and is determined by the van der Waals non-bonding interactions between molecules in the crystal. In 2-hexanone this amounts to a few hundred Kelvin in magnitude depending upon the precise details of the crystal structure. The latter has yet to be determined before more detailed conclusions may be drawn.

4. CH_3 Groups Bonded to sp^2 Carbon Atoms

We have embarked upon a series of investigations designed to study methyl groups bonded to sp^2 carbon atoms in an attempt to obtain data relevant to the study of the term V_{attr} in the potential energy. A few representative samples are discussed here, the results of the experiments having been summarised in the table.

In Tiglic acid, $CH_3CH=C(CH_3)COOH$, there are two non-equivalent methyl groups. The barrier heights associated with each have been accurately determined by NMR and INS tunnelling spectroscopy. In this molecule both CH_3 groups are bonded to sp^2 carbon atoms; in contrast to the Ketones, however, the functional group to which the CH_3 is bonded is -ene. The observed barriers here fall within the range observed for the low barrier sites in the Ketones [3].

The temperature dependence of T_1 has been measured for Acetyl Acetone. We observe two minima, one at 11 K and another at 40 K. This provides an example of a molecule for which the ENOL-tautomer is the stable structure in preference to the KETO-tautomer as illustrated below, where the hydrogen bond is assumed to be asymmetric.

```
              O - H....O                          O         O
ENOL:         |        ||            KETO:        ||        ||
       CH3  - C = CH - C - CH3             CH3 -  C - CH2 - C - CH3
```

This explains the presence of two minima in T_1; they are assigned to the two chemically distinct CH_3 groups in the molecule. The predicted barrier heights are comparatively low, both -ene and carbonyl groups are represented here. One exception to this rule appears to be Acetophenone, in which the predicted barrier height is relatively high. This may reflect further factors not considered in this discussion, for example the proximity of the CH_3 to the phenyl ring.

5. The Dynamics of CH_3

Using the measured tunnel splittings and barrier heights in Table 1,the reorientation rate and hence T_1 has been calculated for 2-hexanone using the model of Clough et al [2]. The continuous line in Fig. 1 represents the results of this calculation. Five identifiable CH_3 groups contribute to the relaxation in the temperature range studied. The theoretical representation of the experimental data is very good. The exercise has been repeated for 4-heptanone and the solid line in Fig. 2 represents the results of the calculations. The values of dipole-dipole coupling constants have been adjusted here to fit the depth of the T_1 minima. We emphasize that apart from this there are no adjustable parameters.

Acknowledgements

The authors express their thanks to Prof. S. Clough for many helpful discussions and to Dr. A. Heidemann and Dr. A. Magerl of the Institut Laue-Langevin, Grenoble, for assistance with the INS measurements. Part of this work is supported by the Science and Engineering Research Council and RMG acknowledges support in the form of a research studentship. AMA thanks the King Abdulaziz University, Jeddah, for a scholarship.

References

1. S. Clough, A. Heidemann, A. J. Horsewill, J. D. Lewis, M. N. J. Paley: J. Phys. C (1981) 14, L525.
2. Ibid.: J. Phys. C (1982) 15, 2495.
3. R. M. Green, A. J. Horsewill: Mol. Phys. (1986) 57, 887.
4. S. Clough, K. Abed, private communication.
5. P. J. McDonald, G. J. Barker, S. Clough, R. M. Green, A. J. Horsewill: Mol. Phys. (1986) 57, 901.
6. J. E. Anderson, W. P. Slichter: J. Phys. Chem. (1965) 69, 3099.
7. E. Lombardi, G. Tarantini, L. Pirola, L. Jansen, R. Ritter: J. Chem. Phys. (1974) 61, 894.
8. O. J. Sovers, C. W. Kern, R. M. Pitzer, M. Karplus: J. Chem. Phys. (1968) 49, 2592.
9. L. C. Allen: Chem. Phys. Letts. (1968) 2, 597.
   R. B. Davidson, L. C. Allen: J. Chem. Phys. (1971) 54, 2828.

NH_4^+ Motions in NH_4MF_3 Cubic Perovskites

R. Navarro[1], *E. Palacios*[1], *J. Bartolomé*[2], *R. Burriel*[2], and *D. González*[2]

[1]Instituto de Ciencia de los Materiales de Aragón (I.C.M.A.), C.S.I.C.-Universidad de Zaragoza, Departamento di Física, E.T.S.I.I., E-50009-Zaragoza, Spain

[2]I.C.M.A., C.S.I.C.-Universidad de Zaragoza, Departamento de Termología, Facultad de Ciencias, E-50009-Zaragoza, Spain

The NH_4^+ rotational contribution to the heat capacity of NH_4MF_3 perovskites (M= Zn, Co, Mn, Cd, Mg) are reviewed and compared in the light of the new structural and spectroscopic data. The influence of NH..F bonding is evidenced.and a

1. INTRODUCTION

In the last decade our group has concentrated on the studies, both theoretical and experimental, of the phase transitions and of the thermal, magnetic and structural properties of fluoroperovskites./1-8/ Particularly, the ammonium compounds NH_4MF_3, with M= Zn, Co, Mn, Cd and Mg have received much of our attention, due to the enhanced anomalies of the heat capacity, $C_{p,m}$, at the transition temperature, T_c, and to their peculiar behaviour in contrast with the non-ammonium isostructural perovskites./9,10/

Recently, previous and new $C_{p,m}$ results have been processed systematically, /9,10/ trying to deduce the different contributions in both the anomalous and the non-anomalous regions. Our efforts have been devoted essentially to the consequences of the NH_4 motion,since the physical properties of NH_4MF_3 compounds are strongly governed by the quantum behaviour of NH_4 rotors, which move under electrostatic and other interactions in the cage formed by the M and F ions (Fig. 1)

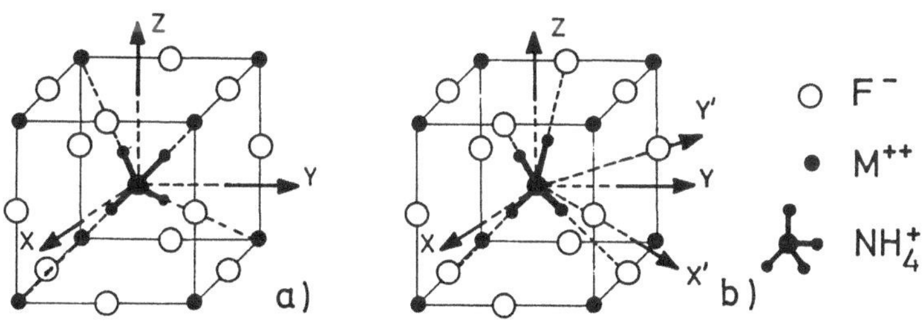

Figure 1. Orientations of maximum (a) and minimum (b) energy of the rotor xyz: lattice fixed axes; x'y'z': molecule fixed axes.

In the present contribution, the $C_{p,m}$ results are reviewed in the light of the known structural and spectroscopic data, pointing out the common trends and analysing the different interactions. Moreover, a criticism of the theoretical model/1/ so far used is done.

2. HEAT CAPACITY AND STRUCTURAL RESULTS

A detailed exposition of the $C_{p,m}$ experimental results and of the evaluation of the different contributions may be found elsewhere/9,10/ and hereafter the discussions are restricted to the relevant points.

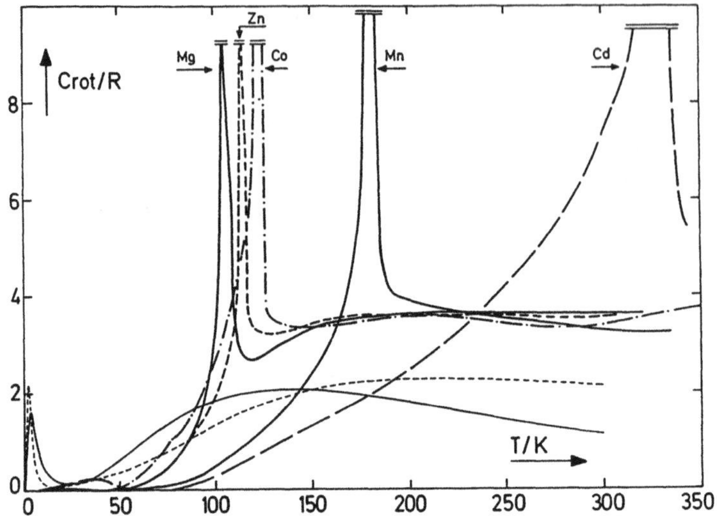

Figure 2. Deduced rotational heat capacities, C_{rot}, after subtraction of lattice and magnetic contributions, of NH_4MF_3. Light continuous and dashed lines: derived C_{rot} from the electrostatic model for M= Cd and Co (low and high electrostatic barrier, respectively)

From the $C_{p,m}(T)$ data the heat-capacity at constant volume, $C_{V,m}(T)$, was deduced. Then, assuming additivity of interactions, the lattice and rotational contribution, $C_{rot}(T)$, were derived, and in Fig. 2 the results have been represented. In the case of Mn salt the magnetic contribution has been subtracted but not in the Co compound, where only the contribution due to the thermal population of higher energy levels/10/ has been discounted.

The relevant results of Fig. 2, together with the structural data, have been summarized in table I, being the main conclusions:

i) The transitions at T_c have first-order character and, upon cooling, the compound reduces its Pm3m symmetry, but the cell volume remains almost constant./3/ This is similar to the structural transitions found in non-ammoniacal cubic perovskites, in which softening of the R_{25} or M_3 modes take place.

ii) The anomalous entropy content, ΔS_{rot}, roughly evaluated by smooth interpolation of $C_{rot}(T)$ for temperatures well above and below T_c, is in all cases close to R ln 3.

iii) The structural data,/3/ below T_c give pseudo-tetragonal symmetry with c>a. However, the presence of weak-ferromagnetism in the Mn and Co salts/1,5/ indicates lower symmetry (P4bm recently reported for NH_4MnF_3 /11/).

iv) Above T_c $C_{rot}(T)$ reaches values between 3.5 and 4 R for all the compounds and below T_c $C_{rot}(T)$ sharply decreases, which indicates that the NH_4^+ are strongly

Table I

Summary of experimental data of NH_4MF_3, M = Mg, Zn, Co, Mn and Cd. For comparison, data of NH_4MnCl_3 have been included. Structural data at room temperature except for the Cd compound (cubic above 331 K)

	Mg	Zn	Co	Mn	Cd	NH_4MnCl_3
T_c(K)	107.5	115.1	124.7	182.1	331.2	265.5
a(nm)	0.4056	0.4118	0.4132	0.4242	0.4401	0.5043
c/a	1.013	1.0170	1.0193	1.0138	1.023	orthorom.
$\Delta S_c/R$	1.05	1.15	1.38	1.29	1.25	0.074

bound. Therefore, the transition separates two regions; above T_c in which $C_{rot}(T)$ is almost independent of the M ion and below T_c where $C_{rot}(T)$ becomes M-dependent.
v) Existing $C_{p,m}$ results below 10 K for the Zn and Co salts deviate from the T^3 law, as shown by their Debye temperatures, $\Theta_D(T)$. For NH_4CoF_3 $\Theta_D(10\ K) = 380\ K$ and $\Theta_D(1.3\ K) = 100\ K$, and for NH_4ZnF_3 $\Theta_D(5\ K) = 400\ K$ and $\Theta_D(2.5\ K) = 340\ K$.

3. ANALYSIS AND DISCUSSION

3.1 Electrostatic point-charge model

While the octopole-octopole interaction plays an important role in ammonium halides,/12/ in our case it will be one order of magnitude lower due to the larger NH_4^+-NH_4^+ distances. Thus, in a good approximation, the neighbouring NH_4^+ may be considered as point charges.

The first analysis of the behaviour of NH_4MF_3 family was done with the electrostatic point charge model/1/ applying the following assumptions: The ions are considered static point charges and for each NH_4^+orientation, ω, gives a pure electrostatic barrier, $V(\omega)$. The NH_4^+ is taken as a rigid tetrahedron with a NH distance of 0.104 nm and an effective charge of $+0.43 \cdot e$ for each proton./1/ The lattice ion motions do not interact directly with the NH_4^+ rotations, since their frequencies are much higher than the characteristic jump frequencies of the rotor. That is not the case of soft modes (i.e. R_{25} or M_3), being the energy reduction of the rotor the triggering mechanism of the transition. On such basis a single-rotor Hamiltonian was established in the form:

$$H = P^2/2I + \hbar^2/2I \cdot V(\omega) \qquad (1)$$

where I is the NH_4 momentum of inertia, P the angular momentum and $\hbar$ the Planck constant. Using the cell dimensions $V(\omega)$ has been evaluated by the Evjen method/1/ for the minimum and maximum energy positions (fig 1a and 1b) and a saddle point. Moreover to obtain the eigenvalues of (1) a shortened expansion of $V(\omega)$ in cubic harmonics was used for the cubic and tetragonal phases:

$$V(\omega) = \beta[\beta_4 V_{44}(\omega) + \beta_6 V_{66}(\omega)]$$
$$V(\omega) = \beta[\beta_{41} V_{41}(\omega) + \beta_{42} V_{42}(\omega) + \beta_{61} V_{61}(\omega) + \beta_{62} V_{62}(\omega)] \qquad (2)$$

where the β and β_{JM} parameters were fixed with the above $V(\omega)$ results.The main results from the model are:

i) The phase transition is of first order and its reorientational contribution to the latent heat is given by the reduction of accessible distinguishable NH_4 orientations by pure rotation.
ii) The rotational hindering barrier, controled by the β parameter diminishes for increasing unit cell size.
iii) Since the crystallographic distortion increases β, the activation energy, E_a, for classical jumps will be higher at $T<T_c$.
iv) Above T_c, $C_{rot}(T)$ is smooth and presents a maximum of 2 R near T=200 K.
v) Below T_c, $C_{rot}(T)$ is also smoothly decaying with temperature.
vi) An anomaly in $C_{rot}(T)$ due to tunnelling motions should appear at temperatures near 10 K.

3.2 Criticism of the model

As mentioned in all studied cases the transition is of first order with an entropy content of $\Delta S/R \sim \ln 3$. The low-symmetry crystal structure of NH_4MnF_3 /11/, recently determined,indicates the formation of a C_s local symmetry for the NH_4 with minimum NF distances. This result is then consistent with the entropy content met in all members of the series and with prediction i).

Table II

Results of the electrostatic point charge model. β= dimensionless strength parameter, E_{ac}/k and E_{at}/k activation energies in the cubic and pseudo-tetragonal phases respectively. Experimental N.M.R. results for E_a and ω_t are also included. (ω_t= average tunnelling frequency)

		Mg	Zn	Co	Mn	Cd	NH_4MnCl_3
		45.	40.	40.	33.	26.	10.
	E_{ac}/k	583.	490.	520.	405.	295.	113.
THEO.	E_{at}/k	649.	650.	600.	471.	369.	202.
	E_{ac}/k	450.	680.				
EXP.	E_{at}/k	700.	825.			2000.	
	$\omega_t(s^{-1})$	$3x10^7$	$8x10^6$			$<10^5$	

In NH_4MnCl_3 the structural phase transition was not accompanied by a sharp anomaly. This was interpreted as caused by its much larger unit cell which reduced β to values near the free rotor limit, thus verifying i) and ii) /4/.

The determination of the jump activation energy E_a from NMR measurement allows the derivation of the rotational barrier heights. The results are included in table II. Indeed, the barrier increases strongly below T_c as predicted in iii)

However the barrier obtained for NH_4CdF_3 (E =2000 K) is in clear conflict with the expected value of E $\sim$300 K following the electrostatic model. As we shall show in paper II this effect is caused by other contributions besides the electrostatic interaction.

From our systematic heat capacity experiments one observes that prediction iv) is fulfilled, though the height at the maximum in C_{rot} is always 1.5 R higher in the experiment. It can be caused by the procedure for obtaining C_{rot} , but also due to stronger rotor-phonon interaction. A more drastic discrepancy is found respect to prediction v). Indeed we always observe a sharp decrease with decreasing temperature which cannot be explained at all by our model. Finally, in no case have we observed the predicted vi) tunnelling maximum, though the excess in C_p observed is probaby due to this cause. Lack of thermodynamic equilibrium may smear out the anomaly.

The tunnelling motion of the NH_4^+ groups has been probed by pulsed NMR /13/ obtaining the average tunnelling frequencies listed in table II for the non-magnetic members of the series. The values were obtained with the classical expression given by O 'Reilly et al. /14/ for the inverse relaxation time in the rotating frame $1/T_{1\rho}$. The largest frequency is found in NH_4MgF_3 while for NH_4CdF_3 it was too small to be resolved.

Concluding, though these compounds were mentioned as good approaches to the electrostatic point model /1,15/, we now find important discrepancies in NH_4CdF_3 (largest cell case) and in general in the low-temperature distorted phase of the series. It seems that hydrogen bonding and repulsive forces should be better accounted for to explain the available data. In order to clarify these aspects new Raman spectroscopic data are being collected. Quasielastic neutron scattering below T_c would also be suitable for the solution of this problem.

The financial support of the Spanish Comisión Asesora de Investigación Científica y Técnica (proj. num. 3380/83), and Spanish-German Integrated Action num. 7B are acknowledged.

1. J. Bartolomé, R. Navarro, D. González and L. J. De Jongh: Chem. Phys. Lett. 48 536 (1977); Physica 92B 23 and 45 (1977)
2. Chr. Steenbergen, L. A. de Graaf, L. Bevaart, J. Bartolomé and L. J. De Jongh: J. Chem. Phys. 70 1450 (1979)
3. R. B. Helmholdt, G.A. Wiegers and J. Bartolomé: J. Phys. C13 5081 (1980)
4. R. Burriel, J. Bartolomé, R. Navarro and D. Gonzalez: In Recent Developments in Condensed Matter Physics, ed. by J. T. Devreese et al, Vol.4 (Plenun Press 1980) p.1
5. J. Bartolomé, R. Burriel, F. Palacio, D. González, R. Navarro, J. A. Rojo, L. J. De Jongh: Physica 115B 190 (1983)
6. R. Burriel, J. Bartolomé, R. Navarro and D. González: Ferroelectrics 54 253 (1984)
7. E. Palacios, J. Bartolomé, R. Navarro, J. García, D. González and H. B. Brom: Ferroelectrics 55 287 (1984)
8. J. Bartolomé, F. Palacio, J.M. Calleja, F. Agulló-Rueda, M. Cardona and R. Migoni: J. Phys. C18 6083 (1985)
9. E. Palacios, R. Navarro, R. Burriel, J. Bartolomé and D. González: J. Chem. Thermodynamics (1986) in press.
10. R. Navarro, R. Burriel, J. Bartolomé and D. González: J. Chem. Thermodynamics (1986) in press and submitted for publication.
11. J. M. Calleja, private communication
12. A. Hüller, J. W. Kane: J. Chem. Phys. 61 3599 (1974)
13. To be published.
14. D.E. O´Reilly, T. Tsang. J. Chem. Phys. 46, 4,1291 (1967).
15. W. Press. Single-Particle Rotations in Molecular Crystals. Springer Tracts in Modern Physics.(1981).

Spectroscopic Raman Study of Trifluoroperovskites of Ammonium Under High Pressure

E. Palacios[1], *J. Bartolomé*[1], *F. Agulló-Rueda*[2], *J.M. Calleja*[2], *M. Cardona*[3], *K. Syassen*[3], *and K. Strößner*[3]

[1]Instituto de Ciencia de los Materiales de Aragón (I.C.M.A.), Universidad de Zaragoza, E-50009 Zaragoza, Spain
[2]Departamento de Optica y Estructura de la Materia, Universidad Autónoma de Madrid, E-Madrid, Spain
[3]Max-Planck-Institut für Festkörperforschung, Heisenbergstr. 1, D-7000 Stuttgart 80, Fed. Rep. of Germany

1. INTRODUCTION

The compounds NH_4MF_3 (M= Mg, Zn, Co, Mn, Cd) have the cubic perovskite structure at high temperature, distorting to quasi-tetragonal symmetry at lower temperatures (fig. 1). In a previous communication /1/, some discrepancies between the electrostatic model and experimental data were discussed, mainly the high barrier (2000 K) to rotation of NH_4 groups in NH_4CdF_3.

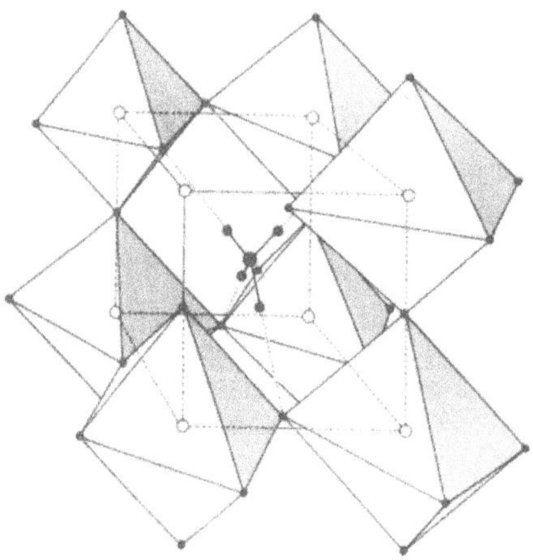

Figure 1. MF_6 octahedra surrounding an NH_4 ion in the $P_{4/mbm}$ low-temperature structure

In ammonium halides NH4X (X= F, Cl, Br, I) the hydrogen bonding (HB) between the protons and the halide ions provides their high rotational barrier /2/. Indeed, such a high barrier and the N-H..F distances involved are indicative of the presence of HB in NH_4MF_3.

The formation of HB is usually found by the softening of the stretching frequencies of the NH bond /3/. So, in NH_4-F HB systems those frequencies (symmetric and asymmetric stretchings) should decrease with decreasing N-F distance /4/, whereas when we compare different N-H--X (X= F, Cl, Br, I) systems, the strength of the HB decreases when X goes from F to I.

On the other hand, the study of internal modes under high pressure allows to study continuously their dependence on N-F distance, thus, avoiding the effect of other different ions which appear when the distance is modified by ionic substitution.

TABLE 1. Cell parameters /nm/ and frequencies /cm^{-1}/ of the internal modes of the ammonium at room temperature. NH_4CdF_3 is cubic above T_c= 331.3 K with a= 0.4394 nm. * This work

	a	ν_1	ν_2	ν_3	ν_4
NH_4MgF_3	0.4056 /5/	3234 *		3305 /6/	1426 /6/
NH_4ZnF_3	0.4116 /7/	3220 /4/	1695 /8/ 1655 /4/	3290 /8/ 3280 /4/	1425 /8/ 1430 /4/
NH_4CoF_3	0.4132 /7/		1685 /8/	3280 /8/	1425 /8/
NH_4MnF_3	0.4242 /7/	3168 /4/	1668 /4/	3245 /4/	1430 /8/ 1428 /6/ 1432 /4/
NH_4CdF_3	a=0.4368 c=0.4447	3074 *		3200 *	1440 /8/
free NH_4^+		3271 /9/		3379 /9/	

2. EXPERIMENTAL

The frequencies of the internal modes of the NH_4 are determined at room temperature and atmospheric pressure and are summarized in Table 1. There are four modes: ν_1 (symmetric stretching, non-degenerate), ν_2 (symmetric bending, twofold degenerate), ν_3 (asymmetric stretching, threefold degenerate), and ν_4 (asymmetric bending, threefold degenerate). The mode ν_1 varies on substitution of the M atom with the ratios $\Delta\nu / \Delta a$ = -2330, -4130, -6180 cm^{-1}/nm, obtained by the differences (Mg:Zn), (Zn:Mn) and (Mn:Cd), respectively. Though the variations in cell parameters are too big to extract any law trend, the sign is always negative. This is contrary to the sign in most HB systems.

The symmetric stretching frequency (ν_1) in NH_4ZnF_3 and NH_4MnF_3 was measured by Raman spectroscopy under high pressures, up to 23 GPa, in an ambient of methanol + ethanol, with a diamond high-pressure cell. The Raman spectra were obtained in the range of 3000 - 4000 cm^{-1} using the λ_1= 488 nm and λ_2 = 476.5 nm lines of a Kr laser. Results are plotted in Fig. 2. The error in the determination of the lines is not greater than 10 cm^{-1}. The data follow roughly a linear dependence with pressure

$$\nu_1 = 3208 + 4.591\ P\ , \text{ for } NH_4ZnF_3 \text{ and}$$
$$\nu_1 = 3174 + 3.414\ P\ , \text{ for } NH_4MnF_3,$$

where ν_1 is in cm^{-1} and P in GPa (1 GPa = 10 Kbar).

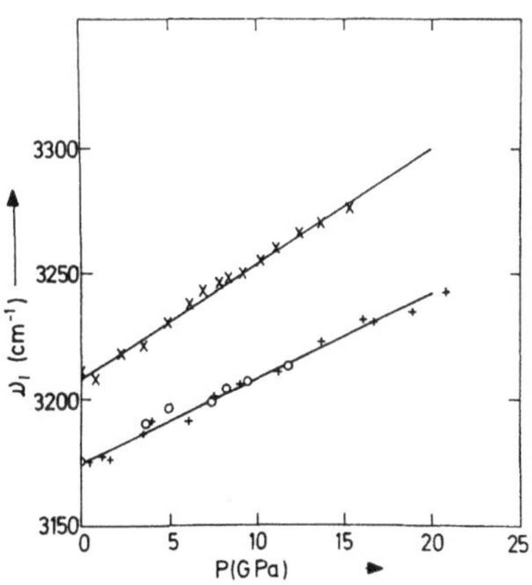

Figure 2. Pressure dependence of the symmetric stretching frequency ν_1 of NH_4ZnF_3 (x) and NH_4MnF_3 (o) measured with λ_1= 488 nm, and (+) with λ_2=476.5 nm

TABLE 2. Frequencies (in cm^{-1}) at room temperature of the internal modes of the ammonium and derivatives of the stretching modes with respect to pressure (in GPa) for ammonium halides.

	ν_1	ν_2	ν_3	ν_4	$d\nu_1/dP$	$d\nu_3/dP$
NH_4F /3/	2870	1729	2810	1849		+6.2 /10/
NH_4Cl /3/	3048		3086			
/11/	3041	1710	3138	1400	-3 /12/	0 /10/
NH_4Br /3/	3038		3124		very	
/11/	3040	1689	3135	1400	small /13/	-1.1 /10/
NH_4I /3	3021		3070		9 /14/	-7.86 /10/

3. DISCUSSION

Usually the dependence of ν_1 on the N-F distance is not the same when that distance is changed by pressure, or when it is modified by ionic substitution. In the case of ammonium halides (Table 2), whereas ν_1 and ν_3 increase on going from NH_4F to NH_4I (although it decreases slightly from NH_4Cl to NH_4I), $d\nu_1/dP$ and $d\nu_3/dP$ are clearly positive in NH_4F, clearly negative in NH_4I, and very small in NH_4Cl and NH_4Br. Such different behaviours might be explained as the result of competition between the HB, which weakens from F to I, and repulsive forces, F^- being the hardest ion and I^- the softest one.

In the NH_4MF_3 family one has N-H--F bonds of varying length with M substitution. One would expect that $d\nu_1/da$ (a= cell parameter), measured under variable applied pressure, to be the same as obtained by ionic substitution, because the HB and repulsive forces are short range and rather independent on the M^{++} ion. Using the data of compressibility at atmospheric pressure of NH_4MnF_3 and NH_4ZnF_3, calculated from phonon dispersion relations /4/ we can see that both estimations (See Table 3) agree fairly well. The discrepancies may be caused by the approximate values of compressibility, the non-linearity of it at high pressures, or the small different effect of Zn^{++} and Mn^{++}.

The positive value of $d\nu_1/dp$ is quite rare in usual HB systems, and it is explained as the effect of repulsive forces in systems where the N-F distance is very small /10/, but that effect is very unusual when the N-F distance is as large as 0.3 nm (like in NH_4BF_4, $(NH_4)_2SiF_6$ or NH_4PF_6). Such a kind of behaviour would mean that there are strong repulsive forces in the NH--F HB system. An empirical model for the H--F interaction potential could be used to fit the experimental frequencies but yielded unrealistic values for the activation energy. So far, an adequate model taking account of all the details has not been found.

In short, the difficulty lies in that, on one side the influence of a repulsive term is shown by the positive slope of ν_1 versus pressure, although the net force

TABLE 3. Comparison of $d\nu_1/dP$ (P in GPa) from pressure experiments with those obtained using compressibility and ν_1 (in cm^{-1}) for several compounds at atmospheric pressure. B is the bulk modulus (in GPa) and a (in nm) is the cubic unit cell size.

	a	$\Delta\nu_1/\Delta a$	B	da/dP	$d\nu_1/dP$ (cal)	$d\nu_1/dP$(ex)
NH_4MnF_3	0.4242	-2330 (a)	79.2	-1.785×10^{-3}	4.16	4.59
NH_4ZnF_3	0.4116	-4130 (b)	102.2	-1.342×10^{-3}	5.54	3.41

would be attractive, as evidenced by stretching frequencies lower than the estimated for NH_4 free ion ones. On the other hand, the fact that the barrier height diminishes with decreasing N-H--F distance indicates the presence of an important compensating repulsive force. The fact that the NH_4 is not a regular tetrahedron even at high temperatures, as evidenced from Raman spectroscopy, may hint to the shortcoming of the so far applied models of rigid rotor. Indeed, the bending of the NH_4 arms implies a positive energy which would reduce the effective barrier to rotations, in which two or several bonds are destroyed and rebound. Further work is needed to explain this interesting hypothesis.

The financial support of the Spanish Comisión Asesora de Investigación Científica y Técnica (proj. no. 1423/82), and Spanish-German Integrated Action no. 7,B are acknowledged.

REFERENCES

1. R. Navarro, E. Palacios, J. Bartolomé, R. Burriel, D. González: This workshop.
2. The electrostatic model is developed in A. Hüller, J.W. Kane: J. Chem. Phys. 61, 3599 (1974).
   Discussions of the model:
   H.D. Hochheimer, M.L. Shand, C.T. Walker, A. Hüller. J. Chem. Phys. 71, 5008 (1979), and
   R.W. Gerling, A. Hüller: J. Chem. Phys. 78, 446 (1983).
3. Plumb, Hornig. J. Chem. Phys. 23, 947 (1955).
4. J. Bartolomé, F. Palacio, J.M. Calleja, F. Agulló Rueda, M. Cardona, R. Migoni: J. Phys. C 18, 6083 (1985).
5. X ray on powdered samples.
6. J.T.P. Dunsmuir, A.P. Lane: Spectroch. Acta 28A, 45 (1972).
7. R.B. Helmholdt, G.A. Wiegers, J. Bartolomé. J. Phys C 13, 5081 (1980).
8. K.C. Patil, E.A. Secco: Canad. J. Chem. 29,1529 (1971).
9. Y. Yamaguchi, H. Schaefer. J. Chem. Phys. 73, 2310 (1980). Theoretical derivation of free NH_4^+ frequencies.
10. S.D. Haman. Aust. J. Chem. 31, 11 (1978).
11. N.E. Schumaker, C.W. Garland. J. Chem. Phys. 53, 392 (1970).
12. A.M. Heyns. J. Phys. Chem. Solids 41, 769 (1980).
13. Y. Ebisuzaki, J. Chem. Phys. 63, 4947 (1975).
14. A.M. Heyns, K.R. Hirsch, W.B. Holzapfel. J. Chem. Phys. 73, 105 (1980).

Part II

Rotational Tunneling II – Coupled Systems

Rotational Tunneling of Coupled Groups Studied by Inelastic Neutron Scattering

A. Heidemann

Institut Laue-Langevin, 156X, F-38042 Grenoble Cedex, France

1. INTRODUCTION

Rotational tunneling of small molecules, molecular groups and ions in solids at low temperatures has been studied extensively in recent years by inelastic neutron scattering [1]. Most of the experimental results could be explained on the basis of a theory emphasizing the extreme single particle character of the tunneling motion [2]. Coupling between different molecular groups, leading to a collective rotational motion, can be expected to be important in cases where groups are located close together. The potential governing the motion of a group consists of terms due to this coupling and terms due to the essentially rigid remainder of the crystal lattice. If the former are small compared with the latter, the spectral structure resulting from coupling is difficult to resolve. However, cooperative phenomena can be observable for systems with low single particle potential barriers and short distances between groups.

More than 27 years ago coupling effects between molecular groups have already been observed by microwave spectroscopy of gaseous systems and treated theoretically. Swalen and Costain discussed the microwave spectrum of acetone in the gas phase [3]. Acetone has two CH_3 groups per molecule coupled via an intramolecular potential hindering the rotational motion. Both groups tunnel around axes which are inclined to each other. Swalen and Costain expanded the rotational potential into a Fourier series :

$$V(\phi_1, \phi_2) = \frac{V_1}{2}(1 - \cos 3\phi_1) + \frac{V_2}{2}(1 - \cos 3\phi_2) + V_{12} \cos 3\phi_1 \cos 3\phi_2 + V'_{12} \sin 3\phi_1 \sin 3\phi_2 + \text{sixfold and higher order terms.} \tag{1}$$

ϕ_1 and ϕ_2 are the rotational coordinates of the two CH_3 groups, V_1 and V_2 the single particle potentials, V_{12} and V'_{12} the strengths of the interaction potentials. A very similar form to eq.(1) was used 25 years later by Clough et al [4] to describe the hindering potential of CH_3 groups in solid lithium acetate at low temperatures.

We will present in this paper a review of coupling phenomena in rotational tunneling in solids at low temperatures observed by inelastic neutron scattering (INS). Theoretical aspects will be discussed in section 2. In section 3 we will describe experimental results.

2. THEORETICAL ASPECTS

2.1 Introduction

In contrast to gases where the molecules rotate freely between collisions, they are subject to an angle-dependent potential in the solid. It can be written down in the following way [1] :

$$W^i(\omega_i \{\omega_j, \vec{R}_j\}) = \sum_{j=1}^{N+M}{}' V_1^{ij}(\omega_i, \vec{R}_i, \vec{R}_j) + \sum_{j=1}^{M}{}' V_2^{ij}(\omega_i, \omega_j, \vec{R}_i, \vec{R}_j) ; \qquad (2)$$

M and N denote the number of molecules and monoatomic units in the crystal, respectively. The prime means that the term i = j should be omitted. The first term stands for the single-particle potential, the second describes the interaction which depends on the orientations ω_i and ω_j of two groups. Quantum mechanically W^i needs to be integrated over the states of all the neighbours of the i-th molecule. But usually approximations are required to solve the problem, for example the molecular field theory. In the following sections we will present some models which are able to describe simple systems.

2.2 Pair of coupled CH_3 groups

We consider two methyl groups which interact strongly but are essentially isolated from other pairs in the crystal. Clough et al.[4] and later Häusler and Hüller [5] have calculated the energy levels of the corresponding Hamiltonian

$$H = -B\left[\frac{\partial^2}{\partial\phi_1^2} + \frac{\partial^2}{\partial\phi_2^2}\right] - V_3(\cos 3\phi_1 + \cos 3\phi_2) - W_3 \cos 3(\phi_1 - \phi_2). \qquad (3)$$

The two CH_3 groups are supposed to rotate around a common axis. B is the rotational constant of one group ($\approx$ 0.6 meV). There are different possibilities to solve the Schrödinger equation with this Hamiltonian. One is to use for the wavefunction ψ a series expansion into products of free rotor states

$$\psi = \frac{1}{2\pi} \sum_{m \neq n} a_m a_n e^{im\phi_1} e^{in\phi_2} \qquad (4)$$

and to diagonalize numerically the Hamiltonian matrix [4]. This method works well for low barrier systems. Häusler et al.[5] developed a method which allows to treat also high barrier systems.

The symmetry species involved are representations of the direct product group $^{(1)}C_3 \times {}^{(2)}C_3$. The spin functions associated with the rotational functions can be written in terms of 2^6 functions like $|\alpha\alpha\alpha\rangle|\alpha\alpha\alpha\rangle$ [6]. The ground state multiplet contains four levels (see Fig. 1).

Only four transitions can be excited by INS as indicated in Fig. 1. Häusler et al. [5] have calculated the frequencies of these transitions for various values of the parameter pair (V_3, W_3). One important result is that in contrast to previous ideas

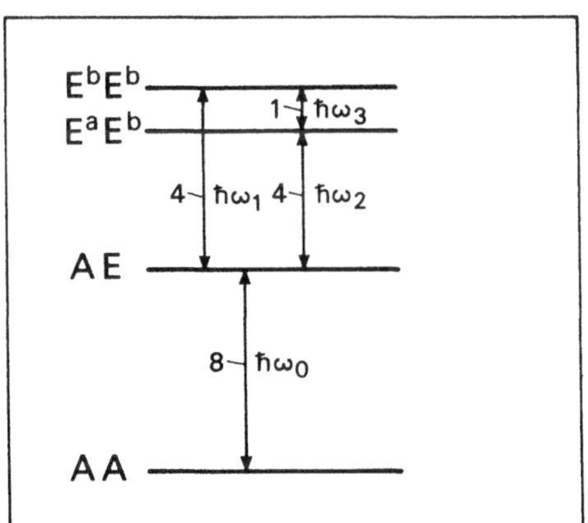

Fig. 1 : The energy levels of two coupled methyl groups in the librational groundstate. The numbers next to the arrows indicate the weights of the corresponding neutron-induced transitions. Using the pocket state formalism we get [7]

$\hbar\omega_0 = -t - g - k$

$\hbar\omega_1 = -t - k$

$\hbar\omega_2 = -t + g$

[2] even in the case of rather strong single-particle potentials coupling effects should be visible. The old argument was that in a deep potential the modulus of the A and E species wave functions is almost the same, therefore neighbouring molecules will not notice the difference and the energy levels should not be influenced. The flaw in this argument is connected with the extreme sensitivity of the tunnel frequency to changes in the potential. Experimentally, however, it is certainly more difficult to observe coupling effects in high barrier cases because of the limit in the energy resolution of neutron spectrometers.

The Hamiltonian (3) is the simplest one can imagine to describe two coupled CH_3 groups. In principle there is no difficulty to include higher order terms in the Fourier series and to generalize it such that it is valid for CH_3 groups with inclined axes. Life becomes very difficult if one uses the same technique to describe coupling between more than two groups.

A complementary approach to solve the problem of a pair of coupled CH_3 groups is to start from pocket states. We will briefly outline the idea and present the essential results [7].

A pair of CH_3 groups can be described by a set of nine pocket states :

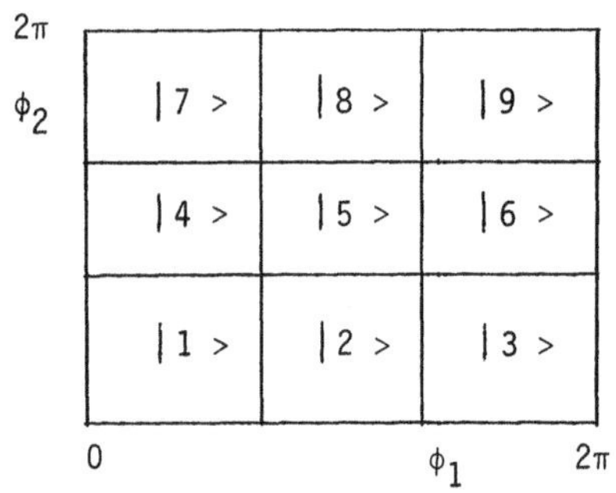

They form a complete set of normalized, non-orthogonal basis vectors of a 9-dimensional Hilbert space in which we can diagonalize the Hamiltonian matrix :

	|1>	|2>	|3>	|4>	|5>	|6>	|7>	|8>	|9>
<1|	D	t	t	t	g	k	t	k	g
<2|	t	D	t	k	t	g	g	t	k
<3|	t	t	D	g	k	t	k	g	t
<4|	t	k	g	D	t	t	t	g	k
<5|	g	t	k	t	D	t	k	t	g
<6|	k	g	t	t	t	D	g	k	t
<7|	t	g	k	t	k	g	D	t	t
<8|	k	t	g	g	t	k	t	D	t
<9|	g	k	t	k	g	t	t	t	D

(5)

The matrix elements t correspond to the rotation of one group only, g and k to simultaneous rotations of <u>two</u> groups in the same or opposite direction respectively. After diagonalization of (4) we obtain splittings of the groundstate multiplet as indicated in Fig. 1. For the limit g = k = 0 (no coupling) we find three equally spaced levels, as it should be.

As a last remark we would like to mention that one can replace the Hamiltonian (3) by an effective spin Hamiltonian [8] :

$$H_{eff} = \omega_t (\sigma_1^2 + \sigma_2^2) + W \vec{\sigma}_1 \cdot \vec{\sigma}_2 \quad ; \qquad (6)$$

ω_t is the tunnel splitting of a single CH_3 group, σ_i are spin matrices for spin 1 particles. Therefore the eigenvalues of σ_z are 0 or ± 1 and correspond to one CH_3 group being in A or E state respectively. W is the coupling strength. In principle this formalism can be extended to cases with interactions between more than two groups [8] :

$$H_{eff} = \omega_t \sum_i \sigma_i^2 + \sum_{ij} W_{ij} \vec{\sigma}_i \cdot \vec{\sigma}_j \qquad (7)$$

2.3 Coupled three-dimensional rotors (CH_4)

Solid methane in phase II can be considered as the model system of N (>> 2) coupled three-dimensional rotors. Yamamoto et al [9] studied solid CH_4 theoretically using a quantum mechanical mean-field approximation based upon the extended James-Keenan (EJK) model with octopole-octopole interactions between nearest neighbour molecules. One of the reasons why a mean-field approximation can be applied to solid CH_4 is that at finite temperature CH_4 is not a pure material but a mixture of A, E and T spin symmetry species. Since each species has a different rotational level structure the translational symmetry of the crystal is lost and each molecule sees a spin-averaged rotational potential. The Hamiltonian of the EJK model is given by [9] :

$$H = B \sum_i (J_i^2 + V_c(\omega_i)) + \frac{1}{2} \sum_{i \neq j} W_{ij}(\omega_i, \omega_j) \qquad (8)$$

with

$$V_c(\omega) = \beta_4 V_4(\omega) + \beta_6 V_6(\omega), \qquad (9)$$

$$W_{ij}(\omega_i, \omega_j) = \sum_{\tau,\nu} C_{\tau\nu}^{ij} U_\tau(\omega_i) U_\nu(\omega_j), \qquad (10)$$

J_i is the angular momentum operator, V_4 and V_6 are expressed in terms of Wigner functions. $U_\tau(\omega)$ are tetrahedral rotator functions [9]. Kobashi et al [10] calculated with this Hamiltonian the collective librational modes of pure A methane. Yamamoto et al [9] predicted on the basis of this Hamiltonian a shift of the tunnel frequency with temperature. The Hamiltonian was, however, not used to calculate rigorously the tunneling groundstate multiplet because of the considerable difficulties involved.

A very simple phenomenological model was invented [11] to describe the low temperature dependence of the tunnel spectrum of CH_4. It is based on the mechanism of inhomogeneous broadening due to the mixture of spin symmetry species. The assumptions are

1) There are nearest neighbour interactions only,
2) Deviations from cubic symmetry are ignored,
3) No distinction is made between T and E molecules,
4) The hindering potential at a given site is produced by a linear superposition of A and E molecules which are distributed binominally.

The results are i) a linear dependence of the tunneling splitting on the concentration c_A of the A molecules and a $\sqrt{c_A(1 - c_A)}$ dependence of the intrinsic linewidth.

2.4 Neutron scattering cross section

The INS cross section of uncoupled three-dimensional rotors was calculated by Hüller [2] for high barrier systems on the basis of the pocket state formalism. Calculations for low barrier systems using series expansions into free rotor states followed later [12, 13, 14]. No detailed calculations are so far available for cases with coupling. However, the structure factors can be expected to be very similar to those of uncoupled groups. One important selection rule will always hold : the total spin change ΔI of the system at one neutron scattering event cannot be greater than one. Therefore it is not possible to excite with neutrons a transition from AA to EE of a pair of CH_3 groups.

3. EXPERIMENTAL RESULTS

3.1 Introduction

In this section we will present results of INS experiments in the field of tunnelling of coupled groups. Most of the experiments have been carried out at the high flux reactor of the ILL in Grenoble with essentially three spectrometers [15] :

1. The multichopper time-of-flight spectrometer IN5 for cold neutrons. For 10 Å neutrons the energy resolution is about 15 μeV.
2. The backscattering spectrometer IN13 for thermal neutrons with an energy resolution and range of 8 μeV and 300 μeV, respectively. High momentum transfers up to 5.5 $Å^{-1}$ can be achieved.
3. The backscattering spectrometer IN10 for cold neutrons with an energy resolution of 0.3 to 1.0 μeV (FWHM) and an energy range from - 15 to 40 μeV.

First, we will present results of low barrier systems (CH_4, lithium and manganese acetate). Then we will mention very recent experiments on high barrier systems.

3.2 Methane

CH_4 in Phase II has an 8-sublattice structure consisting of 6 ordered sites (symmetry D_{2d}) and two disordered sites (symmetry O_h)[16]. The ordered molecules represent a model system of many coupled three-dimensional quantum rotors. INS studies with high energy resolution performed at low temperatures between 0.05 and 5 K revealed coupling effects in the temperature dependence of the spectral shape [11] (see Fig. 2). Two important observations were made :

1. The tunnel frequencies shift to higher values with decreasing temperatures in agreement with the prediction of Yamamoto et al. [9].
2. The intrinsic line width of the peak corresponding to the A → T transition decreases with decreasing temperature and reaches a finite value if extrapolated to T = 0K.

The frequency and intrinsic line width of the A → T transition are plotted in Figs. 3 and 4 as a function of C_A and $\sqrt{C_A(1 - C_A)}$ respectively. The straight line in Fig. 3 is the result of a fit with the model of inhomogeneous broadening. The straight line in Fig. 4 was obtained from the same model with the parameters used in Fig. 3. The experimental values of the line width are always higher than the calculated ones, indicating that another additional line-broadening mechanism must be present. About its nature one can so far only speculate : oxygen impurities, T-state splitting, surface effects, non-crystallinity ? Or is it related to so far not understood collective phenomena ?

3.3 Lithium acetate (LIAC)

Lithium acetate ($LiCOOCH_3 \cdot 2H_2O$) is the most thoroughly studied molecular crystal showing cooperative effects in the tunneling spectrum. The crystal structure is

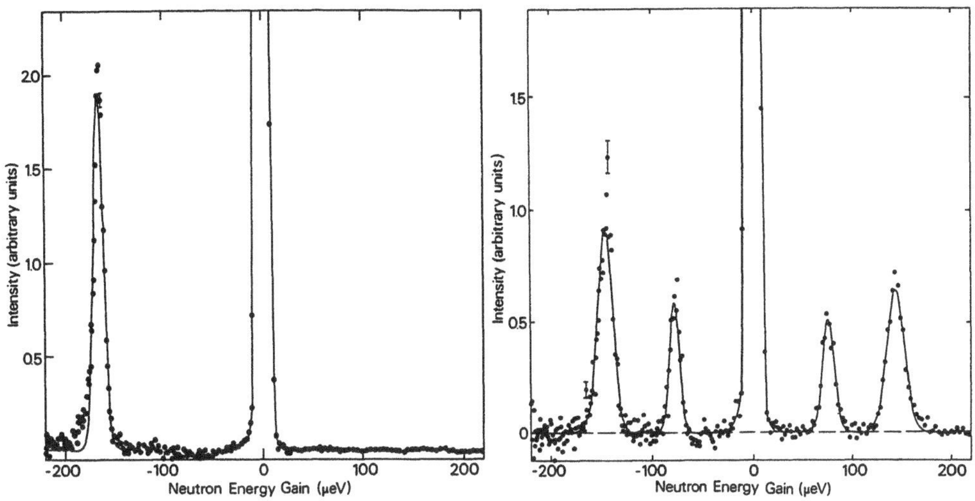

Fig. 2 : INS spectra of O_2 doped CH_4.
(a) Spin temperature $T_s < 0.2$ K (b) $T_s = 5$ K [12]

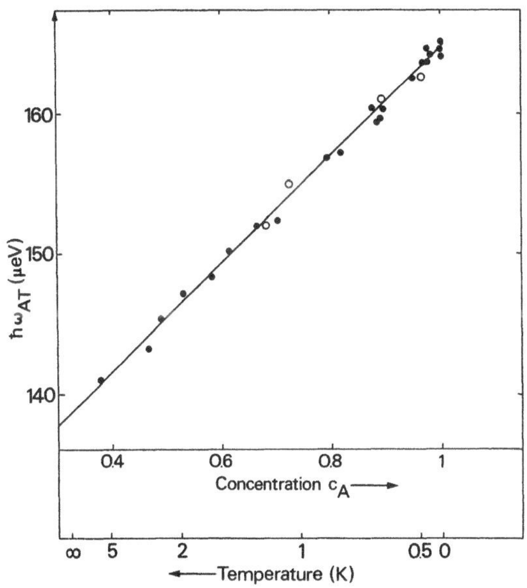

Fig. 3 : Frequency of the A-T peak as a function of the concentration c_A of the A spin species. The straight line represents a fit with the model of inhomogeneous broadening [12].

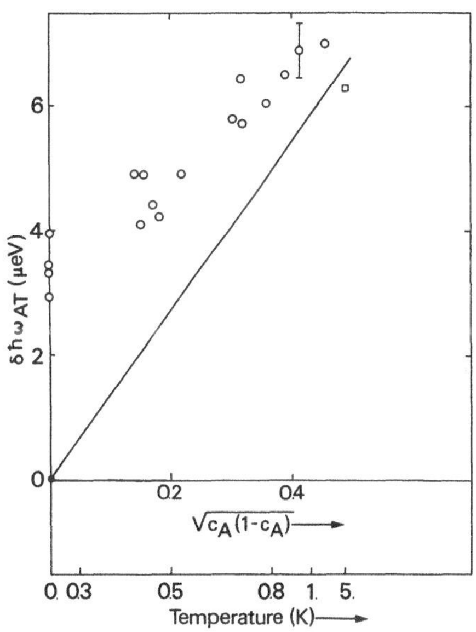

Fig. 4 : Linewidth of the A-T peak as a function of $[c_A(1-c_A)]^{1/2}$. The straight line is calculated with the model of inhomogeneous broadening with the parameters used in Fig. 3 [12].

orthorhombic with space group Cmmm [17]. A projection on the bc plane is shown in Fig. 5. The CH_3 groups occur in pairs (distance 2.5 Å) with a common axis of rotation. The distance between nearest pair axes is 6.6 Å. A typical INS spectrum of LIAC is shown in Fig. 6 [4]. Good agreement between the theory (see section 2.2) and experiment is found if one uses the parameter pair

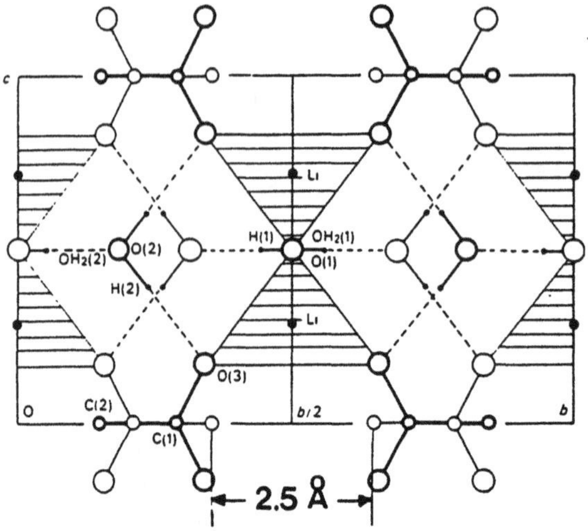

Fig. 5 : Projection of the structure of LIAC on the bc plane [17].

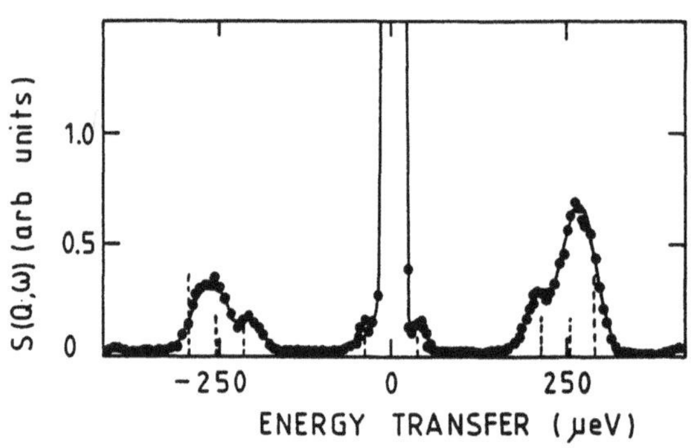

Fig. 6 : INS spectrum of LIAC at a temperature of 2.6 K. The dashed lines indicate the result of the model calculation with the Hamiltonian (3) [4].

$$V_3 = 2.96\ B, \qquad W_3 = -\ 6.58\ B.$$

This corresponds to two CH_3 groups in a staggered orientation. In LIAC the interaction term W_3 dominates the hindering potential.

Recent pressure-dependent measurements [18] on LIAC showed that the average tunnel frequency ω_0 and the average splitting $\Delta\omega = \frac{1}{2}(\omega_3 + \omega_0 - \omega_1)$ (see Fig. 1) shift to lower values with increasing pressure. A comparison of the pressure-dependent data with the model of section 2.2 leads to the conclusion that V_3 increases more rapidly with pressure than the absolute value of W_3. This clearly indicates that the pressure dependence of the crystal structure of LIAC cannot be described by assuming equal and linear scaling of all atom distances as a function of pressure.

The lowest lying librational peak of LIAC is shifted in agreement with the model calculation to higher frequencies with increasing pressure [18]. However, the absolute values of the frequency of the first librational peak differ by a factor 1.5 between theory and experiment. Apparently, it is not possible to describe with the Hamiltonian (3) simultaneously the tunneling and librational spectra. In principle, improvements are possible by adding higher order Fourier components to the hindering potential.

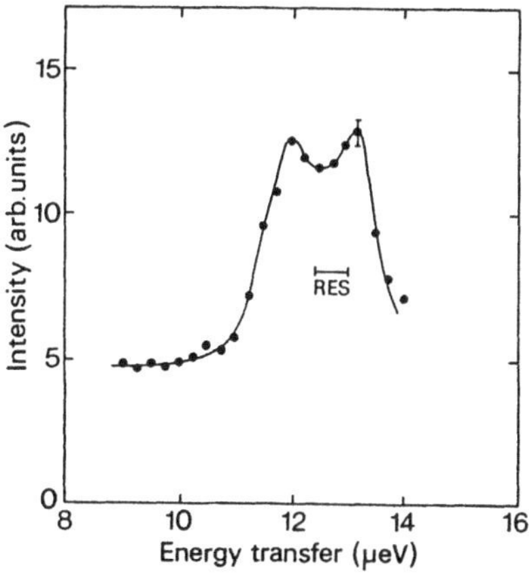

Fig. 7 : INS spectrum of fully deuterated lithium acetate [19].

The isotope effect can be used with advantage to determine the shape of the rotational potential. We have plotted in Fig. 7 the tunneling spectrum of fully deuterated LIAC [19]. A double peak structure is observed with lines at 12 and 13 μeV and with an intrinsic line width of about 0.5 to 1 μeV. The frequencies are by about a factor 2 smaller than the values calculated by scaling the ones measured for protonated methyl groups in LIAC. This again might tell us, that our model Hamiltonian (3) is not good enough. Another question is whether it is legitimate to scale the Hamiltonian to determine the isotope effect in a case where coupling is producing the bulk part of the hindering potential.

As we have seen in the case of CH_4, information about coupling effects can also be obtained from the temperature dependence of the INS spectra. Recently, some interesting results were obtained with LIAC [20]. The shape of the main tunneling peak at 250 μeV changes with temperature, i.e. the outer line grows at the expense of the inner ones with decreasing temperature, confirming nicely the level scheme produced by Hamiltonian (3). The main tunneling peak shifts to higher frequencies with decreasing temperature. This can only be explained by the assumption that also interactions with more distant CH_3 groups have to be taken into account. Friedrich et al [21] measured the spin conversion time τ_c of LIAC via a neutron transmission experiment and found that τ_c varies with temperature like T^{-7}. Therefore it takes more than 100 hours to reach spin temperatures less than 2.5 K. Several unsuccessful attempts were made recently to speed up the conversion rate in LIAC. Also a level crossing experiment with γ-irradiated LIAC in an external magnetic field of such a size that the Zeeman splitting of the free radicals matched the tunnel splitting did not show any influence on the INS spectrum. The reason for these negative results is probably that the spin diffusion process is too slow in LIAC [22].

3.4 Manganese acetate (MNAC)

Manganese acetate ($Mn(COOCH_3)_2.4H_2O$) contains magnetic Mn^{2+}ions. Therefore spin conversion of the CH_3 groups is rapid. The structure of MNAC is monoclinic with six molecules per unit cell. The CH_3 groups sit on three inequivalent sites and occur in pairs forming a cluster of six particles [23]. INS measurements yielded the following results [24] : three tunneling frequencies (1.2, 50, 137 μeV at 5 K) are observed and assigned to the three inequivalent sites. Only the temperature dependence of the high-frequency tunneling peak carries the signature of coupling. The lineshape and average line frequency are strongly temperature dependent for $0.05\ K < T < 10\ K$, very similar to the results in LIAC and CH_4. A qualitative description of the results is again possible with the model Hamiltonian (3). Quantitative differences, however, show that coupling between more than two groups has to be considered. One - maybe - very important observation is that - like for CH_4 -

the intrinsic linewidth if extrapolated to T = 0K remains finite. Again no satisfactory explanation could be found.

As a final remark we would like to mention that the tunneling spectrum of MNAC is extremely sensitive to hydrostatic pressure [25] : the outer peak is shifted from 137 μeV at zero pressure to about 70 μeV at only 1.2 kbar !

3.5 One-dimensional high barrier systems

As outlined in section 2 the observation of cooperative effects in rotational tunneling becomes more difficult with increasing barrier height to rotation. One criterion for coupling is that the tunneling spectrum has to be sensitive to temperature changes in the region $\hbar\omega_t \approx k_BT$, if conversion takes place. Unfortunately, most of the molecular crystals with CH_3 groups have very long conversion times. One example is nitromethane, the tunneling peak of which has a finite linewidth [26]. It is, however, not known whether it is due to coupling. Cobalt acetate, which converts fast because of its magnetic cobalt ions, has a tunnel peak at 26 μeV with an intrinsic width of about 2 μeV. It could be shown recently [27] that the peak shape is independent of temperature down to 100 mK. Therefore coupling effects can be excluded.

A clear-cut case for coupled tunneling motion in a high barrier system is that of CD_3 groups in LIAC (see section 3.3).

As a last example we will present INS results on dimethyl-tin-dichloride (DMTC) [28]. This compound has a crystal structure with layers of tin and chlorine atoms and of methyl groups. The axes of the latter are inclined to this plane. The layer of the CH_3 groups consists of pairs arranged side by side on inclined rotation axes and separated by d_{CC} = 3.24 Å. The closest distance between CH_3 groups of different pairs is 4.61 Å.

The INS experiments revealed a tunneling peak at 49 μeV with an intrinsic width of 2.8 μeV (assuming a Gaussian line shape) and a weak shoulder at about 0.7 μeV on the wing of the elastic peak. In the meV range a dominant peak occurs at 5.5 meV and some more features above 10 meV. The model Hamiltonian (3) could be used again to describe well <u>all</u> the observations. The weak shoulder at 0.7 μeV is explained by the E_aE_b to E_aE_a transition. The following parameter pair was obtained :

$$V_3 = 27.0\ \mathrm{B}, \qquad W_3 = 16.8\ \mathrm{B}.$$

As both model parameters are positive, the case of frustrated coupling is realised where the interaction potential counteracts the single particle potential. The Hamiltonian (3) is strictly speaking only valid for pairs with a common axis. Therefore it is a bit surprising that the agreement between theory and experiment is so good.

Until now we have discussed exclusively cases of <u>inter</u>molecular coupling. The early results with infrared spectroscopy were obtained in gases with rotations and <u>intra</u>molecular coupling effects [3]. One case where intramolecular coupling might be responsible for the shape of the INS tunneling spectra is that of metal hexamine halides [29].

4. CONCLUSIONS

A review has been given on the theoretical description and the experimental observations (by INS) of coupling effects in the field of rotational tunneling in solids. Microscopic theories are available only for isolated pairs of one-dimensional rotors. Coupling of $N > 2$ groups can so far only be described by a mean field theory which takes in addition local field fluctuations into account. Experimentally only cases with intermolecular interactions have been studied. Cases where intramolecular interactions play an important role do certainly exist but have not yet been well identified. Non zero intrinsic linewidths in the very low temperature spectra of CH_4 and MNAC are so far not understood. Is there a link to long-range collective excitations ? These and related questions might be part of the future activity in the field of rotational tunneling in solids at low temperatures.

ACKNOWLEDGEMENT

The author wishes to thank S. Clough, A. Hüller, M. Prager and W. Press for numerous discussions.

REFERENCES

1. W. Press, Springer Tracts in Modern Physics, 92, Berlin, Heidelberg, New York (1981)
2. A. Hüller, Phys. Rev. B 16, 1844 (1977)
3. J.D. Swalen, C.C. Costain, J. Chem. Phys. 31, 1562 (1959)
4. S. Clough, A. Heidemann, A.H. Horsewill, M.N.J. Paley, Z. Phys. B 55, 1 (1984)
5. W. Häusler, A. Hüller, Z. Phys. B 59, 177 (1985)
6. P.S. Allen, Faraday Symp. Chem. Soc. 13, 133 (1978)
7. W. Häusler, Thesis (1985) Erlangen
8. A. Hüller, A. Würger, to be published
9. T. Yamamoto, Y. Kataoka, K. Okada, J. Chem. Phys. 66, 2701 (1977)
10. K. Kobashi, R.D. Etters, T. Yamamoto, J. Phys. C 17, 13 (1984)
11. A. Heidemann, K.J. Lushington, J.A. Morrison, K. Neumaier, W. Press, J. Chem. Phys. 81, 5799 (1984)
12. Y. Ozaki, Y. Kataoka, T. Yamamoto, J. Chem. Phys. 73, 3442 (1980)
13. A. Hüller, W. Press, Phys. Rev. B 24, 17 (1981)
14. A. Heidemann, I. Anderson, B. Jeffryes, B. Alefeld, Z. Phys. B 49, 123 (1982)
15. B. Maier, Neutron Beam Facilities at the HFR Grenoble (1986)
16. W. Press, J. Chem. Phys. 56, 2597 (1972)
17. J.L. Galigné, M. Mouvet, J. Falgueirettes, Acta Cryst. B 26, 368 (1970)
18. A. Heidemann, J. Eckert, L. Passell, W. Häusler, to be published in Z. Phys. B
19. A. Heidemann, H. Friedrich, to be published
20. A. Heidemann, K. Abel, C.J. Barker, S. Clough, to be published
21. H. Friedrich, unpublished
22. A. Heidemann, H. Friedrich, R. Scherm, ILL Experimental Report 9-03-418 (1986)
23. E.F. Bertaud, D. Tranqui, P. Burlet, J.M. Moreau, Acta Cryst. B 30, 2234 (1974)
24. A. Heidemann, S. Clough, P.J. McDonald, A.J. Horsewill, K. Neumaier, Z. Phys. B 58, 141 (1985)
25. A. Heidemann, C. Vettier, unpublished
26. D. Cavagnat, A. Magerl, private communication
27. A. Heidemann, H. Friedrich, ILL experimental report 9-03-456 (1986)
28. M. Prager, A. Heidemann, W. Häusler, Z. Phys. B 64, 447 (1986)
29. H. Blank, J.K. Cockcroft, G.J. Kearley, these proceedings.

Study of the Dipolar Proton Spin System in Lithium Acetate with NMR Dispersion Spectroscopy

E. Crits and L. Van Gerven

Center for Nuclear Magnetism and Radiospectroscopy,
Laboratorium voor Vaste Stof-Fysica en Magnetisme,
Katholieke Universiteit Leuven, B-3030 Leuven, Belgium

1. Introduction

Lithium acetate contains almost freely rotating methyl groups [1]. These CH_3-rotors lie in coaxial pairs, with their planes of rotation separated by 2.5 Å along the rotation axis, a not so large distance compared to the intra-methyl proton distance of 1.8 Å. Hence intermethyl dipolar interactions between the proton spins of two adjacent methyl groups become important. They enhance the Zeeman- and dipolar spin-lattice relaxation rates [1]. This effect of the intermethyl interaction is most pronounced in the angular dependence of T_1, T_{1D} and in the NMR spectrum of lithium acetate single crystals [2,3,4,5].

As in γ-picoline, dipolar nuclear spin polarization can be created after a rapid change in temperature [2,6]. This nuclear polarization effect is also influenced by the intermethyl interactions since the transitions between the E_a and E_b tunnel levels are destroying the dipolar polarization. Unlike in γ-picoline, the dipolar polarization effect in lithium acetate has not received much attention until now.

Because this Haupt effect, as it is often called, strongly alters the NMR dispersion line shape χ' (H_s), it is possible to study its evolution as a function of time, angle (in single crystals) and temperature. For this purpose we have developed a special type of NMR dispersion spectrometer (NMRDS) which provides very fast and quasi-absolute line shape determinations [7,8]. It is based on the absolute relation - without any unknown parameter, except η - between χ' and the frequency shift $\Delta\omega_o$ of any simple, but very stable HF oscillator:

$$\chi' = -\frac{2}{\eta}\frac{\Delta\omega_o}{\omega_o} \qquad (1)$$

where χ' = the real part of the complex HF susceptibility of a sample introduced into the coil of the LC-resonator of the oscillator
ω_o = oscillator frequency
η = filling factor of the coil.

Because of its absolute character, it is easy to compare line shapes measured under different conditions. So, by studying the evolution of the NMR dispersion spectrum in this rather original way, we can get information about the dipolar polarization behaviour as a function of time, angle and temperature.

Together with the result obtained from pulse experiments like T_1 and T_{1D} measurements, this will provide us with some valuable insight into the role of the intra- and intermethyl interactions in lithium acetate.

2. Experimental

The experimental set-up is described by CRITS et al. [8]. The samples we have been using are powder crystals of $CH_3COOLi.2H_2O$ and single crystals of

$CH_3COOLi.2H_2O$ and $CH_3COOLi.2D_2O$. After each temperature jump we follow the evolution of the dipolar polarization as function of time. For the single crystals we also measured the angular dependence of the evolution of the dipolar polarization.

In the presence of a strong dipolar polarization - as it is the case here - the NMR dispersion, as well as the NMR absorption line, consist of a dipolar and a Zeeman contribution, according to Provotorov's theory [9]. Using the fact that the dipolar term in the dispersion line is even and the Zeeman term odd, we can disentangle them easily and study the evolution of the dipolar line as a function of time. This evolution can then be studied as a function of angle and temperature. Important parameters of the dipolar resp. Zeeman dispersion lines are the peak-to-peak values χ'_{Dm} resp. χ'_{Zm} or the corresponding frequency shifts $(\Delta\omega_o/2\pi)_{Dm}$ resp. $(\Delta\omega_o/2\pi)_{Zm}$.

3. Results

In fig. 1 some dipolar dispersion lines are shown for different values of the angle between the crystal b-axis and the magnetic field. The magnitude and the sign of the dipolar dispersion depend on the direction of the temperature jump. The Haupt effect becomes very pronounced at temperatures lower than 40 K. We obtain a maximum polarization around a temperature of 15 K, as can be seen from fig. 2. At 4.2 K the dipolar dispersion becomes very small in comparison with the Zeeman dispersion so that it is very difficult to observe the dipolar contribution.

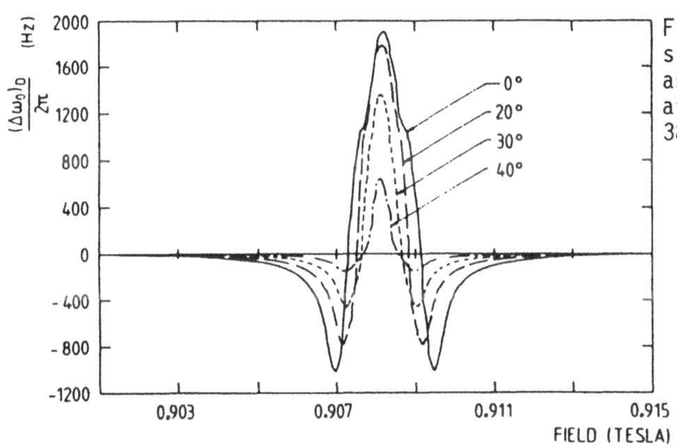

Fig. 1.: Dipolar lines of a single crystal of $CH_3COOLi.2D_2O$ as a function of angle. T = 10 K, after a jump from 6.6 K. $\omega_o/2\pi$ = 38.2 MHz

The temperature dependence of the Haupt effect can be explained as follows. The $E \rightarrow A$ conversion transitions caused by intramethyl interactions generate dipolar polarization. The strength of this polarization is determined by the population difference between the E and A levels, which is *higher* at *lower* temperatures. On the other hand the strength of the polarization is also determined by the spin conversion rate which is *lower* at *lower* temperatures.

Moreover, we have to consider the *depolarizing* transitions between the E_a and E_b levels due to the intermethyl interactions, which will become more effective at *lower* temperatures. This is confirmed by the fact that by lowering the temperature to 4.2 K the dipolar spin-lattice relaxation rate increases while the conversion rate decreases. The combination of these three effects produces a maximum in polarization strength around 15 K [10]. This also explains the very small dipolar dispersion at liquid helium temperatures.

In fig. 3 the dependence of the dipolar polarization on the angle between the crystal b-axis and the magnetic field is shown. This dependence has a $(1 - 3\cos^2\theta)^2$ behaviour and does not depend on the deuteration of the crystal

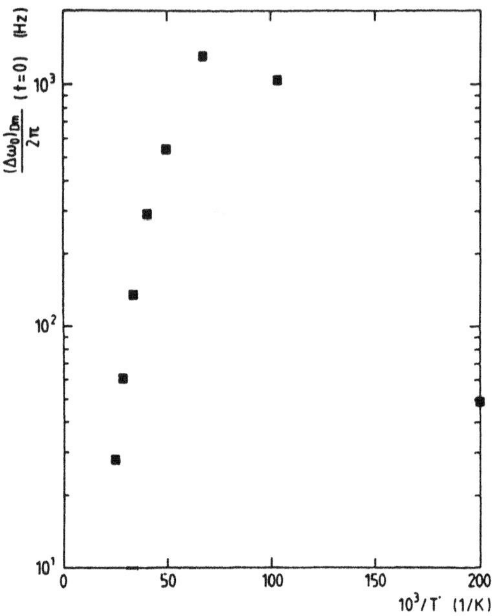

Fig. 2.: p-p value of the dipolar dispersion line of $CH_3COOLi.2H_2O$ powder as a function of temperature. The p-p value is measured at 5 min - which we take as t = 0 - after the temperature jump which is always 5 K. $\omega_o/2\pi$ = 38.2 MHz

water protons [3,6,11]. We have also measured the angular dependence of the decay time of the dipolar polarization, or conversion time τ_{con} at different temperatures. This is shown in fig. 4. The conversion time becomes minimum in the neighbourhood of the "magic angle". This is a very unexpected result because normally one would expect no angular dependence if one considers only the ground rotational levels

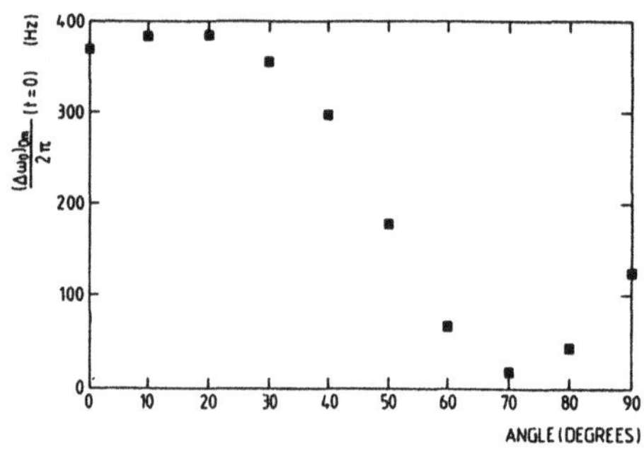

Fig. 3.: P-p value of the dipolar dispersion line of a single crystal of $CH_3COOLi.2D_2O$ as a function of angle taken 5 min - which we take as t = 0 - after a temperature jump from 10 K to 7 K. $\omega_o/2\pi$ = 38.2 MHz

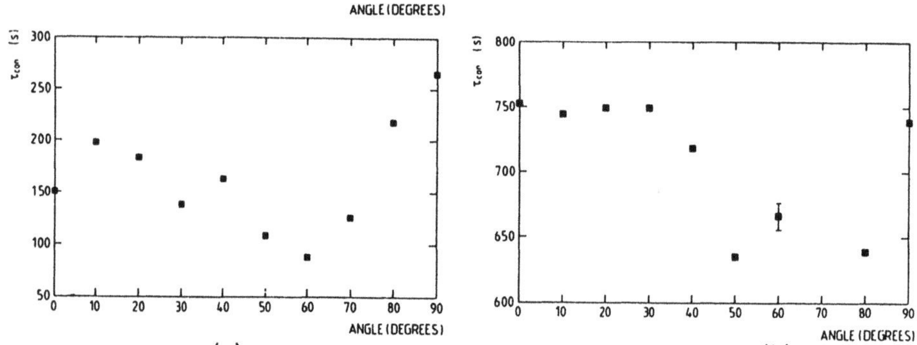

(a) (b)

Fig. 4.: Nuclear spin conversion time in a single crystal of $CH_3COOLi.2D_2O$ as a function of angle after a temperature jump (a) from 35 K to 30 K; (b) from 10 K to 7 K. $\omega_o/2\pi$ = 38.2 MHz

[3,12]. The relative difference between the minimum and maximum conversion time decreases with decreasing temperature. This indicates that one probably has to take into account the higher librational levels also. In any case, we do not have a consistent explanation for this angular dependence yet.

Finally, we also performed some NMRDS measurements on the Li^7 nuclear spins in lithium acetate. In fig. 5 spectra recorded after a temperature jump are shown for different angles. These spectra are the superposition of three dipolar dispersion lines which are shifted due to the quadrupolar interaction [13]. We observe the same angular dependence of the dipolar polarization and of the decay times as for the proton NMRDS, although the absolute values of the decay times appear to be somewhat larger. A possible explanation for the observed Li^7 spectra and their evolution might be that dipolar order created by and in the methyl proton system can be transferred to other spin systems. This is maybe the case as well for the protons of the crystal water, if present. Because the latter also contribute to the NMR proton dispersion signal, this could account for the small difference in the dipolar spectrum between the deuterated and non-deuterated samples.

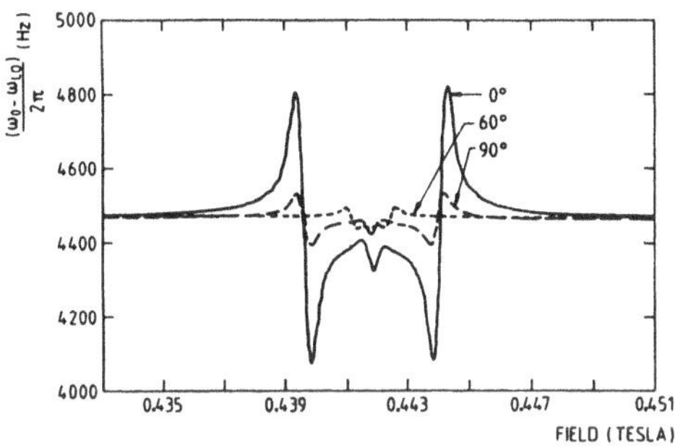

Fig. 5.: Recorded dispersion lines of Li^7 in a single crystal of $CH_3COOLi.2D_2O$ as a function of angle. T = 13 K, after a jump from 20 K.
$\omega_o/2\pi \cong 7.3$ MHz
$\omega_{LO}/2\pi$= local oscillator frequency

4. References

1. P.S. Allen and P. Branson, J. Phys. C 11, L121 (1978)
2. P.S. Allen, Faraday Symp. of Chem. Soc. 13, 133 (1978)
3. B. Van Riet, Doctor's thesis, Leuven (1983)
4. G. Ślósarek and N. Pislewski, Act. Phys. Pol. A65, 315 (1984)
5. G. Ślósarek and N. Pislewski, Act. Phys. Pol. A68, 717 (1985)
6. J. Haupt, Phys. Lett. 38A, 389 (1972)
7. B. Van Riet and L. Van Gerven, J. Phys. E 15, 558 (1982)
8. E. Crits and L. Van Gerven, "XXIInd Congress Ampère on Magnetic Resonance and Related Phenomena", Eds. K.A. Müller, R. Kind and J. Roos, Zürich, 590 (1984)
9. M. Goldman, Spin Temperature and Nuclear Magnetic Resonance in Solids (Oxford Press, London 1970)
10. P. Beckmann, S. Clough, J.W. Hennel and J.R. Hill, J. Phys. C 10, 729 (1977)
11. S. Clough, Phys. Lett. 42A, 371 (1973)
12. M. Punkkinen and S. Clough, J. Phys. C 7, 3403 (1974)
13. R.V. Pound, Phys. Rev. 79, 685 (1950)

Tunnelling of Ammine Ligands in Nickel Hexammine Iodide

G.J. Kearley, H. Blank, and J.K. Cockcroft

Institut Laue-Langevin, 156X, F-38042 Grenoble Cedex, France

The temperature-dependence of the INS spectrum of $Ni(NH_3)_6I_2$ shows clear evidence of coupling between the tunnelling levels of the NH_3 groups. A 3-parameter model is presented which reproduces the main spectral characteristics of this complex system reasonably well.

1. Introduction

The low-temperature phases of the metal hexammine salts, $M^{2+}(NH_3)_6X_2$ (where M is a divalent transition-metal ion and X is a univalent anion), are of general interest by virtue of the wide range of tunnel splittings displayed by the ammine ligands. It may reasonably be assumed that the geometry of the hexammine cluster is little changed by substitution at the central metal atom within the cluster or substitution of one spherical anion for another, so that there is the opportunity to study tunnelling of the NH_3 groups in progressively different environments.

Details of the arrangements of the ammine ligands within the $Ni(NH_3)_6^{2+}$ cluster are not known for the low-temperature phases of these salts (see reference 1 and references therein) and the arrangements of these ligands within the illustration of the cluster (Figure 1) are arbitrary. In fact, much of the initial interest in the tunnelling spectra was aimed at deriving some structural information [2]. However, the N-N distance within the octahedral hexammine cluster is 2.14Å for the room-temperature phase of $Ni(NH_3)_6I_2$ (Figure 1), and with such a small separation, interactions between neighbouring NH_3 groups are expected to be important [3].

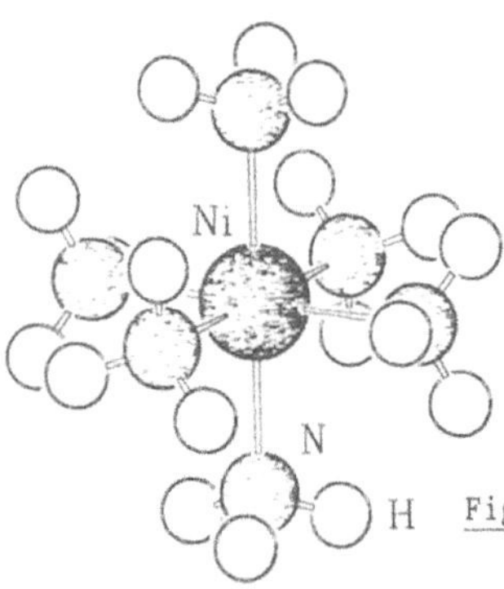

Figure 1: Schematic illustration of the metal hexammine ion.

In this communication we present our preliminary analysis of the tunnelling spectrum of $Ni(NH_3)_6I_2$ which may be considered as an intermediate case, with tunnel-splittings falling in the range 40 - 80 μeV, this region being conveniently accessible to time-of-flight spectrometers.

2. Experimental

Nickel hexammine iodide was prepared by dissolving the nickel hexaquohydroxide in concentrated ammonia solution. The nickel hexammine iodide was then precipitated by addition of ammonium iodide solution, filtered and washed with concentrated ammonia solution. The sample was recrystallised from dilute ammonia solution and dried over KOH in an atmosphere of ammonia (most metal hexammine salts lose ammonia rapidly when exposed to air).

The INS spectra were recorded using the IN5 multichopper time-of-flight spectrometer at the Institut Laue-Langevin, Grenoble, France. Samples, giving a 10% scattering probability, were sealed in thin-walled aluminium containers, with temperature-control being obtained using a standard liquid-helium cryostat. Energy-resolution of 6.0 μeV (FWHM) was achieved by using an incident neutron beam of long wavelength (13 Å, 0.48 meV). In order to maintain a reasonable count-rate a rather high neutron pulse-rate was used, this being possible by virtue of the low sample-temperatures which limit energy-loss processes within the sample to less than about 1.5 meV. Each spectrum required about 12 hours data collection.

Data treatment and fitting programs, using standard algorithms, were locally written. The calculated spectrum was generated from a set of Gaussian functions, with energy-dependent widths, convoluted with the measured instrumental resolution function. The use of a high neutron pulse-rate required that a flat background be subtracted from the time-of-flight spectra before conversion to the energy scale.

3. Results and Discussion

The tunnelling spectrum of $Ni(NH_3)_6I_2$ is comprised of three well-resolved bands at energy-transfers of 42, 67 and 82 μeV (Figure 2), the observation of more than one peak being attributable either to crystallographic differences between NH_3 groups, coupling, or a combination of these. Casual inspection of the spectrum at 1.5 K and 12.0 K (Figure 2) reveals considerable sensitivity of the band-intensities to changes in temperature, and it is clear that there exists a ladder of states rather than a collection of insulated 2-level systems.

Two important questions arise: Firstly, which species play an important role in determining the hindering potential of the NH_3 groups: the iodide ions, nearest-neighbour NH_3 ligands, all ligands sharing a common metal atom, or are these components of similar importance? Secondly, accepting the presence of ammine:ammine interactions, do the spectra give any evidence of inequivalences in the ammine sites? Unfortunately, the hexammine system is somewhat intractible for precise calculations [4], but we may however, use the temperature-dependence of the spectra to discern the relative importance of various interactions.

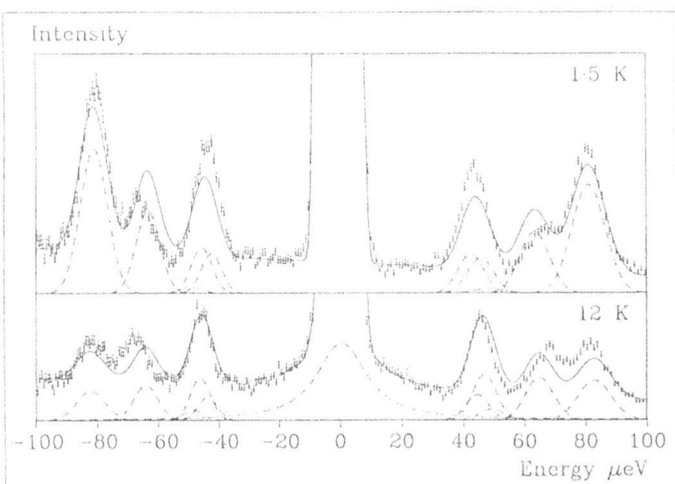

Figure 2: Tunnelling spectra of $Ni(NH_3)_6I_2$. The solid line represents the fitted function, with the components being shown below.

A simple model which was found to give a reasonable fit to the observed data considers the ligands within the hexammine cluster to be indistinguishable with the hinderance potential of any NH_3 group being dependent on the tunnel-states, either A or E, of its four nearest neighbours. Each ligand may then have any one of six different surroundings as shown at the top of Figure 3. At temperatures much higher than the tunnelling energies the A and E states are equally populated and the probabilities of the six different surroundings would then be 1:4:4:2:4:1, respectively. This situation would give rise to a spectrum comprised of six peaks in energy-loss and energy-gain with relative intensities 1:4:4:2:4:1 working from high energy-transfer inwards (the AAEE and AEAE features could be reversed). A comparison of the spectrum predicted by this model with the basic three-band structure of the observed spectra requires that the difference between the hindering potential of some of the six environments is very slight in some cases.

More careful consideration of the transitions within the octahedral ion reveals that the nearest-neigbour constraint allows the energy-level diagram for the entire octahedron to be described with only three parameters. Figure 3 shows the transitions between the energy levels of the octahedral system in terms of the tunnelling within one of the six environments, a,b,c,d,e,f, detailed above, along with constraints between these transitions. The degeneracies of the levels are easily calculated and we may also derive the populations by a simple Boltzmann distribution at the sample-temperature. The basic spectrum is therefore described, not only for peak positions, but also for peak intensities, in terms of only three parameters. A single peak-width parameter is used which provides for a linear increase of peak-width with increasing energy-transfer [5].

At 12.0 K the populations of the A and E states are very similar, and the results of refinement with populations appropriate to this temperature are shown by the solid line in Figure 2, with the separate components being shown below with a broken line. At much lower temperatures the population of predominantly A-state environments increases at the expense of E-states, with a corresponding increase in intensity of the high energy-transfer features of the spectrum. This general behaviour is correctly reproduced by the calculated spectrum (solid line, Figure 2), which is based on the same parameters as the 12.0 K spectrum but with populations modified to suit the lower sample-temperature. Details of the redistribution are illustrated by the separate components shown by the broken line, with the energy-transfer data and assignments being collected in Figure 3.

The spectrum at a sample temperature of 12 K (Figure 2) shows the presence of appreciable quasielastic broadening, which has been found to increase in intensity with increasing temperature, but with a width that is almost temperature independent. This observation is consistent with a very low activation energy. At 19 K a phase transition occurs in which all the ammine ligands become crystallographically equivalent and undergo diffusive rotational motions. This restricts the temperature-range over which the tunnelling transitions can be studied in these salts.

4. Conclusions

The data presented here do not permit the form of the potential to be derived and we cannot entirely exclude the possibility of a ladder of low-energy states arising from a complex potential for 'isolated' NH_3 groups. However, observed spectra at two very different temperatures are reproduced reasonably well by the spectra calculated from the simple coupled model, and we may conclude that this model approximates the resultant of a very complex system rather well. An important component of the spectrum, correponding to tunnelling of a ligand with all its neighbours in the ground-state, was not observed due to its low multiplicity, and thus low intensity. This peak is expected to fall at energy transfers of 121 μeV (outside the spectral range presented here), and it is anticipated that a study using a sample-temperature of ca. 100 mK will be undertaken in the near future to positively identify this component.

A more detailed description and analysis of this system will appear elsewhere, but we may nevertheless offer some response to the questions raised at the

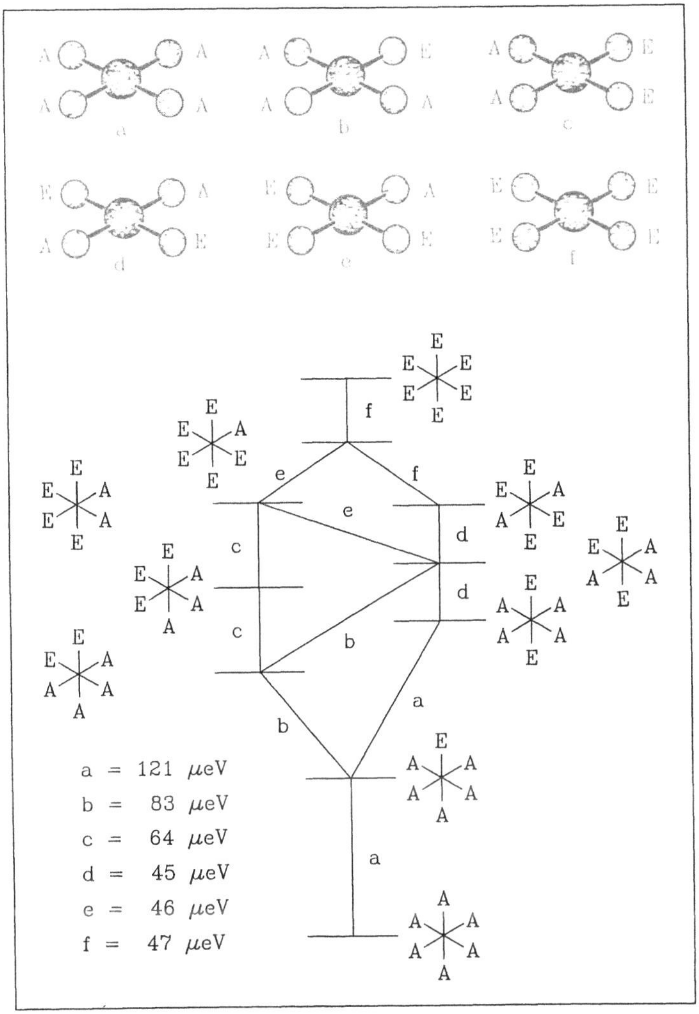

Figure 3: Energy-level scheme for the tunnel-states of the six NH_3 groups in the hexammine cluster. The six distinguishable environments for any NH_3 group, based only on the states of the four nearest neighbours, are detailed above. The constraints implied by the arrangement of energy-levels are: d=2b-a, e=2c-b, and f=4(c-b)+a. No data are available above 100 μeV at present.

beginning of this communication. Firstly, since the data are fitted reasonably by our model, we may deduce that the dominant species for the hindering potential of a given NH_3 group are the nearest-neighbour ligands. Secondly, since the peak widths seem to be well accommodated by an inhomogeneous broadening term [5], within the limits of our data it seems that crystal-field effects are swamped by the interactions between ligands.

1. A. R. Bates, S. R. Hughes and D. J. Somerford: J. Phys. C: Solid State Phys., 16, 2847 (1983).
2. J. Eckert and W. Press: J. Chem. Phys., 73, 451 (1980).
3. M. Prager, A. Heidemann and W. Hausler: Z. Phys. B, 64, (1986).
4. W. Hausler and A. Huller: Z. Phys. B, 59, 177 (1985).
5. G. J. Kearley, J. K. Cockcroft, A. N. Fitch and B. E. F. Fender: J. C. S. Chem. Comm. In press.

The Free Rotor, NH_3, in Metal Hexammine Ions

H. Blank, J.K. Cockcroft, and G.J. Kearley

Institut Laue-Langevin, 156X, F-38042 Grenoble Cedex, France

The low-energy INS spectrum of $Co(NH_3)_6(PF_6)_2$ shows clear evidence of almost free rotation of the NH_3 groups. A slightly modified V_6 hindrance potential is shown to fit the observed data reasonably well.

1. Introduction

The barrier to rotation of the NH_3 groups in metal hexammine salts is expected to be rather low (see previous contribution), particularly when the anion is large, and highly polarizable. For example, the tunnelling features observed in the inelastic neutron scattering (INS) spectrum of $Ni(NH_3)_6I_2$ [1] and $Ni(NH_3)_6Br_2$ [2] bear little resemblance to each other, the iodide showing three features centred around 60 μeV, whilst the bromide has only a single peak at 8 μeV.

In this communication, we examine the low-temperature motions of ammine ligands in $Co(NH_3)_6(PF_6)_2$, in which the characteristics of the PF_6^- anion appear to allow almost free rotation of the NH_3 groups. Examples of almost free rotation in solids are somewhat scarce [3-6] and to our knowledge the present case is not only the first example for NH_3 groups but also for such low barriers in ionic salts.

2. Experimental

We were unable to prepare the hexafluorophosphate directly but found that this salt could successfully be prepared via the nitrate derivative.

Cobalt(II) nitrate was dissolved in dilute ammonia solution to provide cobalt(II) hexammine nitrate. An excess of NH_4PF_6 solution was added in order to precipitate cobalt(II) hexammine hexafluorophosphate, which was then filtered and washed with dilute ammonia solution. The product was dried and stored over KOH in an atmosphere of ammonia.

The INS spectra were collected using the IN5 spectrometer at the Institut Laue-Langevin, France, with an incident wavelength of 3.9 Å (5.38 meV), with a momentum transfer of 2.5 $Å^{-1}$. Energy resolution at zero energy transfer was 0.23 meV. Sample containment is described in the previous contribution. The illustrated spectrum required two hours continuous measuring time, with data reduction being achieved using standard algorithms.

3. Results and discussion

Figure 1 shows low-energy-transfer INS spectrum of $Co(NH_3)_6(PF_6)_2$ to contain at least five neutron energy-loss features at -0.542, -1.09, -1.32, -1.88, and -2.41 meV. Both the energy-transfer and high intensity of the dominant feature at 0.542 meV strongly suggest that its presence can be attributed to almost free rotation of the NH_3 groups [5]. Clearly, the sample-temperature of 1.5 K limits the number of observable transitions between rotational levels, and it would be unreasonable to suggest that all the observed spectral features have this origin. Presumably, at least some of the peaks arise from whole-body motions of the hexammine cluster.

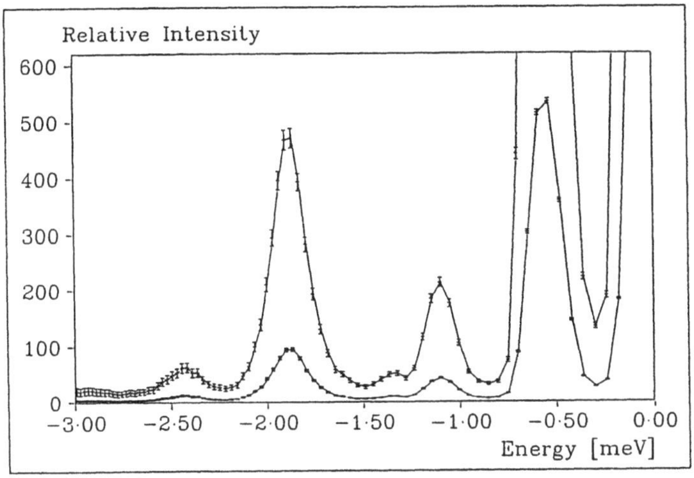

Figure 1: The INS spectrum of $Co(NH_3)_6(PF_6)_2$. The lower spectrum has been reduced by a factor of 5.

It is easiest to proceed if we first consider the free-rotor levels which would give the following transitions: 0.655 (0>1), 1.96 (1>2), 2.62 (0>2), 5.26 (1>3), and 5.90 (0>3). Comparison of the observed 0>1 transition with the free-rotor value, 0.542 and 0.655 meV, respectively, would suggest that the free-rotor levels are only slightly perturbed in the hexammine cluster, so that we may tentatively assign the peaks at 1.88 and 2.41 meV to the 1>2 and 0>2 transitions, respectively.

The observation transitions only between the three lowest levels do not allow us to enter a detailed discussion of the form of the potential, but we may, however, use these observations to determine which of the various plausible potentials are most likely, and consequently derive a barrier height. Table I presents the values for the energy levels calculated for a 3-fold, 6-fold, 9-fold, 12-fold, and 24-fold potential, based on the above peak assignments and requiring exact agreement for the 0>1 transition. Inspection of Table I shows that the

Table I: Calculated transition energies between rotational levels under various hindrance potentials. Observed energy-transfers are included for comparison.

potential-form	barrier height [meV]	0>1	1>2	0>2	1>3	0>3
V_3	2.74	0.542	2.385	2.927	5.424	5.966
V_6	13.55	0.542	1.457	1.999	2.62	3.162
V_9	31.50	0.542	1.567	2.109	3.89	4.431
V_{12}	56.60	0.542	1.595	2.137	4.127	4.669
V_{24}	231.80	0.542	1.620	2.162	4.30	4.837
$0.88\ V_6 + 0.12\ V_3$	10.60	0.543	1.891	2.433	3.421	3.963
all experimental values [meV]: 0.542 1.09 1.32 1.88 2.41						

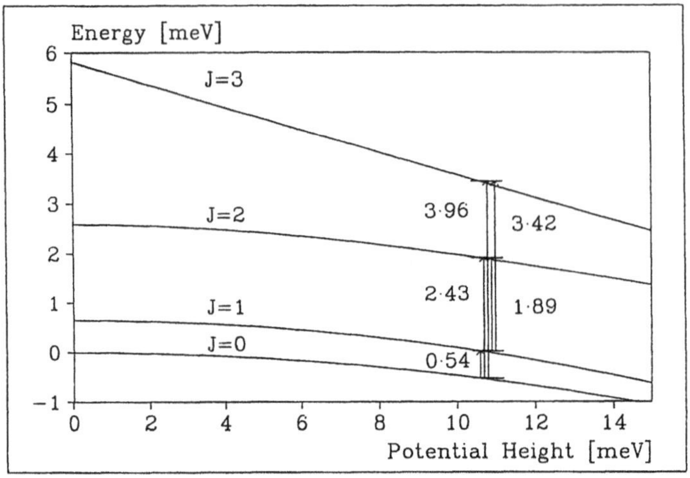

Figure 2: Variation of the energy levels of the one-dimensional rotor, NH_3, as a function of barrier height. A potential of the form:
$V=0.12(V_3/2(\cos 3\Phi)) + 0.88(V_6/2(\cos 6\phi))$
was used. The arrows represent the observed transitions.

3-fold potential can be eliminated since it does not provide even an approximate match with any selection of the observed features. Not suprisingly, as higher symmetries are approached, the energy levels become less perturbed by the magnitude of the hindrance potential, and very loose agreement can be obtained between our initial assignments and any of the high-symmetry potentials.

If we admit additional peaks to our initial assignment, a reasonable agreement between the peaks not only at 0.542, 1.88 and 2.41, but also at 1.31 meV can be found with the V_6 potential (Table I). We may speculate that the form of the potential may be described by a slightly modified V_6 potential, and we have used an additional V_3 component to this end [7]. The best agreement was achieved using a combined sixfold:threefold potential (0.88:0.12), the resulting potential being illustrated in Figure 2.

4. Conclusions

There is little doubt that the main spectral feature and at least some of the other peaks arise from transitions between the rotational levels of the ammine ligands. Positive identification of all the observed peaks is rather difficult, since as the temperature is increased the spectrum collapses rapidly until at 20 K a phase transition leads to classical rotational diffusion of the ligands. Nevertheless, some progress in this respect may be made if spectra of the deuterated material, and spectra as a function of pressure, become available.

The central question however, remains unanswered: how do we reconcile the observation of almost free rotation in this material with the strong NH_3:NH_3 coupling observed in the iodide analogue (previous contribution), and further, with the very small tunnel splittings found for some of the other members of this series? It seems that the answer to this question will have to await a much clearer understanding of the subtle structural differences between these salts.

1. J.Eckert, W.Press: J. Chem. Phys. 73, 451 (1980)
2. W.Press, M.Prager: J. Chem. Phys. 67, 5752 (1977)
3. W.Schott: Z.f.Physik 231, 243 (1970)
4. H.Kapulla, W.Gläser: IAEA, Vienna, 841 (1972)
5. B.Alefeld, A.Kollmar, B.A.Dasannacharya: J. Chem. Phys. 63, 4415 (1975)
6. A.E.Zweers, H.B.Brom, W.J.Huiskamp: Phys. Lett. 47A, 347 (1983)
7. M.Prager, K.-H.Duprée, W.Müller-Warmuth: Z. Phys. B51, 309 (1983)

Tunnel Spectroscopy by Dipolar-Driven NMR

K.J. Abed, G.J. Barker, S. Clough, and A.J. Horsewill

Department of Physics, University of Nottingham,
Nottingham, NG7 2RD, UK

At low static fields nmr transitions are driven at least in part by the dipole-dipole interactions being modulated by the applied rf field. The result is that the $\Delta m = 2$ transition is comparable in intensity with the main $\Delta m = 1$ transition. More importantly, transitions which are symmetry forbidden at high field may become allowed at low field. This applies to the tunnelling sidebands of CH_3 and NH_4 groups observed at low temperatures, where the transitions involve a change in proton permutation symmetry. By carrying out nmr in fields of the order of 0-0.02 T tunnel sidebands are easily observed. This has opened a new window of tunnel spectroscopy for tunnel frequencies from 40 kHz - 1 MHz. Several examples will be shown. To maintain reasonable sensitivity, a field cycling procedure is used in which the initial polarisation and final measurement are carried out at high fields. Apart from measuring tunnel frequencies, a number of interesting and unexpected features have been observed. The small tunnel frequencies accessible to this technique are characteristic of CH_3 groups terminating aliphatic chains. It therefore provides a sensitive probe of the environment of CH_3 groups in a very wide range of materials.

1. INTRODUCTION

This paper reports on a new method [1](dipole-dipole driven nuclear magnetic resonance spectroscopy) of measuring small tunnel frequencies (40 kHz to 1 MHz) which is proving very productive, though so far it has only been applied to methyl groups. These small tunnel frequencies are characteristic of the high hindering barriers found where CH_3 groups are attached to sp^3 hybridized atoms, for example as terminating groups of aliphatic chains. Barrier heights in such molecules lie in a fairly narrow range and are probably dominated by intramolecular interactions, which are more accessible to calculation than the inter-molecular interactions usually most important when the CH_3 is bonded to a sp^2 atom. Many molecules studied contain pairs of methyl groups. Examples are $S(CH_3)_2$, $O(CH_3)_2$, $RCH(CH_3)_2$ etc. In these cases, the interaction between methyl groups must be an important, probably dominant, part of the total hindering potential experienced by each group. It is therefore very likely that their motions are to some degree correlated, and the question arises whether the present range of experimental techniques can study this correlation. We report on the new technique from this point of view. Many molecules contain only single or separated methyl groups. It is a very common feature of the crystalline environment though, for methyl groups of different molecules to cluster together. Thus the dynamical coupling of methyl groups probably occurs also in these materials.

The correlation of the rotational motion of a methyl group with the motion of its surroundings, is an essential part of the transition with increasing temperature, from coherent quantum motion to thermally activated hopping motion[2]. At sufficiently low temperature, with few excited thermal phonons, methyl rotation is described with a simple Hamiltonian.

$$\mathscr{H} = -\frac{\hbar^2}{2I}\frac{\partial^2}{\partial\varphi^2} + V\cos 3\varphi \qquad (1)$$

and a rotational symmetry condition on the eigenfunctions

$$\psi(\varphi + 2\pi) = \psi(\varphi) \tag{2}$$

which restricts the Bloch wave solutions of (1) with the general form $(-\pi \leqslant k \leqslant \pi)$

$$\psi(\varphi) = \sum_n u_n \exp(ikn\varphi) \tag{3}$$

to values of $k = 0$ and $\pm 2\pi/3$ so that $\exp(3ik)=1$.

This only works at low temperature though because (2) is too restrictive. The reason is that we are dealing with a large system of a methyl group plus the crystal lattice in which it is embedded, and the assumption that this can be separated into a one-dimensional Hamiltonian (1) and a phonon bath is by no means trivial. This separation cannot be done in an arbitrary way and its validity depends on the meaning of φ. Although this coordinate depends mainly on the CH_3 rotational coordinate, it also incorporates the lattice coordinates to a degree which increases with increasing temperature. In this way the correlation of the environmental motions with the methyl rotation is built in implicitly into (1). Since φ is not a pure angle (2) does not apply and neither does the restriction of the k values to 0 and $\pm 2\pi/3$. Thus $\varphi = 0$ and $\varphi = 2\pi$ are fundamentally distinguishable, though they are not always easily distinguished. At low temperatures only k values near 0 and $\pm 2\pi/3$ are thermally excited, so that it appears as if (2) applies. At higher temperatures, a broader range of k values is excited, and the discrete tunnel splittings observed at low temperatures evolve into the broadened bands at higher temperatures. As the restrictions on k disappear, so the correlation of rotation with nuclear spin, which gives rise to nuclear spin symmetry species, also disappears. Thus the identification of $k = 0$ with the A nuclear spin state and $k = \pm 2\pi/3$ with the E state is no longer possible when k is smeared over the whole range from $-\pi$ to π. The 'rotational' state (the inverted commas recall that φ is not a pure rotation angle) has become uncorrelated with the proton permutation symmetry and hence with the nuclear spin symmetry.

When two coupled methyl groups are concerned, the Hamiltonian becomes two-dimensional and the Bloch waves analagous to (3) propagate in a plane. With increasing temperature two kinds of change can be anticipated, namely (a) an increase in the participation of lattice motions as in the one-dimensional case (b) an increase in the participation of each methyl group in the rotations of the other.

2. COUPLED ROTATIONS IN THE QUANTUM REGIME

It is a simple matter to discuss the ground torsional state of a two-dimensional version of (1) and (2) using nuclear spin permutation operators P_x and P_y in a nuclear spin Hamiltonian. The idea here is to use the correlation of nuclear spin and rotational functions to simulate the effect of tunnelling by means of nuclear spin operators. Thus for two uncoupled methyl groups the spin Hamiltonian which simulates the effects of tunnelling rotation has the form

$$\mathscr{H} = -\frac{1}{3} h\nu_t^x (P_x + P_x^{-1}) - \frac{1}{3} h\nu_t^y (P_y + P_y^{-1}) \tag{4}$$

with ν_t^x and ν_t^y the two tunnel frequencies.

The basis functions are the 64 (i.e. 2^6) products of the nuclear spin functions α, β for the six protons. Symmetry-adapted combinations can then be written as products $X_{xm} Y_{ym}$ where X and Y are the symmetry species A, E^a or E^b for the two groups and xm, ym take the values $\pm 3/2$, $\pm 1/2$ for A and $\pm 1/2$ for E, with the suffixes x and y labelling the two groups. The effect of P_x is to cyclically permute the first three spin functions, and P_y similarly permutes the last three. $P_x\, \alpha\alpha\beta\alpha\beta\alpha = \beta\alpha\alpha\alpha\beta\alpha$; $P_y\, \alpha\alpha\beta\alpha\beta\alpha = \alpha\alpha\beta\alpha\alpha\beta$. The eigenvalues of $P_x + P_x^{-1}$ are 2 for A functions and -1 for E functions.

A coupling term is introduced into (4) by adding $\mathscr{H}_c$

$$\mathscr{H}_c = -\frac{1}{3}h\nu_c(P_xP_y + P_x^{-1}P_y^{-1}) . \tag{5}$$

The meaning of this is that correlated rotations in which the two groups both turn in a clockwise direction, or both in an anticlockwise direction, are favoured over those in which one turns clockwise and the other anticlockwise. The parameter ν_c is a measure of the importance of correlated rotations.

The eigenfunctions and eigenvalues are now

$A_x A_y$	$-\frac{2}{3}h(\nu_t^x + \nu_t^y) - \frac{2}{3}h\nu_c$
$A_x E_y^a,\ A_x E_y^b$	$-\frac{1}{3}h(2\nu_t^x - \nu_t^y) + \frac{1}{3}h\nu_c$
$E_x^a A_y,\ E_x^b A_y$	$-\frac{1}{3}h(-\nu_t^x + 2\nu_t^y) + \frac{1}{3}h\nu_c$
$E_x^a E_y^a,\ E_x^b E_y^b$	$\frac{1}{3}h(\nu_t^x + \nu_t^y) + \frac{1}{3}h\nu_c$
$E_x^a E_y^b,\ E_x^b E_y^a$	$\frac{1}{3}h(\nu_t^x + \nu_t^y) - \frac{2}{3}h\nu_c$.

If we assume that because of the large intra methyl group dipole-dipole interaction, transitions occur in which the state of one methyl group may be changed, we find that the allowed transition frequencies are ν_c, ν_t^x, $\nu_t^x \pm \nu_c$, ν_t^y and $\nu_t^y \pm \nu_c$. (When we consider dipole-dipole driven nmr in the next section, the nuclear Zeeman energy has to be added to the above eigenvalues). This type of Hamiltonian therefore has a fairly clear signature. It is the one which has been used to discuss the pairs of groups in lithium acetate[3] with the assumption in that case that $\nu_t^x = \nu_t^y$ and $\nu_t^{xy} \gg \nu_c$. If ν_c is the dominant term then one expects to see the frequencies ν_c, $\nu_c \pm \nu_t^x$, $\nu_c \pm \nu_b^y$.

3. DIPOLE-DIPOLE DRIVEN NMR

The new technique can be regarded simply as nuclear magnetic resonance at very low field ($<$ 0.03 T). The point of going to low field is that the selection rules are very different than at high fields and forbidden transitions become allowed. The important ones are those in which the methyl group changes its symmetry species, since these transitions occur at fields dependent on the tunnel splitting. At high fields these are very weak[4] and can only be detected when the tunnel splitting lies in a narrow range of 40 - 100 kHz. By contrast, at low fields the transitions are strongly allowed and a frequency window from 40 kHz to 1 MHz is opened, and perhaps beyond. This is a window in which very many methyl tunnel frequencies lie.

An explanation for the change in selection rules can be given by assuming that a different representation is needed at low field. The conventional discussion of nmr has the quantization axis z along the static external field. The high-frequency oscillating field applied perpendicular to this is then used as a time-dependent perturbation to cause transitions, the flip-flop transitions of the nuclear spins due to the dipole-dipole interactions being invoked to cause the phase randomization. At low field, this application of time-dependent perturbation theory becomes incorrect, because the oscillating field has a frequency sufficiently low for the dipole-dipole flip-flop transitions to adapt adiabatically to it, at least in part. The r.f. field now has the role of switching energy between the nuclear Zeeman and dipolar reservoirs, rather than pumping energy into the nuclear Zeeman reservoir. In this case it is better to take the quantization axis along the total instantaneous magnetic field, so including the applied oscillating field into the main Hamiltonian. The consequence of this is to transfer the time-dependence from the oscillating field to a part of the dipole-dipole interaction. This latter can then be used as the time-dependent perturbation. Because of its complexity, it has many matrix elements and gives rise to a rich spectrum. The particularly important features are that both $\Delta m = 1$ and $\Delta m = 2$ transitions (single nuclear spin flips and double flips) are observed, with sidebands due to molecular tunnelling.

In order to have adequate sensitivity, a magnetic field cycling system is used which enables the initial growth of nuclear magnetization and the final measurement of that magnetization to take place in a strong magnetic field, while the actual experiment occurs in a weak field. The sequence is as follows: (1) The nuclear magnetization is reduced to zero at 0.6 T to provide a well-defined starting condition. This is followed by a preparation period (typically 4 minutes) at 0.6 T to allow magnetization to grow. (2) The field is switched rapidly (4 sec) to low field B and an oscillating field of frequency ω is switched on perpendicular to B. This lasts for a time t. (3) The field is switched back to 0.6 T and the nuclear magnetization $M(B, \omega, t)$ is measured by observing the transient following a single 90° pulse. By incrementing the magnetic field B, a spectrum at fixed ω and t is obtained.

4. TUNNELLING SPECTRA BY LOW-FIELD NMR

About twenty new tunnel splittings have been measured so far. These include a series of ketones and a series of n-alkanes, with a number of molecules having pairs of methyl groups. The aim has been to search for systematic features of tunnelling spectra and of the dependence of tunnel frequency on molecular structure. Although it is too soon to be sure that very general features have been identified, a number of things are emerging which appear to be of some interest and to have a bearing on the possible coupling of groups.

The first observation is that a relatively high proportion of molecules with methyl groups on sp^3 atoms exhibit tunnel splittings in the accessible region, typically between 40 kHz and 800 kHz. In some cases, for example the n-alkanes, there is remarkable consistency of tunnel frequency from one molecule to another. This indicates that most of the hindering potential is intra-molecular in origin. The second notable observation is that nearly all the molecules studied appear to have in addition a small tunnel frequency in the range 40 - 100 kHz. This gives rise to side peaks which are imperfectly resolved from the main nmr peaks, and for this reason they are not always as easily measured as the higher frequencies. The most typical form of spectrum is therefore one exhibiting two sets of sidepeaks on each of the $\Delta m = 1$ and $\Delta m = 2$ main peaks. A third observation is that at least in some cases, the high-frequency sidepeaks themselves have low-frequency satellites. This is expected from the analysis of section 2 as will be explained below.

When the nuclear Zeeman energy is added to the rotational energies given in section 2 and the fields calculated at which transitions of energy $\hbar\omega$ occur, ω being the angular frequency of the oscillating field, then one finds the fields to be given by $B = \omega/\gamma$ and $B = \omega/(2\gamma)$ for the main $\Delta m = 1$ and $\Delta m = 2$ transitions in which the spin symmetry state does not change. For uncoupled pairs of methyl groups (i.e. $\nu_c = 0$) there are additional side peaks at $B = (\omega \pm \omega_t^a)/\gamma$ and $(\omega \pm \omega_t^a)/(2\gamma)$ and at $(\omega \pm \omega_t^b)/\gamma$ and $(\omega \pm \omega_t^b)/(2\gamma)$. For coupled pairs there are also satellites at $B = (\omega \pm \omega_c)/\gamma$, $B = (\omega \pm \omega_c)/2\gamma$, $B = (\omega \pm \omega_t^a \pm \omega_c)/\gamma$, $(\omega \pm \omega_t^b \pm \omega_c)/(2\gamma)$ etc. These last are satellites of satellites and may be used to identify coupled pairs.

Such satellites of satellites are not frequently seen with strong intensity. In favourable cases (for example in undecane) they show up quite well. In other cases they are certainly much weaker than the main satellites. The evidence in favour of coupling is therefore not conclusive, though it is otherwise hard to explain the very frequent observation of two different tunnel frequencies in the ratio of approximately 10:1.

Our somewhat tentative interpretation is therefore as follows. A methyl group which terminates an aliphatic chain experiences a hindering potential whose height is equivalent to about 2000K and this means that the tunnel frequency is a few hundred kHz. The crystal structure favours the clustering of methyl groups and the mutual intermolecular interactions contribute to the hindering potentials to the extent of about 200K. This can increase or decrease the single methyl group tunnel frequency, but it also has the effect of introducing a degree of correlation of the rotations of the interacting groups. This manifests itself by the appearance of small tunnel frequencies of the order of 50 kHz. This relatively small value is associated with the fact that since both groups move, the rotation represents a greater displacement through the coordinate space than when a single group rotates. It is of course possible that more than two groups are involved in collective rotations, but since the small tunnel frequencies are imperfectly resolved, little can be deduced about that.

Although the mechanical coupling of methyl groups through a mutual contact or proximity is the most obvious coupling mechanism, it is also possible that well-separated methyl groups may be coupled where the rotations of both are coupled to the same molecular vibrational mode.

Three typical examples of tunnel spectra are shown in the figures, for dimethyl ether, 5-nonanone and n-undecane. The first has close pairs of methyl groups, the others terminal groups. Inter molecular interactions between methyl groups are probably significant in all cases. The spectra consist of negative peaks on a fairly flat background. The deepest peak, and the one occurring at the highest field is in all cases the $\Delta m = 1$ transition at $B = \omega/\gamma$. At half this field is the $\Delta m = 2$ peak. Each of these is flanked by satellites, the spacing of the $\Delta m = 2$ satellites being half that of the $\Delta m = 1$ satellites.

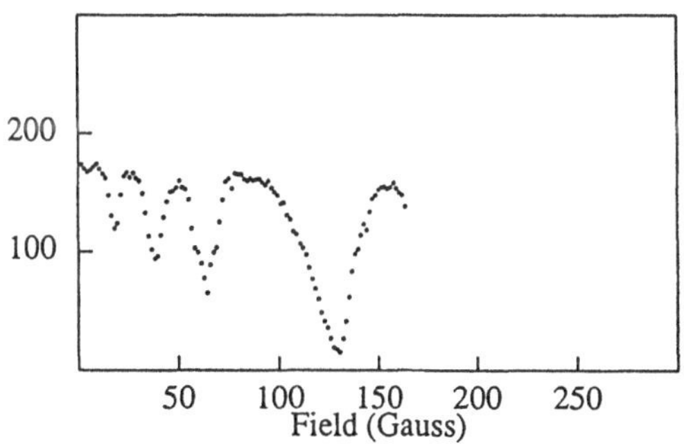

Figure 1. A low-field spectrum of dimethyl ether showing evidence of two tunnel frequencies 390 kHz and 45 kHz.

Figure 1 shows a spectrum of dimethyl ether with $\omega/2\pi = 550$ kHz and $t = 25$ sec. The $\Delta m = 1$ and 2 peaks are near 130 and 65 Gauss. Near 40 Gauss is a low-field satellite of the $\Delta m = 1$ peak (separation about 90 Gauss). Near 20 Gauss is the corresponding $\Delta m = 2$ satellite (separation $\sim$ 45 Gauss). The high-field $\Delta m = 2$ satellite near 110 Gauss can be seen on the side of the main $\Delta m = 1$ peak. The large tunnel frequency responsible for all these peaks is 390 kHz. The small tunnel frequency is responsible for shoulders on both the $\Delta m = 2$ and $\Delta m = 1$ peaks, most clearly seen on the former.

Figure 2 shows a spectrum of 5-nonanone with $\omega/2\pi = 600$ kHz and $t = 20$ sec. The $\Delta m = 1$ and 2 peaks are near 140 and 70 Gauss. Two pairs of satellites of the $\Delta m = 2$ peak can be seen as can one of the corresponding low-field $\Delta m = 1$

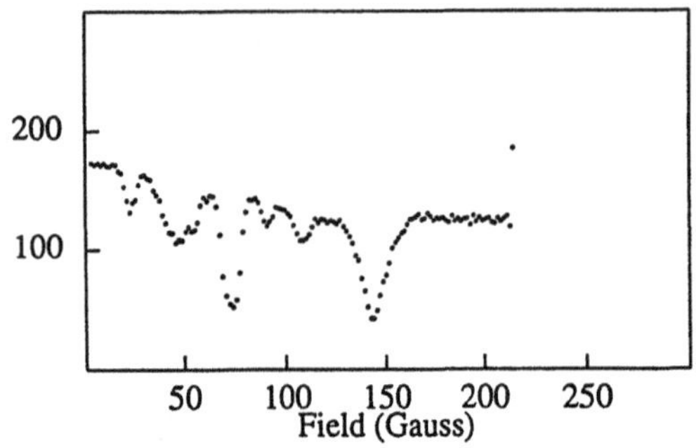

Figure 2. A low-field spectrum of 5-nonanone. Three tunnel frequencies, 386 kHz, 150 kHz and 36 kHz are measured.

satellites. Shoulders on the main $\Delta m = 1$ peak reveal the existence of a small tunnel splitting. The three tunnel frequencies are 386 kHz, 150 kHz and 36 kHz.

Figure 3 is a spectrum of undecane and is interesting because of the evidence of a $\Delta m = 1$ satellite at the sum of the large and small tunnel frequencies. This can be seen near 125 Gauss. As discussed previously, this is an indication of the effects of coupling.

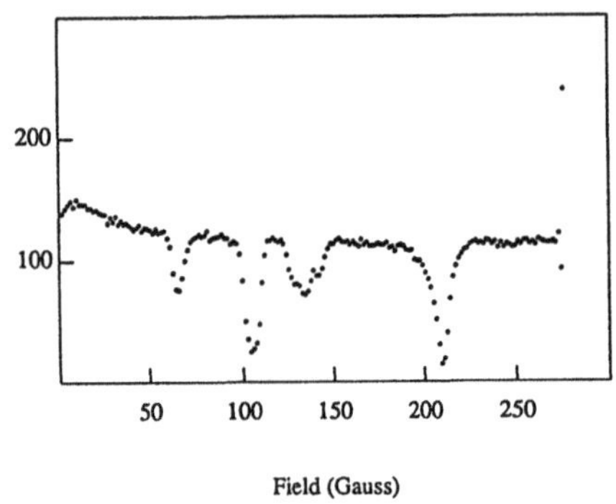

Figure 3. A low-field spectrum of n-undecane. Tunnel frequencies 342 kHz and 44 kHz are observed and a satellite displaced from the $\Delta m = 1$ peak by the sum of these frequencies is observed.

A number of normal alkanes have been studied. In all cases large and small tunnel frequencies are detected. An interesting feature is that there is a systematic difference between those with an even number of carbon atoms and those with an odd number. The latter have large tunnel frequencies some 10% larger than the former. This reflects a difference in crystal structures, the even numbers being triclinic and the odd are orthorhombic. In all cases the molecules are parallel with the methyl groups lying in sheets. Pairs of sheets belonging to different sets of molecules are in close proximity. The strongest interactions are probably between groups in different sheets rather than between groups in the same sheet. It is therefore likely that these systems may be treated as interacting pairs of methyl groups.

5. LOW-FIELD RELAXATION AND DYNAMIC POLARIZATION

The large amount of data which we have collected reveals three unexpected features two of which can be seen in figure 3. The first three data points at the high field in figure 3 exhibit wide variation, until after a few measurements, the data settle to the more or less uniform level. This initial effect depends on the previous history and it can be amplified by spending a long time at high field before starting the experiment. It demonstrates that there exists some other energy reservoir, not contained in our analysis, with which the nuclear Zeeman system is in contact during the sweep from high to low field. This could be due to motions associated with lattice defects. During the first few measurements the temperature of this reservoir assumes a value such that the net transfer of energy with the Zeeman reservoir, during a measurement sequence, becomes equal to zero. For the remainder of the experiment the reservoir has no influence on the data.

The second feature visible in figure 3 is the rise at low field of the background level. In fact the background level at fields above 50 Gauss falls with increasing t due to relaxation of the magnetization towards its very small equilibrium value. At low field this decline does not occur. It therefore appears that relaxation is slower at very low field (less than 50 Gauss) than above. This is the opposite of what is normally expected. No satisfactory explanation of this phenomenon has been found so far.

The third interesting feature which is occasionally observed is that the negative peaks are sometimes quite strongly asymmetric, even to the extent of reversing sign on the low-field side. This is what is expected if the population of the tunnel levels deviate strongly from the thermal equilibrium values. This is conveniently described by introducing a tunnel temperature. The results indicate that this tunnel temperature may become negative, meaning a greater population in the E level than in the A. It is not surprising that some disturbance of this temperature results from the measurement sequence, but the size of the effect in some samples is surprising. It demonstrates that low-field nmr spectroscopy has considerable potential for studying dynamic spin symmetry conversion and low-field relaxation processes, and is not limited only to the measurement of tunnel frequencies.

REFERENCES

1. S. Clough, A. J. Horsewill, P. J.McDonald and F. O Zelaya: Phys. Rev. Letters 55, 1794, 1985
2. S. Clough, P. J. McDonald and F. O. Zelaya: J.Phys.C : Solid St. Phys. 17, 4413, 1984
3. S. Clough, A. Heidemann, A. J. Horsewill and M.N.J. Paley: Z.Phys.B - Condensed Matter 55, 1, (1984)
4. J. A. Ripmeester, S. K. Garg and D. W. Davidson: J. Chem. Phys. 67, 2275, (1977)

Zeeman-Tunnel Resonance Line Shapes in NMRRR Investigations of Rotational Tunneling

G. Vandemaele, A. Buekenhoudt, and L. Van Gerven

Center for Nuclear Magnetism and Radiospectroscopy,
Laboratorium voor Vaste Stof-Fysika en Magnetisme,
Katholieke Universiteit Leuven, B-3030 Leuven, Belgium

A new magnet system with a very fast switching power supply (switching over a range of 1.5 T with a rate of 50 T/s) is used to study rotational tunneling in copper acetate and in some ammonium compounds at low temperatures. These solids contain CH_3- respectively NH_4^+- groups, whose librational levels split up, due to motional hindering, into A and E, respectively A, 3T and E tunnel levels, according to the threefold, respectively tetrahedral symmetry of their surroundings. The tunnel splitting $h\nu_t$ of the ground librational level is studied by field-cycling [1]. This experimental technique enables us to study the spin-surroundings relaxation at variable fields B_r ($\equiv 2\pi\nu_o/\gamma$) by means of one measuring field B_m, resonant to the spectrometer frequency $\nu_s (\equiv \gamma B_m/2\pi)$, in which the saturation and measuring pulses are applied. Anomalies in the recovered magnetization M_r, measured as a function of B_r, are expected at Zeeman-tunnel resonance, where $\nu_o = \nu_t$ or $\nu_t/2$ (hence $B_r = B_t (\equiv 2\pi\nu_t/\gamma)$ or $B_r = B_t/2$). This of course is a NMR Relaxation Resonance (NMRRR) phenomenon.

Essential in this method is the switching from B_m to B_r and back, with a switching rate of 50 T/s, during a time t_s. This switching time t_s should be kept as short as possible (at least shorter than τ_1) to prevent the spin-lattice relaxation masking the Zeeman-tunnel relaxation. When $B_r \cong B_t$, $B_t/2$, we have to switch through the Zeeman-tunnel resonance region. In order to avoid distortion of the Zeeman-tunnel resonance line shape - which normally should be a peak representing the tunnel frequency distribution - the switching rate during this passage should be much larger than the Zeeman-tunnel interaction rate τ_{cro}^{-1}. If this is not the case, two types of distortion can occur, dependent on whether the tunnel reservoir remains in thermal equilibrium with the lattice, or not. This depends on the tunnel-lattice relaxation time τ_{TL}, whether it is much shorter, or not, than the time between two successive passages through the Zeeman-tunnel resonance region.

The performed NMRRR research includes the A-E tunnel splitting in $Cu(CH_3COO)_2.H_2O$ and the T-T tunnel splitting in NH_4ClO_4 at different temperatures, and the A-T tunnel splitting in $(NH_4)_2TeCl_6$ at 4.2 K.

We found the tunnel frequency of copper acetate to be 68 MHz, rather than 65 MHz, as was proposed by COPPENS et al. [1, 2] from their first field-cycling measurements with a slow switching electromagnet (0.03 T/s). For NH_4ClO_4 we measured a tunnel frequency of 45 MHz, equal to the value published by PUNKKINEN et al. [3]. For $(NH_4)_2TeCl_6$, an anomaly in the recovered magnetization is found in the region of 27 MHz, as was seen by PRAGER et al. [4] who suggested a value for ν_t of 24 MHz.

Let us now turn to the Zeeman-tunnel resonance line shapes. Comparison of the observed experimental curves (figures 1, 2 and 3) shows striking differences in the way the $M_r(B_r)$ anomaly near Zeeman-tunnel resonance appears. The expected peak-like shape seems indeed to be distorted by the fact that the *switching time is not negligibly short.*

Because of the very short switching time of our new magnet system (of the order of ms), non-adiabatic effects due to Zeeman-lattice relaxation (showing long τ_1's of the order of 10 to 1000 s) cannot occur during the switching. How-

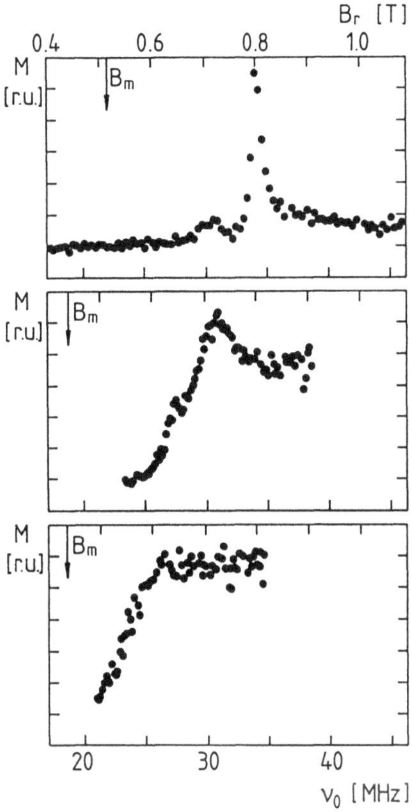

Figure 1. The recovered proton magnetization of $Cu(CH_3COO)_2.H_2O$ as a function of B_r at $T = 4.2$ K (a), 31.8 K (b) and 36.2 K (c) using field cycling: $B_m = 0.5165$ T (a) or 0.4345 T (b) and (c); t_r (the relaxation interval at B_r) = 60 ms.

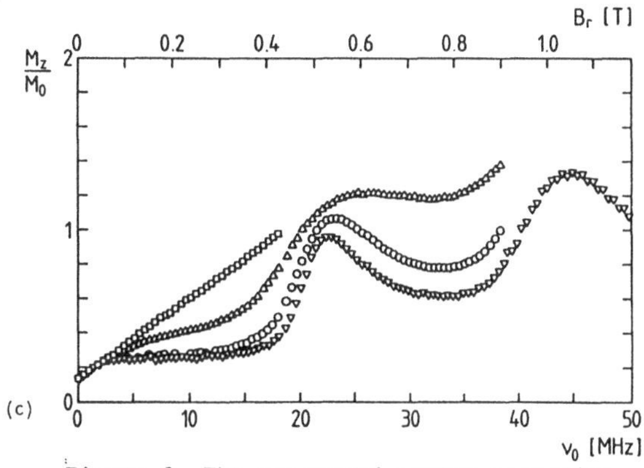

Figure 2. The recovered proton magnetization of NH_4ClO_4 relative to the equilibrium magnetization M_o at B_m as a function of B_r at $T = 4.2$ K, using field-cycling: $B_m = 0.434$ T; $t_r = 100$ ms (▽), 250 ms (O), 850 ms (Δ), 4 s (□).

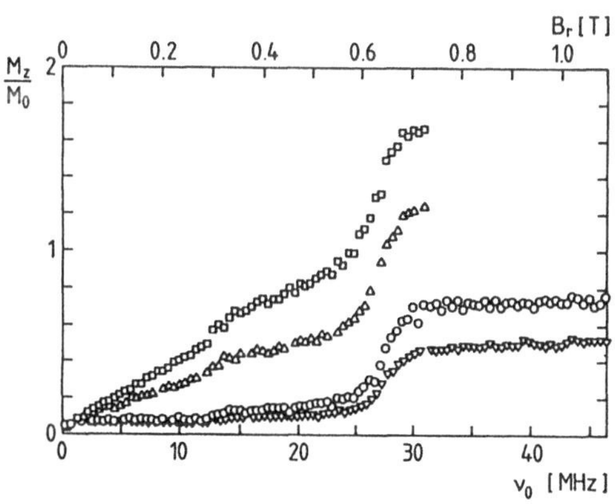

Figure 3. The recovered proton magnetization of $(NH_4)_2TeCl_6$ relative to the equilibrium magnetization M_o at B_m, as a function of B_r at $T = 4.2$ K, using field-cycling: $B_m = 0.2864$ T; $t_r = 60$ ms (▽), 150 ms (O), 1 s (Δ), 2 s (□).

ever, despite the very fast switching, we are not always able to avoid the Zeeman-tunnel relaxation - which is governed by very short characteristic times τ_{cro} - and the corresponding non-adiabatic effects. The latter bring about a step-like $M_r(B_r)$ anomaly which is the integral of the real resonance peak in the case where the tunnel-lattice relaxation time τ_{TL} is very short. In the case of long tunnel-lattice relaxation times, saturation of the tunnel reservoir occurs. For copper acetate, the latter has been proved to exist by a field-cycling experiment with $B_m < B_t/2$ and $B_r > B_t/2$. After each passage through the $B_t/2$ resonance, the recovered magnetization at the constant B_r becomes smaller, indicating a gradual heating of the tunnel reservoir during each passage, however fast. In copper acetate, we also observe a change in the shape of the tunnel lines with temperature, which makes the detection of the tunnel frequency and the study of its temperature dependence very difficult. From 4.2 K up to 22 K, the $B_t/2$ anomaly looks like a sharp, nearly symmetric (double) peak (figure 1.a). We should note that we measured for the first time by means of field-cycling the splitting of the Zeeman-tunnel resonance line, which was detected before with other techniques and ascribed to the existence of two structurally different methyl groups [2]. From 22 K on, the broadening of the peak becomes highly asymmetric and gets a higher tail at the high field side (figure 1.b). The recovered magnetization in this tail increases gradually as the temperature rises, until the Zeeman-tunnel resonance anomaly becomes a step function around 36 K (figure 1.c). This proves that, at higher temperatures, efficient Zeeman-tunnel contact is still present, but saturation of the tunnel reservoir is no longer possible because of the shortening of the tunnel-lattice relaxation time (which is equal to the spin conversion time τ_{con} as we deal with an A-E spin conversion) with increasing temperature [5]. Consequently both Zeeman-tunnel resonances at ν_t and $\nu_t/2$ can be detected in one field-cycling scan.

Similarly to copper acetate at 36 K, for NH_4ClO_4 and $(NH_4)_2TeCl_6$ both resonances are detected in one scan (with $B_m < B_t/2$, B_t). However, at 4.2 K, two peaks are found for perchlorate,whereas for ammoniumhexachlorotellurate, the $M_r(B_r)$ scan consists of two steps (figures 2 and 3). One could explain the steps in $(NH_4)_2TeCl_6$ in the same way as the ones in copper acetate at 36 K if one assumes that the spin-restricted A-T conversion is characterized by a short conversion time τ_{con}, even at 4.2 K, which is very unlikely. For NH_4ClO_4, one would expect at 4.2 K a short tunnel-lattice relaxation time because no real spin conversion is involved. Hence no tunnel saturation should be present. This is proved to be the case by repeatedly switching from $B_m < B_t/2$ to $B_r < B_t/2$, where no decrease in the recovered magnetization is detected. But why then the decrease in M_r at the relaxation fields $B_r > B_t/2$ in figure 2? This can only be explained by the fact that we are indeed able to switch adiabatically through Zeeman-tunnel resonance in NH_4ClO_4, in contrast to the other substances. This is most likely due to the fact that in NH_4ClO_4, for an unknown reason, τ_{cro} is longer than the passage time while in $Cu(CH_3COO)_2.H_2O$ and in $(NH_4)_2TeCl_6$, τ_{cro} would be shorter. Consequently, we deduced the tunnel frequency $\nu_t(TT)$ of perchlorate from the peak value in the $M_r(B_r)$ scan. The line shape of NH_4ClO_4 does not change over the whole studied temperature region from 4.2 K to 16 K. The measured temperature dependence of the tunnel frequency $\nu_t(TT)$ is presented in figure 4.

Further investigations are necessary. It would be particularly interesting to measure the different values of τ_{cro}. For $Cu(CH_3COO)_2.H_2O$ this has turned out to be not feasible up to now, since the value of τ_{cro} seems to be extremely short. It would be worthwhile to try to measure τ_{cro} in NH_4ClO_4 for which, according to our present indications, a longer value is expected.

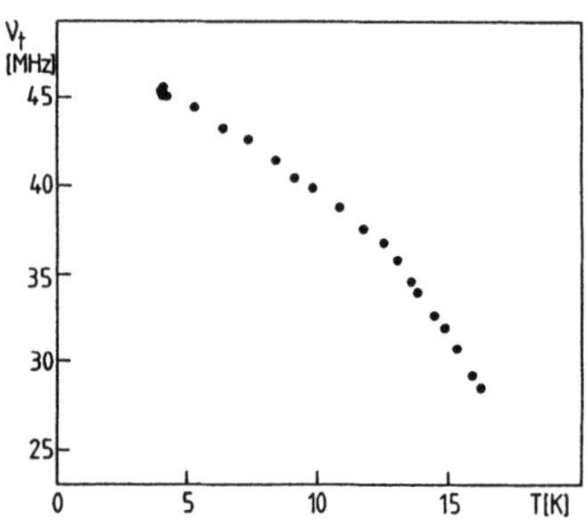

Figure 4. The temperature dependence of the TT-tunnel splitting in NH_4ClO_4, studied in the temperature range from 4.2 K to 16 K.

References

1. Coppens P., Van Gerven L., Clough S. and Horsewill A.J., J. Phys. C 16, 567 (1983)
2. Coppens P., Doctor's thesis, Leuven (1984)
3. Punkkinen M., Ylinen E.E. and Bunkov Y.M., J. Phys. C 10, 4149 (1977)
4. Prager M., Raaen A.M. and Svare I., J. Phys. C 16, L181 (1983)
5. Vandemaele G., Coppens P. and Van Gerven L., Phys. Rev. Lett. 56, 1202 (1986)

Coupled Tunneling of Methyl Groups in Solids as Studied by Proton NMR Method

S. Takeda and H. Chihara

Department of Chemistry, Faculty of Science, Osaka University, Toyonaka, Osaka 560, Japan

Rotational tunneling frequencies or spectra of methyl groups in several solids were measured by NMR methods, especially by NMR field cycling technique. The temperature dependences of tunneling frequencies were also observed. A comparison between the simple tunneling spectrum of isolated methyl group of 2,5-dimethylpyrazine and complicated spectrum of durène, tetramethylpyrazine and hexamethylbenzene indicates that there is coupling between closely located methyl groups of durene, tetramethylpyrazine and hexamethylbenzene molecules.

1. Introduction

Coupled motion of atoms, radicals or molecules in solids is one of very interesting problems of molecular dynamics. Energy splitting of the torsional levels of methyl group due to tunneling effect is very sensitive to change of potential barrier hindering rotation of a methyl group. Thus, mechanical coupling between closely located methyl groups plays more prominent rôle in the rotational tunneling than in the classical hopping rotation. Features of tunnel rotation of a single methyl group have been reasonably understood but there remain many unsolved problems for coupled rotational tunneling of methyl groups.

In this paper, we discuss the phenomena which are believed to result from the coupling effects between the methyl groups on the basis of our previously published results, *i.e.* the tunneling frequencies or spectra of hexamethylbenzene /1/, tetramethylsilane /2/, durene and tetramethylpyrazine /3/, which were investigated by NMR field cycling technique /3/, and some new experimental results.

2. Tunnel Spectra of a Single Methyl Group of 2,5-Dimethylpyrazine

The field cycling spectra of 2,5-dimethylpyrazine (2,5-DMP) in the stable phase are shown in Fig.1. The details of NMR field cycling technique were described elsewhere /3/, which is used to search the magnetization peaks due to so-called tunneling-assisted relaxation. As shown in Fig.1, only two peaks of magnetization, the positions of which are $\nu_{0b}(A) = 3.5 \pm 0.3$ MHz (peak A) and $\nu_{0b}(B) = 7.5 \pm 0.4$ MHz (peak B), were observed. The peak B is clearer than the peak A. These two peaks make a pair corresponding to the condition that $\nu_{0b}(A) : \nu_{0b}(B) = 1 : 2$. This relation is well understood by the prediction of spin-lattice relaxation theory for a single

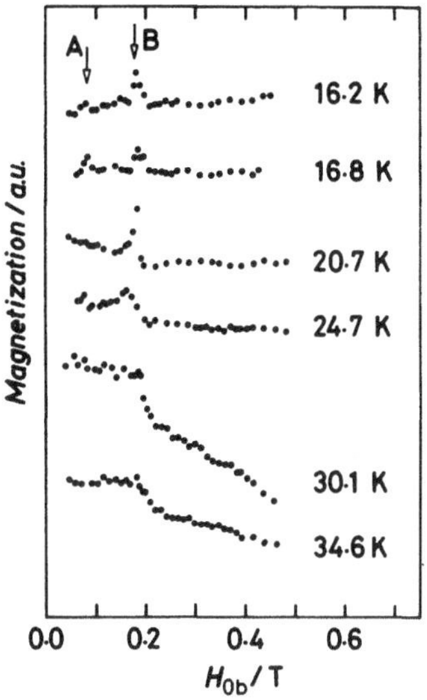

Fig.1 Field cycling spectra of 2,5-DMP. FID signal was observed at H_{0a} = 0.484 T (ν_{0a} = 20.6 MHz). It always took 7 s to change the external field from H_{0a} to another value H_{0b} where the magnetization was allowed to recover for 321 s at 16.2 and 16.8 K, 255 s at 20.7 K, 170 s at 24.7 K, 105 s at 30.1 K and 40 s at 34.6 K. The field cycling runs were conducted from high to low magnetic field at any particular temperature except for 24.7 K at which the runs were conducted from low to high magnetic field.

methyl group, namely $\nu_{0b}(A) = \nu_t/2$ and $\nu_{0b}(B) = \nu_t$. Here ν_t is the tunneling frequency. We thus obtain ν_t = 7.5 MHz for 2,5-DMP at the temperatures shown in Fig.1. The identification of only one pair of magnetization peaks (only one kind of tunneling splitting) is in agreement with the fact that there is only one kind of methyl group in this crystal /4/. Therefore, the methyl group of 2,5-DMP is described as an isolated methyl group.

The tunneling frequency did not change up to 34.6 K as shown in Fig.1 and an evidence of decrease of the tunneling frequency in the higher temperature region was observed as described below. Below the temperature at which the maximum of the spin-lattice relaxation rate was observed, the spin-lattice relaxation was nonexponential, due to the coupling between the Zeeman- and the tunnel-system and the relative weights of magnetization of fast and slowly relaxed components changed as the temperature increased /5/. This result is quantitatively consistent with the picture that the decrease of tunneling frequency, Δ_t, is described by

$$\Delta_t \simeq a \exp(-\hbar\omega_p/kT) \qquad (1)$$

as Hewson claimed /6/ and $\hbar\omega_p$ = 2 ~ 3 kJ/mol (160 ~ 250 cm^{-1}) /5/. Here ω_p is the frequency of phonon which is coupled with methyl rotation.

3. Tunnel Spectra of Two Closely Located Methyl Groups

The field cycling spectra of durene are shown in Fig.2(a). The spectra are rather complicated compared with those of 2,5-DMP; six peaks of magnetization, the positions of which are at 9.2 ± 0.4 MHz (A), 12.2 ± 0.3 MHz (B), 15.6 ± 0.5 MHz (C), 17.8 ± 0.5 MHz (D), 25.3 ± 0.6 MHz (E) and 29.3 ± 0.7 MHz (F), were identified between 15 and 27 K. The positions of these peaks did not change up to 27 K. When we consider the pairwise relation about the positions of magnetization peaks as in the case of 2,5-DMP, we were able to identify three pairs of magnetization peaks, *i.e.* three different sets, A-D, B-E and C-F. Similarly, for tetramethylpyrazine, we found three pairs of magnetization peaks /3/.

The crystallographic data indicates that there are only two kinds of crystallographically inequivalent methyl groups in both durene and tetramethylpyrazine /7/. Therefore, if there is no coupling between two closely located methyl groups in a molecule (these are crystallographically inequivalent), only two pairs of magnetization peaks must be observed. The observation of three pairs of magnetization peaks demonstrate strong coupling between the methyl groups of durene and tetramethylpyrazine. This coupling between methyl groups of durene and tetramethylpyrazine is much stronger than that in acetone /8/. The tunnel splittings of torsional ground state of durene, which were deduced from the field cycling spectra, are shown in Fig.2(b).

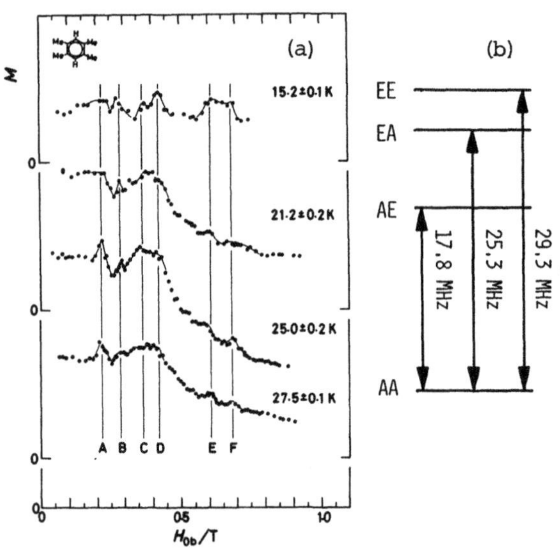

Fig.2(a) Field cycling spectra of durene. FID signal was observed at H_{0a} = 0.458 T (ν_{0a} = 19.5 MHz). The field cycling runs were conducted from low to high magnetic field at any particular temperature to make the base line of the spectra smooth. The time allowed for magnetization recovery at H_{0b} was chosen as 608 s (15.2 K), 153 s (21.2 K), 61.5 s (25.0 K) and 27.4 s (27.5 K).

Fig.2(b) The tunnel level scheme of torsional ground state of durene.

4. Tunneling of Methyl Groups Tied in a Ring; Hexamethylbenzene

The two pairs of magnetization peaks were identified in the field cycling spectra of hexamethylbenzene (HMB) in the low-temperature limit (4.2 K) /1/. The frequency of one pair of peaks (ν_t(B) = 7.9 MHz at 4.2 K) decreased more rapidly and became zero at about 30 K, while the other pair of peaks (ν_t(A) = 9.6 MHz at 4.2 K) remained invariant at these temperatures as shown in Fig. 3. At a higher temperature, the tunneling frequency ν_t(A) seems to decrease and becomes about 2 MHz at 60 K, which was deduced from the measurements of temperature dependence of $T_{1\rho}$ (spin-lattice relaxation time in the rotating frame) /9/. The temperature dependences of tunneling frequencies, ν_t(A) and ν_t(B), are very different from one another although the

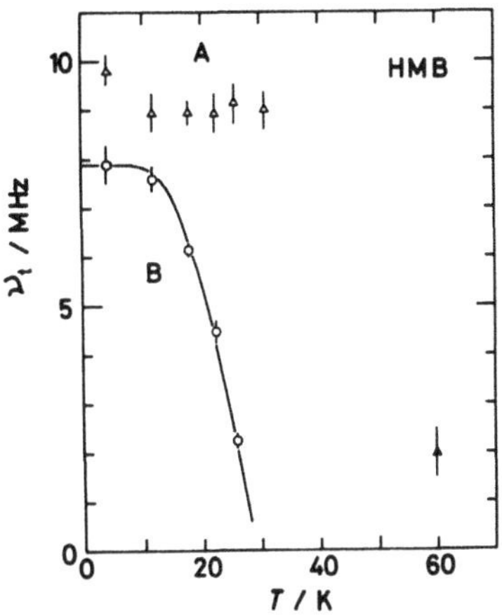

Fig.3 Temperature dependence of tunneling frequencies, $\nu_t(A)$; △ and $\nu_t(B)$; ○ , of HMB. The solid curve was calculated with $\Delta_t = (7.9 \times 10^6 - \nu_t) = 1 \times 10^8 \exp(-\hbar\omega_p/kT)$ where $\omega_p = 51\ \mathrm{cm}^{-1}$. ▲ was estimated by temperature dependence of $T_{1\rho}$ /9/.

tunneling frequencies, 9.6 and 7.9 MHz at 4.2 K, are not so different. This fact cannot be reasonably understood by the picture of isolated independent methyl groups and is probably rationalized when we consider the effect of mechanical coupling between the six methyl groups which are tied in a ring.

The strong coupling between neighboring methyl groups in HMB molecule is confirmed to exist by comparing the hindering barrier of methyl rotation of HMB and that of 1,3,5-trichloro-2,4,6-trimethylbenzene (TCTMB). The temperature variation of spin-lattice relaxation rate, T_1^{-1}, of HMB and TCTMB /10/ is shown in Fig. 4. The high temperature activation energy is 7.9 kJ/mol for HMB and it is 2.4 kJ/mol for TCTMB. The large reduction of the hindering barrier of the methyl rotation due to alternate replacement of the methyl groups by chlorine atoms (which have nearly

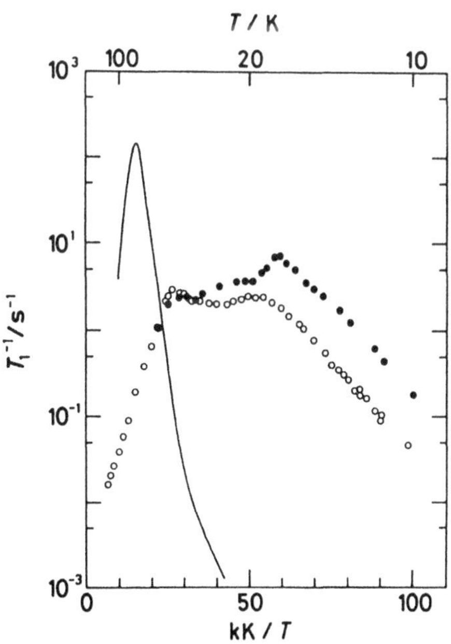

Fig.4 Temperature dependence of spin-lattice relaxation rate, T_1^{-1}, of TCTMB, ○; observed at 15.1 MHz and ●; observed at 6.95 MHz by 180°-τ -90° method. High and low temperature maxima of T_1^{-1} are attributed to the relaxation due to intramethyl and intermethyl dipole interaction, respectively. The nonexponential relaxation due to coupling between Zeeman and rotational polarization system was observed at high temperature and therefore $T_1^{-1} = (\tau_{1/2}/\ln 2)^{-1}$ was plotted in this figure. Here $\tau_{1/2}$ is half-recovery time of magnetization. The appropriate analysis led us to the result that the correlation time was described by $\tau_c^{-1} = 1.0 \times 10^{13} \exp(-2.4\ \mathrm{kJ\ mol^{-1}}/RT) + 2.5 \times 10^{10} \exp(-0.8\ \mathrm{kJ\ mol^{-1}}/RT)$ and the tunnel frequency was estimated to be 830 MHz at 36 K /10/. The solid curve is T_1^{-1} of HMB measured at 10 MHz /12/.

the same van der Waals radius as that of methyl group and are spherical in contrast with the methyl group) indicates that there is strong mechanical coupling between the closely located methyl groups in HMB molecule.

The decrease of tunneling frequency $\nu_t(B)$, $\Delta_t(B)$, can be described by the formula (1) and the phonon frequency, $\omega_p = 50 \pm 5\ cm^{-1}$, was derived by fitting (1) to the observed $\Delta_t(B)$. The phonon frequency of the librational mode about the six-fold axis of HMB is $50 \sim 60\ cm^{-1}$ /11/. The tunneling mode B of methyl groups of HMB molecule is also coupled to this librational phonon.

References

1. S. Takeda, G. Soda and H. Chihara: Solid State Commun. 36, 445 (1980)
2. S. Takeda and H. Chihara: J. Phys. Soc. Jpn. 51, 642 (1982), *ibid.* 54, 3690 (1985)
3. S. Takeda and H. Chihara: J. Magn. Reson. 54, 285 (1983)
4. It was confirmed by ^{14}N-NQR measurement (T. Eguchi and M. Kishita: J. Phys. Soc. Jpn. 53, 3640 (1984)) that there was only one kind of ^{14}N in the stable phase of 2,5-DMP. This fact indicates that there is also only one kind of methyl group in this phase.
5. S. Takeda and H. Chihara: in preparation for publication
6. A.C. Hewson: J. Phys. C 15, 3841 (1982)
7. D.T. Cromer, A.J. Ihde and H.L. Ritter: J. Am. Chem. Soc. 73, 5587 (1951), D.T. Cromer: J. Chem. Phys. 61, 245 (1957), C.H. Stam: Acta Crystallogr. Sect. B 28, 2630 (1972)
8. S. Clough, A.J. Horsewill and P.J. McDonald: J. Phys. C 17, 1115 (1984)
9. K. Morimoto: J. Phys. Soc. Jpn. 50, 2404 (1981)
10. S. Takeda and H. Chihara: in preparation for publication
11. Von D. Bougeard, P. Bleckmann and B. Schrader: Ber. Bunsenges. Phys. Chem. 77, 1059 (1973), J.J. Rush: J. Chem. Phys. 47, 3936 (1967)
12. S. Takeda and H. Chihara: unpublished result. The result is almost consistent with the published T_1^{-1} of HMB (P.S. Allen and A. Cowking: J. Chem. Phys. 47, 4286 (1967), H.W. Bernard, J.H. Tanner and J.G. Aston: J. Chem. Phys. 50, 5016 (1969) and reference /9/).

Tunneling of CH_3 Groups in Molecular Crystals of the $X(CH_3)_n$ Type

J. Peternelj[1], *I. Jencic*[1], *B. Cvikl*[2], *and M.M. Pintar*[3]

[1]Institute "Jozef Stefan", University E. Kardelj, Ljubljana, Yugoslavia
[2]Faculty of Engineering, University of Maribor, and "Jozef Stefan" Institute, University E. Kardelj, Ljubljana, Yugoslavia,
[3]Department of Physics, University of Waterloo, Waterloo, Ontario, Canada N2L 3G1

A system of coupled CH_3-type torsional oscillators embedded in a crystal field of definite symmetry and interacting with an externally applied d.c. magnetic field is considered. Interaction among the neighboring CH_3 groups is assumed to be the result of electrostatic interaction of the respective charge distributions. To obtain an approximate but nevertheless reasonably accurate description of the lowest lying energy eigenstates we examine the minima of the potential energy of the system. Next we construct Gaussian wave packets peaked at each of these minima and form linear superpositions in order to respect the symmetries of the problem under consideration.

1. Introduction

The experimental evidence concerning the tunneling of CH_3 groups in solid lattices [1], indicates that the inclusion of the interaction among neighboring groups into the corresponding Hamiltonian might be necessary for the proper interpretation of the energy spectra in the low-temperature limit. In order to investigate this point, we assume that this interaction is contained in the electrostatic energy of the respective charge distributions as represented by the molecular groups in question. Performing the multipole expansion of this energy, it turns out that the first non-vanishing term which depends on the angular variables of both interacting groups, is the octopole-octopole interaction. All other non-zero terms which are of lower order than the octopole-octopole term are conveniently lumped together and appear in the Hamiltonian only as an addition to the "external" potential. The problem is thus reduced to the solution of the resulting Schroedinger equation, which in general cannot be obtained in analytic form. In what follows, we will describe an approximate solution of the low-temperature tunneling spectrum of interacting CH_3 groups for a few specific examples.

2. The Hamiltonian of the system

For the purpose of this calculation the Hamiltonian can be written as

$$H = H_Z + H_{OT} + H_{TT}, \qquad (2\text{-}1)$$

where H_Z is the Zeeman energy, while H_{OT} and H_{TT} represent the torsional Hamiltonian and the torsion-torsion interaction, respectively. For a system of N CH_3 groups we have

$$H_{OT} = \sum_{n=1}^{N} \left\{ - \frac{\hbar^2}{2J} \frac{d^2}{d\gamma_n^2} + V_3 \left(1 - \cos 3(\gamma_n - \delta_n)\right) \right\} \qquad (2\text{-}2)$$

where J denotes the moment of inertia of the CH_3 group corresponding to the rotation around its symmetry axis as measured by the Euler angle $\gamma_n . \delta_n$ is a constant phase factor, which is left unspecified for the time being. As already mentioned, H_{TT} is obtained from the multipole expansion of the electrostatic energy of two rigid charge distributions. If $(\alpha_n, \beta_n, \gamma_n)$ are the Euler angles [2] determining the orientation of the n-th group with respect to the lattice fixed coordinate system H_{TT} can be approximated by

$$H_{TT} = \frac{1}{2} V_3 \sum_{n \neq n'} \Big(g^{(+)}_{n,n'}(\alpha_n,\beta_n;\alpha_{n'},\beta_{n'}) \cos 3(\gamma_n+\gamma_{n'}) + \\ + g^{(-)}_{n,n'}(\alpha_n,\beta_n;\alpha_{n'},\beta_{n'}) \cos 3(\gamma_n-\gamma_{n'}) \Big) \quad . \tag{2-3}$$

The coupling constants $g^{(\pm)}_{n,n'}$ will be defined later on, when we discuss specific examples.

3. The Schroedinger equation

The solutions of the Schroedinger equation corresponding to the Hamiltonian (2-1), can be written as a product of spin functions and functions depending only on the angular variables $\vec{\gamma} \equiv (\gamma_1, \ldots, \gamma_N)$. Correspondingly, the energy is written as a sum of Zeeman energy and torsional energy E_T. The latter is determined by

$$\left(-\frac{\hbar^2}{2J} \nabla^2 + V(\vec{\gamma})\right) \Phi(\vec{\gamma}) = E_T \, \Phi(\vec{\gamma}) \quad , \tag{3-1}$$

where $\nabla \equiv \left(\frac{\partial}{\partial\gamma_1}, \ldots, \frac{\partial}{\partial\gamma_N}\right)$ and the potential $V(\vec{\gamma})$ is defined as

$$V(\vec{\gamma}) = \sum_{n=1}^{N} V_3 \big(1 - \cos 3(\gamma_n - \delta_n)\big) + H_{TT} \quad . \tag{3-2}$$

In the small oscillations approximation, we write $\vec{\gamma} = \vec{\gamma}_c + \vec{\eta}$, where $\vec{\gamma}_c$ is the solution of

$$\nabla V(\vec{\gamma}_c) = 0 \quad . \tag{3-3}$$

Taking $\vec{\eta} \equiv \vec{\gamma} - \vec{\gamma}_c$, to be small, the ground state wave function $\Phi_0(\vec{\gamma})$ centered at $\vec{\gamma}_c$ is

$$\Phi_0(\vec{\gamma}) = \frac{(\det M)^{1/4}}{\pi^{N/4}} \, e^{-\frac{1}{2}\sum_{n,m} M_{n,m}(\gamma_n-\gamma_{cn})(\gamma_m-\gamma_{cm})} \quad . \tag{3-4}$$

The matrix M is defined as

$$M_{n,m} = \sum_{\alpha=1}^{N} \left(\frac{J\omega_\alpha}{\hbar}\right) u_n^{(\alpha)} u_m^{(\alpha)} \quad , \tag{3-5}$$

and it is assumed that $\omega_\alpha > 0$, for all α. ω_α^2 and $u^{(\alpha)} \equiv (u_1^{(\alpha)}, \ldots u_N^{(\alpha)})$ are the eigenvalues and the eigenvectors respectively of the matrix $U_{n,m}$, which is defined by

$$U_{n,m} = \frac{1}{J} \left(\frac{\partial^2 V}{\partial\gamma_n \partial\gamma_m}\right)_{\vec{\gamma}_c} \quad . \tag{3-6}$$

4. The tunneling approximation for the case of multiple minima

Let us suppose that $\vec{\gamma}_c$ represents a clasically stable minimum of the potential $V(\vec{\gamma})$. Due to the assumed symmetry of the CH_3 groups, it is clear that $\vec{\gamma}_c+(2\pi/3)\vec{\sigma}$, where $\vec{\sigma}\equiv(\sigma_1,\dots,\sigma_N)$, $\sigma_i=0,\pm1$, also represents a stable minimum of the potential $V(\vec{\gamma})$. To account for the resulting tunneling we form the appropriate linear superposition of the wave functions (3-4) peaked at various minima $\vec{\gamma}_c+(2\pi/3)\vec{\sigma}$, as suggested by the symmetry of the system under consideration. Therefore we write

$$\tilde{\Phi}_{0,\vec{\alpha}} = \sum_{\vec{\beta}} \left(\prod_{i=1}^{N} S_{\alpha i,\beta i} \right) \Phi_{0,\vec{\beta}} \,, \tag{4-1}$$

where $\vec{\alpha}\equiv(\alpha_1,\dots\alpha_N)$, $\vec{\beta}\equiv(\beta_1,\dots,\beta_N)$, $\alpha_i,\beta_i=1,2,3$, and there is a one to one correspondence between $\vec{\beta}$ and $\vec{\gamma}_c+(2\pi/3)\vec{\sigma}$. The matrix $S_{\alpha,\beta}$ is constructed from the characters of the irreducible representations of the point group C_3. In the low-temperature limit the functions (4-1) diagonalize approximately the Hamiltonian H_T defined by (3-1). The tunneling splitting between the ground state and the first excited state, $\omega_T(\alpha_i=1;\ i=1,\dots N;\ \alpha_k=2,3;\ k\neq i)$ and which in the zero coupling limit $g^{(+)}\to 0$, corresponds to the energy difference between A and E levels, is in this approximation given by

$$\hbar\omega_T(\alpha_i=1;i=1,\dots,N;\alpha_k=2,3;k\neq 1) = 3e^{-\frac{\pi^2}{9}M_{kk}}\Bigg\{\frac{\hbar^2}{2J}\frac{\pi^2}{9}(M^2)_{kk} - \frac{2}{9}J\omega_0^2\cos 3(\gamma_{ck}-\delta_k)e^{-\frac{9}{4}M^{-1}_{kk}} + \frac{4}{9}\sum_m\Big[g^{(+)}_{km}\cos 3(\gamma_{ck}+\gamma_{cm})e^{-\frac{9}{4}(M^{-1}_{kk}+M^{-1}_{mm}+2M^{-1}_{km})} + g^{(-)}_{km}\cos 3(\gamma_{ck}-\gamma_{cm})e^{-\frac{9}{4}(M^{-1}_{kk}+M^{-1}_{mm}-2M^{-1}_{km})}\Big]\Bigg\} \tag{4-2}$$

where $\omega_0^2\equiv 9V_3/J$.

5. Discussion

The result (4-2) shows that the set of the first excited states in general forms a band of N doubly degenerate levels. The width of this band is determined, of course, by the strength of the torsional-torsional interaction, and what is more interesting, also by the type of the classical minimum $\vec{\gamma}_c$. For example, if we choose a rectangular lattice of CH_3 groups with their symmetry axes parallel, then with a suitable choice of the phase factors δ_n, the equilibrium orientation of the groups is as shown in Fig. 1. In this case the width of the first excited state vanishes, at least in the first order in the exponential factors $\exp(-(\pi^2/9)M_{kk})$. It appears in general, that those minima which more or less preserve the original translational symmetry of the lattice, generate energy spectra with very narrow bands. On the other hand, if the classical solution breaks the translational symmetry of the Hamiltonian, than (4-2) shows that we obtain a nonvanishing band width already in the first order in the exponential factors $\exp(-(\pi^2/9)M_{kk})$.

For a threedimensional lattice, the coupling constants $g^{(\pm)}$ are in general both different from zero and of the same order of magnitude. However, if we consider an isolated cluster $X(CH_3)_4$ with the tetrahedral arrangement of the respective CH_3 groups, we obtain $g^{(-)}=0$, and

$$g^{(+)} = \frac{8935}{10368}\left(\frac{e^2/R}{V_3}\right)\left(\frac{d}{R}\right)^6 \tag{5-1}$$

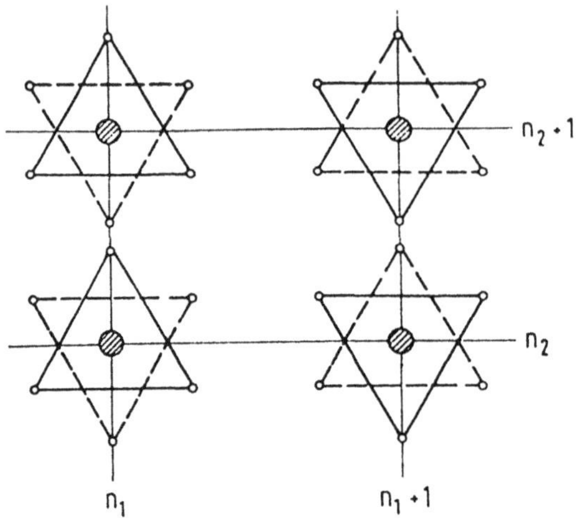

Fig. 1

Equilibrium orientation of the CH_3 groups in the case of a rectangular lattice with the symmetry axes of all the groups being parallel.

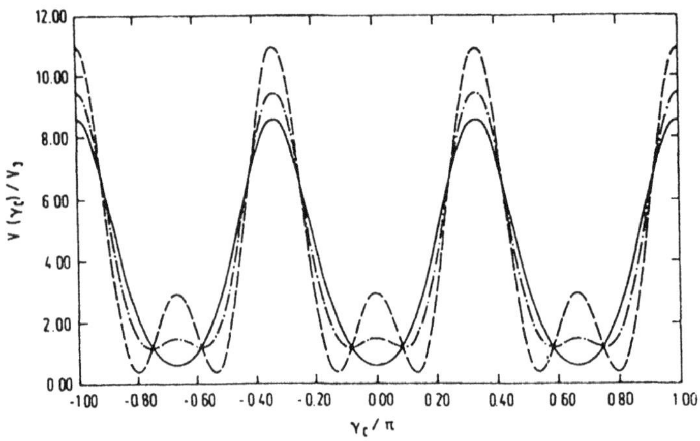

Fig. 2 The angular dependence of the potential $V(\vec{\gamma})$, when all four CH_3 groups have identical orientation. The solid line corresponds to $g^{(+)}=1/4$, and the dashed line to $g^{(+)}=1/2$.

where d is the proton-proton distance within the CH_3 groups, while R is the distance between the centers of the equilateral triangles formed by the protons belonging to two different CH_3 groups. If we take $d = 1.7$ Å, $e^2 = 0.04\ e_0^2$ [3], R = 3 Å [4] and $(V_3/k_B) = 400$ K, we obtain $g^{(+)} = 0.16 < 1/6$. On the other hand, for R = 2.5 Å, (5-1) yields $g^{(+)} = 0.58 > 1/6$. Here we have two distinct situations corresponding to $g^{(+)} < 1/6$ and $g^{(+)} > 1/6$, respectively. When $g^{(+)}$ is less than 1/6 the tunneling spectrum is the same as in the case of isolated CH_3 groups (except for the magnitude of the splitting), while in the case when $g^{(+)} > 1/6$, the corresponding low temperature energy scheme is distinctly different. This result could be anticipated also by examining the $g^{(+)}$ dependence of the potential $V(\vec{\gamma})$ as shown in Fig. 2, and subjected to the constraint $\gamma_1 = \gamma_2 = \gamma_3 = \gamma_4$.

References.

1. K.R. Sridharan, W.T. Sobol and M.M. Pintar, J. Chem. Phys., 82, 4886 (1985).
2. H. Goldstein, Classical Mechanics, 2nd. ed. (Addison-Wesley, 1980) p. 146.
3. L.C. Snyder and H. Bosch, Molecular Wave Functions and Properties: Tabulations from SCF Calculations, (John Wiley & Sons, 1972).
4. G.W. Smith, J.Chem. Phys., 42, 4229 (1965).

Part III

Rotational Tunneling III – Surfaces, Isotope Effect, Matrix Isolation

Hindered Rotation and Tunnelling of Adsorbed Species

S.J.M. Collinson, G.M. Hunt, M.V. Smalley, M.L. Statham, and R.K. Thomas

Physical Chemistry Laboratory, South Parks Road, Oxford, OX1 3QZ, UK

1 Introduction

Hindered rotation and rotational tunnelling spectra can be used to study adsorbed systems in two ways. Firstly, they can give direct information about intermolecular forces. This is particularly important for adsorbed systems because, although there are few that are suitable for such measurements, we do not yet understand molecule-surface forces even for such a simple model adsorbent as graphite. Secondly, the spectra are sensitive to the surroundings of a molecule and can therefore probe its environment. In bulk solids this is not usually at all interesting because direct information about the structure will usually be available from diffraction. However, for adsorbed systems we are often not able to determine the structure of the surface, what site a molecule occupies when there is a choice, or even the structural relation between adjacent molecules. We give examples of both types of application in this paper.

2 Methane on Graphite

Methane on graphite is in many ways a model system for the study of physisorbed systems. It has been well studied theoretically [1,2], thermodynamically [3,4], by neutron diffraction [5,6,7], and by neutron inelastic scattering [8,9]. As for other small molecules such as the inert gas, the structure is determined by the competition between adatom-adatom and adatom-surface forces. The van der Waals diameter of methane is slightly smaller than the commensurate (C) $\sqrt{3} \times \sqrt{3}$ R 30° structure normally adopted by these small molecules on graphite. At coverages well below a monolayer methane adopts this C structure but as the monolayer nears completion it undergoes a transition to an incommensurate (I) structure. The exact nature of the C-I phase transition is still not completely understood from either a theoretical or experimental standpoint.

The rotational tunnelling spectrum of CH_4 in its C phase on graphite has been analysed to give quantitative information about the reorientational potential of the molecule [8]. The tunnelling splitting for CH_4 is shown schematically in Figure 1. In a tetrahedral potential there are three levels but in the trigonal field of the surface and surrounding molecules the degeneracy of the T levels is split and we have the four levels and five possible transitions in the incoherent neutron scattering spectrum, shown on the right of Figure 1. These have all been observed and can be fitted by two parameters from which two barriers to rotation have been derived. Neutron diffraction shows that CH_4 adopts a tripod down configuration on the surface. The barrier heights are 202 cm^{-1} for rotation about the CH bond normal to the surface and 172 cm^{-1} about one of the three equivalent CH bonds that point towards the surface. The interesting feature is that for the barrier heights to be this way round the molecule-molecule interactions must be comparable with the molecule-surface interactions.

The importance of the molecule-molecule interactions is shown by the completely different spectra obtained from CH_4 when it is incorporated into a layer of krypton on graphite [10]. Krypton forms the same C structure as methane at coverages below

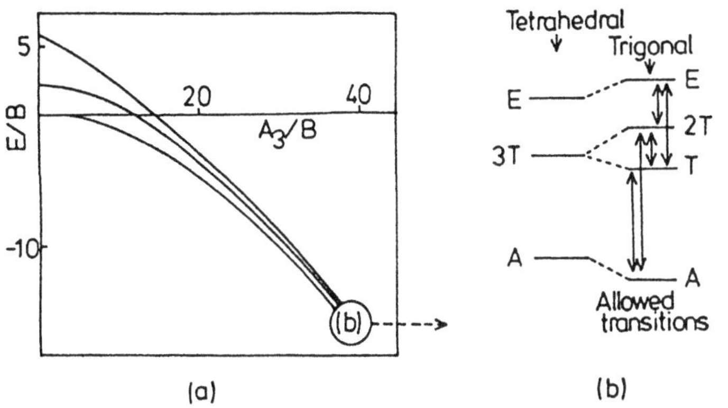

Fig. 1 (a) Correlation of free rotational levels on the left with tunnelling levels of the ground librational state of CH_4 on the right. (b) Expanded diagram of the ground state levels with allowed transitions

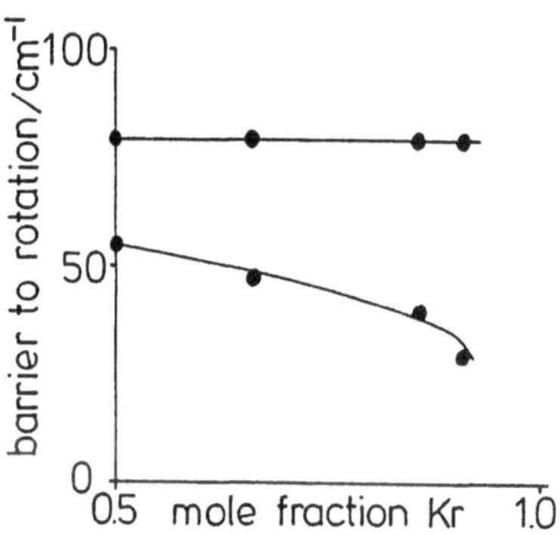

Fig. 2 Variation of the two barriers to rotation of CH_4 in mixed CH_4/Kr layers on graphite

a monolayer and it is possible to incorporate CH_4 into this layer up to a mole fraction of about 0.3. It is found that the spectra of CH_4 in such a mixed layer are characteristic of only weakly hindered rotation. Both barriers to rotation are very much reduced. Figure 2 shows how they vary with the mole fraction of Kr in the layer. The barrier height for rotation about each of the three equivalent CH bonds is 79 cm^{-1} and does not vary significantly with Kr concentration. This is consistent with it being caused only by surface-molecule interactions. The unique barrier is 55 cm^{-1} when the layer is saturated with CH_4 but drops to about 30 cm^{-1} at 90% Kr and looks as though it would drop further at higher Kr concentrations. An atom-atom potential calculation of the two barriers to rotation for a CH_4 molecule surrounded by six Kr atoms gives 72 cm^{-1} and $\sim$ 0 cm^{-1}. The first agrees closely with the experimental result but the second does not. This may be because we have not reached the limiting value experimentally or it may be that we have made false assumptions about the potential.

In the two examples considered above we have quoted extremely accurate values of the barrier heights. However, it should be realized that this accuracy is limited by the assumptions about the shape of the potential. For CH_4 the potential was taken to be a two parameter one, such that rotation about any one of the CH bonds consisted of a single Fourier component. Using the pocket state method of HÜLLER [11] we could fit the observed spectrum exactly. However, we have not yet been able to account quantitatively for the spectra of CH_3D on graphite. This is a much more difficult problem to solve, both experimentally and theoretically, but at present the discrepancy seems likely to have its origin in the assumptions about the potential. For CH_4 in the krypton layer the barriers were too low to permit use of the pocket state treatment and we used the method of SMITH [12] with free rotor wavefunctions as our starting basis set. We assumed the same shape of potential as for CH_4 alone but were not able to fit the observed spectra with anything like the same precision. For a CH_4 molecule, rotating about an axis normal to the surface and surrounded by six kryptons, it is plausible that a sixfold term in the potential should be added to the lowest Fourier component, the threefold term. This could then be the cause of the discrepancy.

For CH_4/Kr mixtures the tunnelling spectrum proved also to be a useful probe of the structure of the system. Although both CH_4 and Kr form C layers and mix in all proportions in the bulk [13], they phase separate on graphite. The tunnelling spectra show that one phase is pure CH_4, the spectrum being identical with that of pure CH_4 on graphite, and the other is CH_4 in Kr up to a limiting fraction of about 0.3, whose spectrum has already been discussed. The most plausible explanation of this separation is that CH_4 and Kr prefer to be adsorbed at different sites. Theoretical calculations indicate that Kr prefers to be above the centre of a carbon hexagon while CH_4 has a slight preference for being directly above a carbon atom. This slight preference allows some CH_4 to be incorporated into the Kr layer but the stronger preference of Kr for the hexagon site prevents any Kr being incorporated into the CH_4 layer. Unfortunately, it has not been possible to determine the adsorption site by any independent experimental means.

The variation of the tunnelling spectrum of CH_4 on graphite with coverage has been a unique probe of the nature of the C-I transition [14]. This transition in general occurs through a series of dislocations which occur because of the mismatch of adsorbate and adsorbent lattices [15]. Depending on the particular system, these structures may be superlattices with a regular arrangement of domain walls. For example, Xe on graphite forms a series of regular hexagonal structures in the region of coverage of the C-I transition [16]. Such static structures will only be observed if there is sufficient energy to lock them to the substrate. In the case of CH_4, PHILLIPS and coworkers [2] have predicted that the variation in energy as a CH_4 molecule moves across the graphite surface is too small to generate domain walls. The incommensurate phase is then predicted to be a "floating" phase, in contrast to what is expected for most other systems. Diffraction has so far been unable to distinguish the two possibilities.

Tunnelling should be particularly sensitive to the differences. The molecule-molecule interactions will increase on going to the I phase and, given the great sensitivity of the tunnelling splitting to the barriers to rotation, there should be a large change in the splitting across the C-I transition. The change is at first sight a surprising one (Figure 3). To a first approximation the spectrum per molecule is independent of coverage in the C phase. On formation of the I phase its intensity falls rapidly with increasing coverage but there is little change in the energy of the transitions. The only possible explanation of this behaviour is that the I phase has a domain structure which includes domains of C phase. At the centre of a domain the spacing of the molecules is determined by the graphite (C phase) but at the edges the molecules are closer together and out of registry with the graphite (heavy domain walls). As the amount of CH_4 is increased the width and the number of walls will also increase, reducing the intensity of the C-phase tunnelling spectrum. Unfortunately, it is not possible to draw any conclusions about the symmetry of the domain wall structure.

3 Hydrogen on Rutile

Macroscopic crystals of the rutile form of TiO_2 have surfaces consisting of the crystal planes (110), (100) and (101) in the approximate ratio 3:1:1. It is usually assumed that this ratio still holds for the microscopic crystals which make up a high surface area powder. Since the unit cell is tetragonal the three planes that make up the surface will have different structures and therefore different adsorption properties. Water may be both chemisorbed and physisorbed on these three surfaces but to differing extents. Thus, without any treatment, all three surfaces are cov-

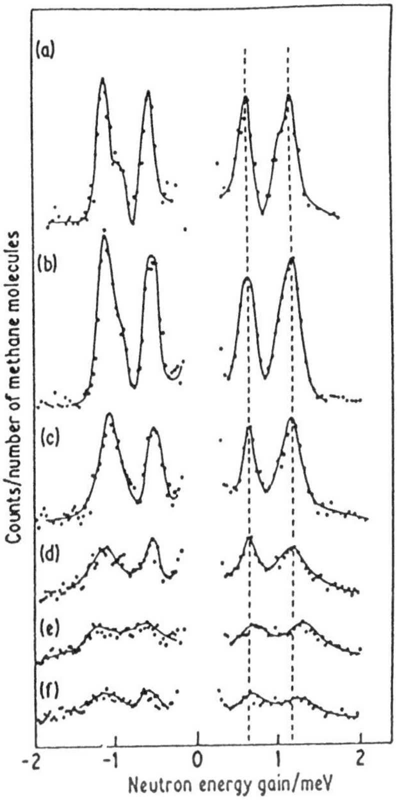

Fig. 3 Tunnelling spectra of CH_4 across the C-I transition at 4.2 K. Coverages (a) 0.46 (b) 0.94 (c) 1.05 (d) 1.13 (e) 1.22 (f) 1.37

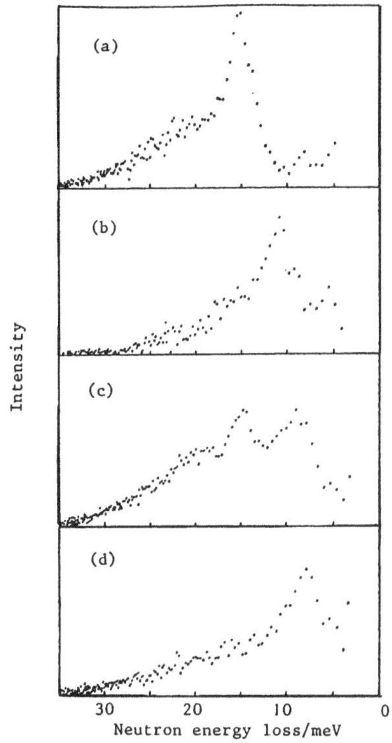

Fig. 4 Neutron energy loss spectra of H_2 on rutile; (a) 0.75 monolayers of H_2 on "wet" rutile, (b) 0.75 monolayers of HD on "wet" rutile, (c) 0.8 monolayers of H_2 on "dry" rutile, (d) 0.5 monolayers of H_2 on "dry" rutile

ered with physisorbed water. Outgassing at room temperature removes physisorbed water from all three faces, outgassing at 200°C removes chemisorbed water from the (100) and (101) faces but leaves chemisorbed water on the (110) face, which is only removed by heating above 200°C [17].

The surface properties above have been deduced largely from infrared spectroscopy. However, comparable conclusions can be drawn independently from the rotational spectrum of H_2 on TiO_2 obtained under a variety of conditions. This is also interesting because TiO_2 may be used to separate ortho and para hydrogen by preferentially adsorbing ortho-H_2, and the separation factor depends on the rotational energy levels on the surface [18].

Four spectra obtained in different circumstances are shown in Figure 4. Figure 4(a) shows the spectrum of H_2 on "wet" rutile, obtained by outgassing at room temperature leaving the whole surface covered with chemisorbed water. There is a sharp line at 15 meV with a broader background perhaps peaking in the 20-25 meV range. The corresponding HD spectrum (Figure 4(b)) has a sharp peak at 11 meV and there is a corresponding large shift in the background scattering. The shift is such that the peak at 11 meV must be a rotational transition. For HD it must be the J=0 to 1 transition and it corresponds to almost completely free rotation. It follows that for H_2 we must also be observing the J=0 to 1 (para to ortho) transition. The J=1 to 2 transition in the more abundant ortho (there is only very slow conversion on this surface) should occur at 29 meV for free rotation. That it does not appear may be partly because features become much weaker as we approach the incident energy of the neutrons (35 meV) and partly because the transition is lost in the broad background scattering. However, we can conclude that H_2 is freely rotating on "wet" rutile.

"Dry" rutile is prepared by outgassing at 200°C and this removes chemisorbed water from the (100) and (101) faces, but not the (110). Figure 4(c) shows a broadish line at 9 meV as well as the two features observed for "wet" rutile. If the coverage is reduced the last two features disappear and we are left only with the line at about 9 meV (Figure 4(d)). These observations together with those of the previous paragraph suggest that the 9 meV transition corresponds to more strongly bound, more hindered, H_2 on the (100) and (101) faces while the other two features are free rotation on the "wet" (110) face. The way that these changes occur with coverage suggest also that the proportion of (110) planes making up the surface is less than the 6o% characteristic of macroscopic crystals and may be more like 40%.

The assignment of the line observed for H_2 on "dry" rutile is not as clear. The shift on replacement of H_2 by HD is so large that the line is lost in the rather broad elastic peak. This suggests that for H_2 we are now observing the J=1 to J=2 line. Without observing any other lines of the spectrum it is not possible to draw definite conclusions about the potential. The simplest potential has the form $C \cdot \cos^2\theta$ where θ is the angle the molecular axis makes with the surface normal. When C is negative the most stable orientation of the molecule is perpendicular to the surface. In a potential of this shape a J=1 to J=2 transition at 8 meV gives a barrier of 2.3 kJ mol^{-1} and predicts that the J=0 to J=1 line occurs at about 9 meV. The two lines therefore approximately coincide and would not be distinguished. No other transitions fall within the energy range of the experiment. This assignment is supported by the value calculated for the separation factor for ortho-para separation. This is found to be 1.26 at 90K, in close agreement with the observed value of 1.3 [18].

4 Hydrogen in Pillared Clays

Smectite clays consist of flat aluminosilicate plates carrying a negative charge. The plates are interleaved by exchangeable hydrated cations. The larger the cation, the further apart the plates are held, and the larger the cationic charge, the more widely spaced are the cations in the interlamellar space. Organic cations of the type $N(CH_3)_4^+$ and $NH_3^+CH_2$-$CH_2NH_3^+$ (ethylene diammonium) may be intercalated into some clays and after gentle heating to remove water a porous material remains, which will sorb gases consisting of small molecules [19]. We have investigated the internal and external surfaces of such materials using hydrogen in the same way as described for rutile. We describe here some experiments on a synthetic laponite intercalated with the ethylene diammonium ion.

Figure 5 shows the spectrum of hydrogen at various coverages, a coverage of one corresponding to complete filling of the internal pores. At a coverage of 0.3 we observe a broad peak at about 9 meV with a suggestion of a shoulder at about 13 meV. As the coverage is increased to 0.6 both peaks increase in intensity but the one at 13 meV increases more. At a coverage of 0.9 the 13 meV peak becomes the strongest peak in the spectrum and there is a suggestion of a peak at about 16 meV. When we saturate the clay with hydrogen the peak at 16 meV becomes more pronounced and there is a new peak at 22 meV.

On the basis of the change of the spectrum with coverage we can identify four sites, two within the pores and two on the external surface. As with TiO_2 the most strongly binding site also has the strongest hindering potential. There are no paramagnetic sites on this clay so that ortho-para conversion is slow and the dominant contribution must therefore be from the J=1 to J=2 transitions. As for H_2 on TiO_2 there are not enough lines to make a clear decision about the potential possible. The one most consistent with the observations is an octahedral field with a negative field parameter giving eight equivalent potential minima. The energy

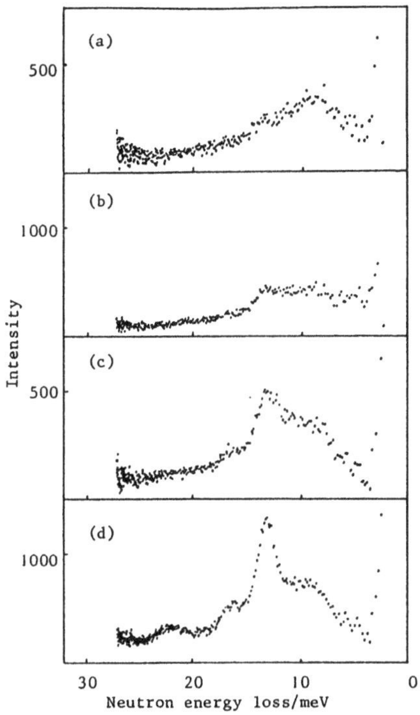

Fig. 5 Neutron energy loss spectra of H_2 in ethylene diammonium laponite at 5 K and coverages of (a) 0.3 (b) 0.6 (c) 0.9 and (d) 1.2

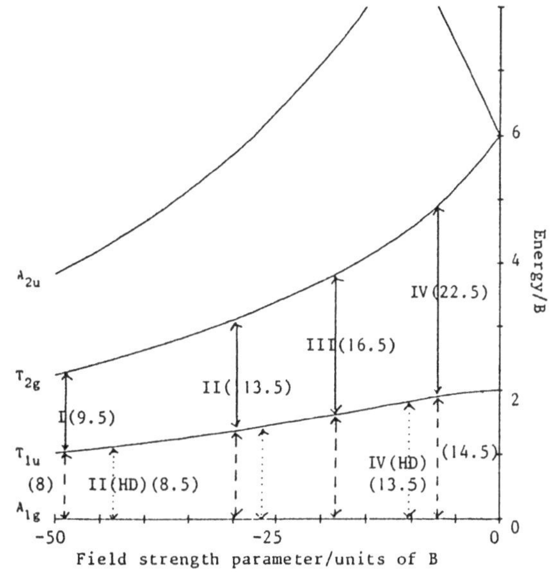

Fig. 6 Energy levels of a linear molecule in an octahedral potential with a negative field parameter

levels have been calculated following SMITH [20] and are shown in Figure 6. The observed transitions for the four sites are marked. There should be corresponding weaker transitions for para hydrogen, marked as dashed lines. Their energies are such that they would not clearly be observed amongst the strong ortho transitions.

However, in neutron energy gain there is a broad peak centred at about 10 meV which corresponds to the overlapping J=1 to J=0 transitions. We can also use Figure 6 to predict the HD transitions. These are marked as dotted lines in Figure 6 and are consistent with an HD spectrum obtained at full coverage, although the resolution is now insufficient to make a clear assignment.

5 References

1. E.S. Severin, D.J. Tildesley: Mol. Phys. 41, 1401 (1980)
2. J.M. Phillips, M.D. Hammerbacher: Phys. Rev. B29, 5865 (1984)
3. A. Thomy, X. Duval: J. Chim. Phys. 67, 286 (1970)
4. H.K. Kim, M.H.W. Chan: Phys. Rev. Lett. 53, 170 (1984)
5. P. Vora, S.K. Sinha, R.K. Crawford: Phys. Rev. Lett. 43, 704 (1979)
6. A. Glachant, J.P. Coulomb, M. Bienfait, J.G. Dash: J. Physique 40, L-543 (1979)
7. T. Rayment, R.K. Thomas, G. Bomchil, J.W. White: Mol. Phys. 43, 601 (1981)
8. M.V. Smalley, A. Hüller, R.K. Thomas, J.W. White: Mol. Phys. 44, 533 (1981)
9. G. Bomchil, A. Hüller, T. Rayment, S.J. Roser, M.V. Smalley, R.K. Thomas, J.W. White: Phil. Trans. R. Soc. B290, 537 (1980)
10. M.P. Allen, C.B. Hovenden, R.P. Humes, S.K. Mehta, R.K. Thomas, M.A. White: Far. Disc. Chem. Soc. 80, 171 (1985)
11. A. Hüller: Phys. Rev. 16, 1844 (1977)
12. D. Smith: J. Chem. Phys. 47, 4680 (1976)

13. S.C. Greer: Phys. Lett. 43A, 73 (1973)
14. R.P. Humes, T. Rayment, M.V. Smalley, R.K. Thomas: to be published
15. P. Bak: Repts. Prog. Phys. 45, 587 (1982)
16. C.W. Mowforth, T. Rayment, R.K. Thomas: Far. Symp. Chem. Soc. 20 (in press)
17. G.D. Parfitt: Prog. Surf. Memb. Sci. 11, 181 (1976)
18. Y. Sandler: J. Phys. Chem. 58, 58 (1954)
19. R.M. Barrer: Zeolites and Clay Minerals (Academic Press, London 1978)
20. D. Smith: J. Chem. Phys. 68, 3222 (1978)

Rotational Tunneling of Matrix-Isolated Tetramethyl Compounds

W. Langel[1], *M. Prager*[2], *O. Schrems*[1], *and E. Knözinger*[1]

[1]Institut für Physikalische Chemie der Universität Siegen, Postfach 101240, D-5900 Siegen, Fed. Rep. of Germany

[2]Institut für Festkörperforschung der Kernforschungsanlage Jülich, Postfach 1913, D-5170 Jülich, Fed. Rep. of Germany

$Sn(CH_3)_4$ was isolated in solid argon, krypton and SF_6, and $Pb(CH_3)_4$ in argon. Tunneling transitions are reported at 68 and 83 μeV for $Sn(CH_3)_4$ and 290 μeV for $Pb(CH_3)_4$ in argon. In Kr and SF_6 matrices no well-defined tunneling transitions were observed, probably due to disorder in the packing site. Hindrance potentials for the methyl group rotation with a high sixfold contribution are found, since the tunneling methyl groups are parallel to 111-planes of the argon lattice and thus face a symmetric environment. Furthermore, the assignments of ν_8, ν_{19} and of a mode at 11 meV are discussed on the basis of far-IR data.

1. INTRODUCTION

The correlation of spectroscopic and diffraction data is a very important cross-check of the understanding of condensed systems. Tunneling spectroscopy is the most sensitive method with respect to the potentials originating from the environment of a molecule. The findings obtained in this way, however, must be interpreted in the light of additional vibrational spectroscopic and diffraction data.

Tetramethyl metal molecules (TMM) have attracted much attention, since rotational tunneling in their methyl groups can be studied conveniently by various methods (INS, NMR) in a homologous series of central atoms (Si, Ge, Sn, Pb) [1,2]. In the bulk phase the inter- and intramolecular interactions hindering the free rotation of the methyl groups cannot be distinguished. In two recent papers [3,4] we presented first results on the isolation of $Sn(CH_3)_4$ embedded in solid argon at cryogenic temperatures. This technique, called matrix isolation, is the most efficient method of reducing the influence of intermolecular interactions. Thus, spectroscopic methods have to be checked whether they are sensitive to the small remaining intermolecular potential.

2. EXPERIMENTAL

The samples were generated by vapor deposition. In contrast to earlier studies of tunneling in isolated molecules by neutron scattering [5,6], this method allows a free choice of the compounds. Samples are generated, however, which are not in thermal equilibrium.

The details of the neutron experiments were described earlier [3,4]. The far-IR and IR experiments were performed with a Fourier transform instrument (Bruker IFS-113v). The energy range of 2.5 to 500 meV could be scanned with a resolution of 150 μeV. Thin films of argon containing 0.5 % TMM were slowly sprayed on to a gold-plated copper mirror kept at 13 K in a closed cycle cryostat (Air Products CSW-202).

3. RESULTS AND DISCUSSION

As was shown earlier [3,4], in $Sn(CH_3)_4$ isolated in argon, one tunneling line at 50 to 80 μeV is obtained depending on the temperature. This splitting is much higher than in the bulk phase (1.7 and 13.3 μeV [2]). The temperature dependence and the line shape indicate that two contributions are present at 68 and 83 μeV [4]. Increasing the temperature from 2 to 25 K strongly shifts the line to lower energies. The activation energy of this shift is around 1.8 meV but slightly different for the two components. A broad transition at 2.1 meV was assigned to the corresponding torsional mode.

The most stable geometry for packing a guest molecule as impurity into the matrix will be the one which induces the smallest perturbation of the surrounding lattice. We suggest that the tetramethyl tetrahedron is embedded in argon in such a way that each triangle formed by CH_3-groups lies in a 111-plane of the fcc argon lattice (fig. 1).

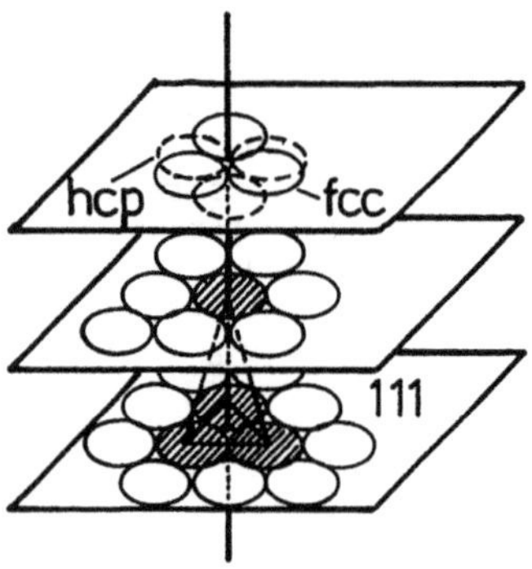

Fig.1 : Site geometry of a TMM-molecule (hatched circles) in an argon lattice (open circles). Dashed circles indicate a hcp stacking fault

The potential quoted earlier, [3,4] for the hindered rotation of the methyl groups of $Sn(CH_3)_4$ in argon have a surprisingly high sixfold contribution. According to our model not only each triangle of CH_3-groups, but also the three H-atoms of each methyl group itself are arranged parallel to a 111-plane. Thus, each CH_3-group is surrounded by a regular hexagon of argon atoms and faces a symmetrical potential.

The respective tunneling splittings reflect the potential which acts on the methyl groups. The widths of the observed tunneling transitions may thus be due to a distribution of non-equivalent hindrance potentials. In the bulk substance two different potentials arise as a result of a deformation of the molecule lowering its symmetry [2]. In the matrix the intermolecular forces are much weaker, and we assume that the tetrahedral symmetry of the molecule itself is maintained. The significant width and especially the occurrence of two overlapping components of the tunneling line of $Sn(CH_3)_4$ in argon can rather be explained by stacking faults in the matrix. They lead to a local hcp-structure (fig. 1). Results obtained by coherent neutron scattering [7] have shown that the structure of the matrices prepared by the method described above tends to be strongly perturbed. The matrices do not have an ideal fcc-lattice. The intensity of the 111-peak increases by a factor of about four after annealing the sample to temperatures above 50 K. These temperatures soften the matrix and allow diffusion and subsequent recrystallization. Temperatures of 30 K and below only induce minor changes in the diffraction pattern.

In the experiments reported here and in [3,4], a tunneling intensity of only a few percent of that of the total incoherent scattering from the protons was observed, indicating that the methyl groups contribute only partly to the observed transitions. After annealing to 30 K a slight increase in the tunneling intensity occurred, which is probably due to a local relaxation. After annealing to more than 50 K the tunneling intensity disappeared. The vibrational spectra [3] show, that clustering and phase separation had taken place.

As stated above, vapor deposition of matrices allows a free choice of the composition of the system to be studied. At first we replaced the central atom Sn by Pb (fig. 2). Immediately after deposition of $Pb(CH_3)_4$ with argon, a broad distribution of intensity is found extending to about 500 μeV. This shows that there is disorder

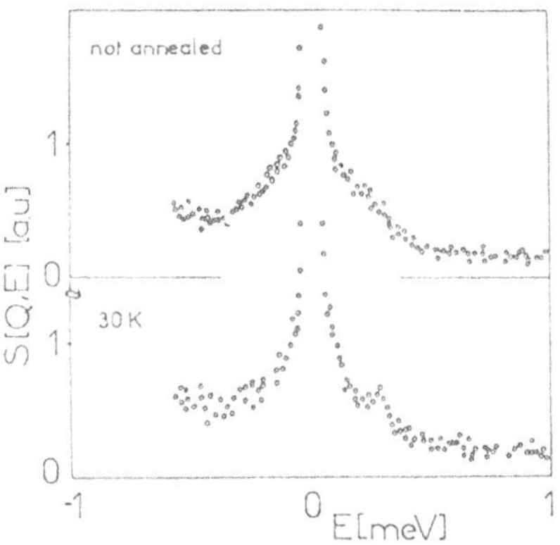

Fig. 2: INS-spectra of $Pb(CH_3)_4$ in argon before (upper trace) and after annealing (lower trace) at 30 K

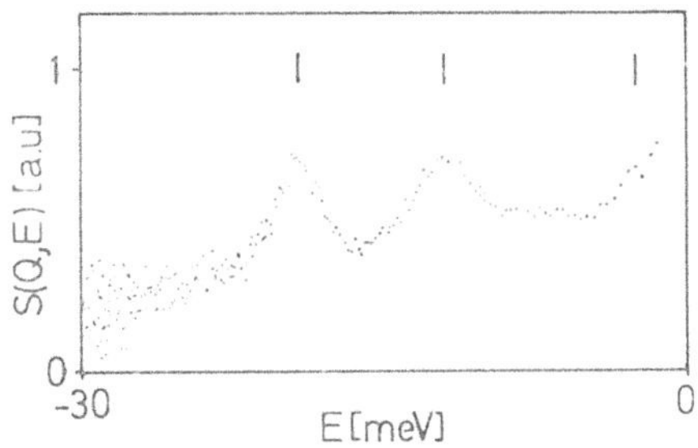

Fig. 3: Low-frequency vibrations of $Pb(CH_3)_4$ in argon (7 K).

in the sample. The upper limit of the tunneling splitting obviously corresponds to $Pb(CH_3)_4$-molecules which are only slightly affected by the matrix cage, if at all. As this limit (500 μeV) is close to the rotational constant of a CH_3-group (655 μeV) with respect to the figure axis, the intramolecular hindrance of the methyl groups in $Pb(CH_3)_4$ is extremely weak.

After annealing to 30 K a small maximum at 290 μeV appears. We assign it to the tunneling transition of the methyl groups in a regular environment of argon atoms. Fig. 3 shows the INS-spectrum of $Pb(CH_3)_4$ up to 30 meV. In this spectral range a weak excitation at 2.8 meV was found which is assigned tentatively to the methyl torsion. This leads to hindrance potentials of the form

$$V_1(\varphi) = -2.6 \text{ meV} \cdot \cos 3\varphi - 17.1 \text{ meV} \cdot \cos 6\varphi$$

$$\text{or } V_2(\varphi) = 9.0 \text{ meV} \cdot \cos 3\varphi + 14.6 \text{ meV} \cdot \cos 6\varphi$$

depending on the relative phase of the threefold and sixfold contributions. The potential parameters for $V_2(\varphi)$ are, as expected, smaller than for the corresponding potential in $Sn(CH_3)_4$ [3], but the ratio of threefold to sixfold term is nearly equal. In contrast to that $V_1(\varphi)$ has a somewhat stronger threefold term as in $Sn(CH_3)_4$, where it is nearly sixfold. The consistence of $V_2(\varphi)$ in $Sn(CH_3)_4$ and $Pb(CH_3)_4$ confirms our interpretation, claiming for a high sixfold contribution to the potential.

In further experiments, $Sn(CH_3)_4$ was isolated in krypton and SF_6. In both cases no well-defined excitations were observed at energies below 1 meV. We assume that the tunneling lines are smeared out owing to a broad distribution of effective hindrance potentials in these matrices. With a transition at 2.1 meV the spectrum of $Sn(CH_3)_4$ in krypton in the range of the CH_3-torsion is very similar to that in argon (fig. 4). In SF_6, however, this transition was broadened and shifted to 1.5 meV. The similarity of the spectra for the torsional mode of the CH_3-group in krypton and argon is surprising, since the tunneling spectra are completely different. Plots of the solutions of the Schrödinger equation for the hindered rotation of a methyl group in a mixed threefold/sixfold potential show, however, that the librational frequency is not very sensitive to the barrier height of $V_1(\varphi)$. On the other hand the tunneling transition strongly depends on the barrier height and will thus easily smear out, when the potential distribution is wide.

The results discussed so far have shown that the influence of intermolecular interactions on matrix-isolated tetramethyl compounds is considerably reduced in

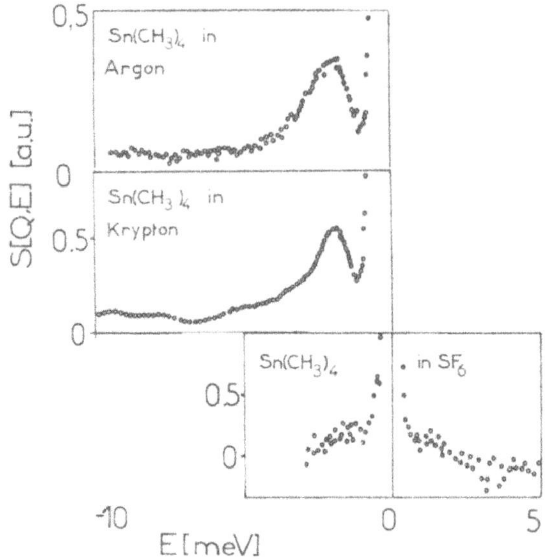

Fig. 4: INS spectrum of $Sn(CH_3)_4$ in argon (upper trace, from [3]), in krypton (middle trace), and SF_6 (lowest trace)

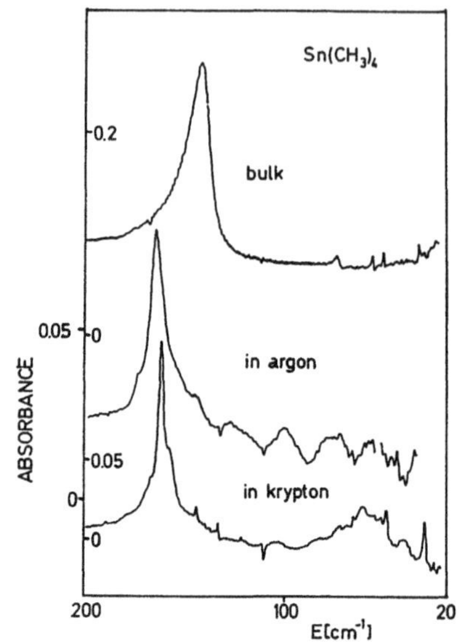

Fig. 5: Far-IR spectra of $Sn(CH_3)_4$ vapor-deposited bulk (upper trace), in argon (middle trace) and krypton (lower trace)

comparison with the bulk phase but it is not negligible. Therefore, we must verify our assumption that the deformation of the molecules lowering the symmetry in the bulk phase [2] is not present in the matrix. The skeletal bending modes $\nu_8(E)$ and $\nu_{19}(F_2)$ are presumably most sensitive to variations of the molecular geometry. They are closely spaced, and only ν_{19} is infrared active [8]. It describes the motion of the central metal atom with respect to the rigid skeleton of four CH_3-groups. As long as these are arranged as a regular tetrahedron, ν_{19} will be threefold degenerate. In the bulk phase it should split into an A- and E-type mode, which are both optically active. We found ν_{19} to appear as a sharp line at about 21 meV in the matrix (fig. 5). The bulk sample shows a much wider line at 17.6 meV with a long tail to higher frequencies, but no splitting into two components. This sample was vapor deposited and not frozen from the liquid, as in earlier experiments [1,2]. We assume that vapor deposition leads to a disordered phase, and that the splitting of the mode is smeared out causing this tail. Tunneling studies of vapor-deposited $Sn(CH_3)_4$ would thus be of great interest.

ν_8 is infrared inactive but should be seen in the INS-spectra. Neither in the bulk nor in the matrix the line at 18 or 21 meV respectively, is split, in agreement with the assumptions made in [8]. Especially in the matrix the transition appears to be very sharp (Fig. 3 and [3]).

The origin of the mode at 11 meV is not yet clear. It is only slightly shifted when $Sn(CH_3)_4$ is replaced by $Pb(CH_3)_4$ or krypton is used instead of argon (Fig. 3 and [3]). As was laid out earlier, the mode could be due to the coupling of the librations of different methyl groups within the same molecule. This coupling should, however, be weaker in $Pb(CH_3)_4$ than in $Sn(CH_3)_4$, and one would expect, therefore, the respective mode to be shifted to lower frequencies. This is not the case.

A second explanation is based on a vibration of the whole molecule against its cage [3]. Since the TMM-molecules have a high mass, such a mode can only have a frequency as high as 11 meV, if it mixes appropriately with an internal mode of the molecule, as e.g. ν_{19}. Thus, in the matrix the tetrahedron of CH_3-groups and the central atom would not only interact with each other but also separately with the cage. This interaction can induce a mixing of ν_{19} and the translations of the whole molecule. A numerical analysis yielded the ratio of frequencies of the two resulting modes as well as the ratio of their amplitudes. It turned out that for $Sn(CH_3)_4$ and for $Pb(CH_3)_4$ the observed frequency and intensity ratios can be fitted under the assumption that the central atom is interacting much stronger with the cage than the methyl group tetrahedron.

4. LITERATURE

1. W. Müller-Warmuth, K.-H. Duprée, M. Prager: Z. Naturforsch. 39a, 66 (1984)
2. M. Prager, K.-H. Duprée, W. Müller-Warmuth: Z. Phys. B 51, 309 (1983)
3. M. Prager, W. Langel: J. Chem. Phys. (in press)
4. M. Prager, W. Langel: Proc. of the 41st Int. Meeting on Phys. Chem., Grenoble, 1986 (in press)
5. Y. Kataoka, W. Press, U. Buchenau, H. Spitzer: In Neutron Inelastic Scattering 1977, Vol. 1, p. 311, Vienna 1978
6. A. Heidemann, J. Havard, K.J. Lushington, J.A. Morrison, W. Press, J. Tomkinson: J. Phys. Soc. Japan 52, 2401 (1983)
7. W. Langel, W. Schuller, S. Popp, E. Knözinger (to be published)
8. W.F. Edgell, C.H. Ward : J. Am. Chem. Soc. 77, 6486 (1955)

Rotational Excitations in $(CH_4)_{1-x}Kr_x$ Solid Mixtures

S. Grondey[1], M. Prager[1], W. Press[2], and A. Heidemann[3]

[1]Institut für Festkörperforschung der Kernforschungsanlage Jülich, Postfach 1913, D-5170 Jülich, Fed. Rep. of Germany

[2]Institut für Experimentalphysik, Universität Kiel, Olshausenstr. 40–60, D-2300 Kiel 1, Fed. Rep. of Germany

[3]Institut Laue-Langevin, 156X, F-38042 Grenoble Cedex, France

Rotational spectra have been determined with inelastic neutron scattering for polycrystalline $(CH_4)_{1-x}Kr_x$ mixtures ($x \lesssim 0.35$) at temperatures T = 1.6,...,77 K /1,2/. The measurements were performed at the TOF spectrometer IN5 of the ILL with an energy resolution (FWHM) between 17 and 160 µeV. The aim of the experiment was to study the influence of spherical impurities (Kr atoms) on the orientational states and excitations of the tetrahedral molecules in solid methane. Solid methane can be described in the most simple approximation as a system of electrostatic octopoles with fixed center-of-mass structure (fcc).

Three principal regions in the phase diagram (see Fig.1) can be distinguished:

① At $x \lesssim 0.2$ and low T perturbed orientational order of phase II;

② At $x \gtrsim 0.2$ and low T frozen-in orientational disorder (phase I);

③ At $T \gtrsim 20...30$ K for all x dynamical disorder of the orientations in phase I.

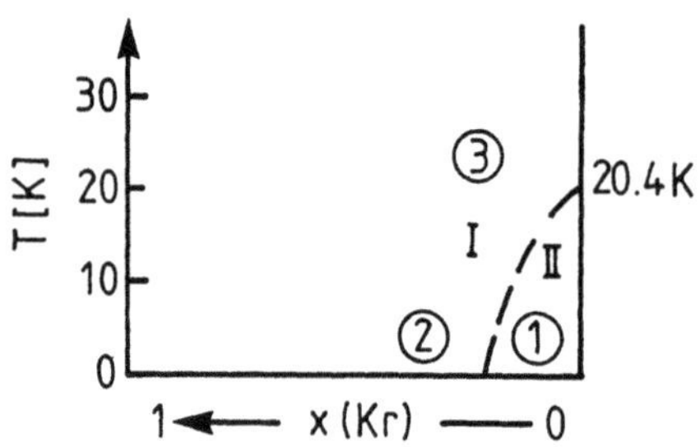

Fig.1 phase diagram (x,T) of the binary mixture $(CH_4)_{1-x}Kr_x$ (schematic). The phases I and II and the regions 1 to 3 of the present investigation are indicated.

1. Perturbed phase II ($x \lesssim 0.2$)

In region 1 of the phase diagram perturbed rotational tunnelling transitions at energy transfers of about 75 and 140 µeV and transitions of almost free rotors (at about 1.08 meV for pure CH_4) /3,4,5/ were observed at $T \approx 4.8$ K (see Fig.2 and 3). With growing Kr content of the samples the scattering intensity is partially shifted to the high energy side of the tunnelling lines. Similarly for the nearly free rotor lines an asymmetry to low energies is found, together with a slight peak shift to lower energies, according to a stronger hindrance of the rotation as the octopole field at the disordered sites no longer cancels by symmetry. Concerning the tunnelling lines, Kr atoms on sites of ordered CH_4 molecules mean a reduction of the rotational potential, a stronger overlap between the orientational wave functions in adjacent potential minima and thus a larger tunnel splitting at neighbouring ordered sites. Distinguishable tunnelling and nearly free rotor scattering contributions were observed up to x = 0.2 /1,2/.

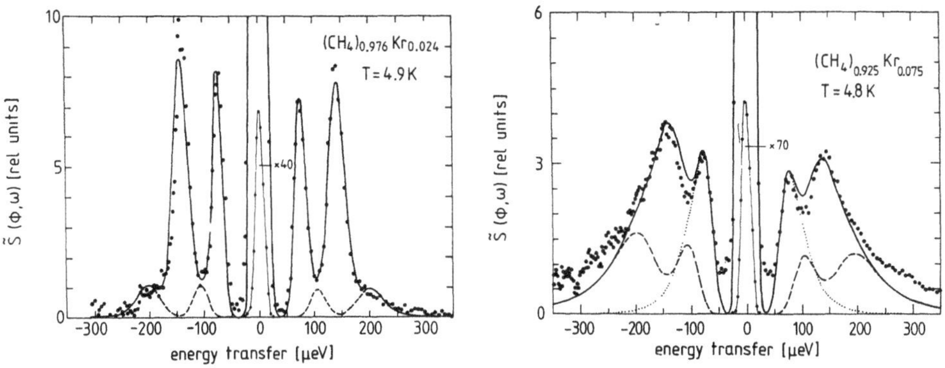

Fig.2 Tunnelling and nearly free rotor lines in phase II of $(CH_4)_{1-x}Kr_x$ solid mixtures (x = 0.024 and 0.075, respectively), T ≈ 4.8 K, wavelength of the incoming neutrons λ_i = 10 Å. $S(\Phi,\omega)$ is the scattering function at constant scattering angle Φ, $\hbar\omega$ the energy change of the neutrons. Thin solid lines are guides to the eyes, only; thick lines result from least squares fits with the model discussed below (solid = total scattering function, dashed = contributions from surroundings with n ≥ 1 (see below), dotted = contributions from the TE and ET transitions).

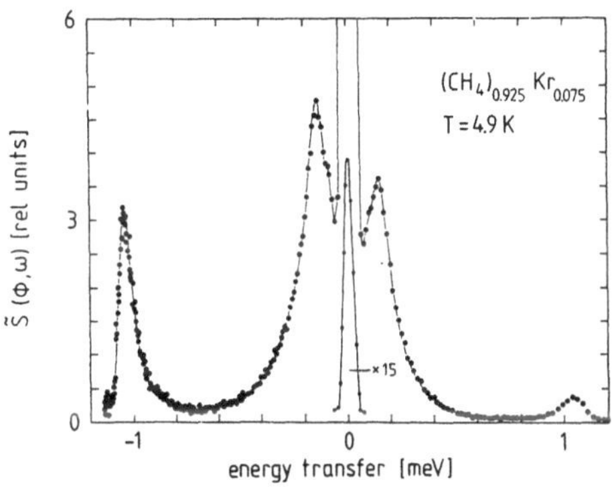

Fig.3 Spectrum for x = 0.075, T = 4.9 K (λ_i = 7 Å), including the transition of the nearly free rotors between the states with rotational quantum numbers J = 0 and J = 1. Thin solid lines are guides to the eyes, only.

To describe the measured tunnelling spectra in a simple way we employed a model which describes the broadened tunnelling lines as a sum of separate lines for local surroundings with n = 0,..,8 Kr atoms replacing orientationally ordered CH_4 next neighbours of an ordered CH_4 molecule. The local rotational potential (molecular field) was assumed to vary linearly with the number n:

$$V_M(n) = V_0/8 \cdot (8 - n) \qquad (1)$$

where V_0 = 37B is the potential at ordered sites in pure CH_4 /6/ The intensities of the sublines were weighted with the binomial probabilities

$$p(n) = \binom{8}{n} \cdot x^n \cdot (1 - x)^{8-n} \qquad (2)$$

for the occurrence of the different surroundings. The intensity distributions of the sublines were calculated as Gaussian distributions of potentials around the values of $V_M(n)$ given in (1), and the potentials were converted to transition energies with exponential laws /6/.

As the thick solid lines in Fig.2 show, the tunnelling spectra at low Kr concentrations (in our spectra up to x = 0.05) could be quite well reproduced with the described model. For x = 0.075 deviations are visible at high energy transfers. They should be due to additional parameters which influence the line positions but are not accounted for by our simple model. As the most important point the local arrangement of the molecules and atoms on nearest neighbour sites, i.e. the local symmetry, has been neglected. Consequently, T-state splittings caused by lowered symmetry have not been considered, as well as interactions with further than nearest neighbours.

2. Orientationally disordered phase I (x $\gtrsim$ 0.2)

For samples with x = 0.27 and 0.35 the measurements at the lowest temperature (T = 1.6 K) show as a new feature only one, surprisingly well-defined peak at an energy transfer $|\hbar\omega| \approx 0.77$ meV (FWHM $\approx$ 0.39 meV). The peak shifts to lower energies with rising temperature, together with a broadening (at T = 4.8 K the position is $|\hbar\omega| \approx 0.63$ meV with FWHM $\approx$ 0.63 meV). Spectra at T = 1.6, 4.8 and 9.9 K are shown in Fig.4 for the sample with x = 0.27.

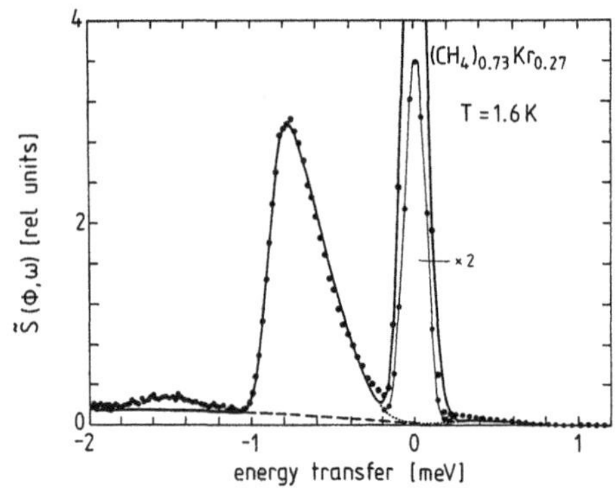

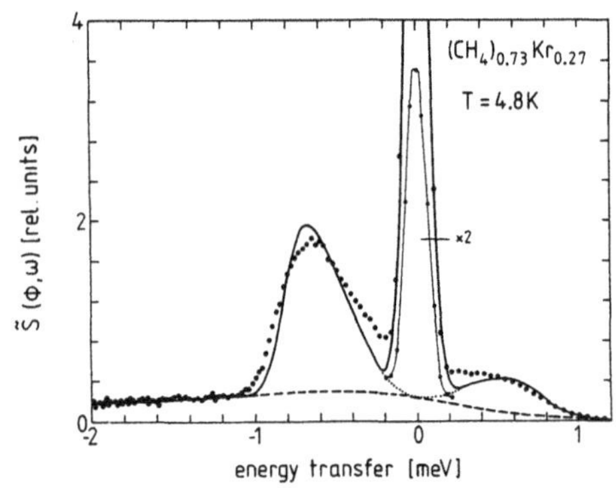

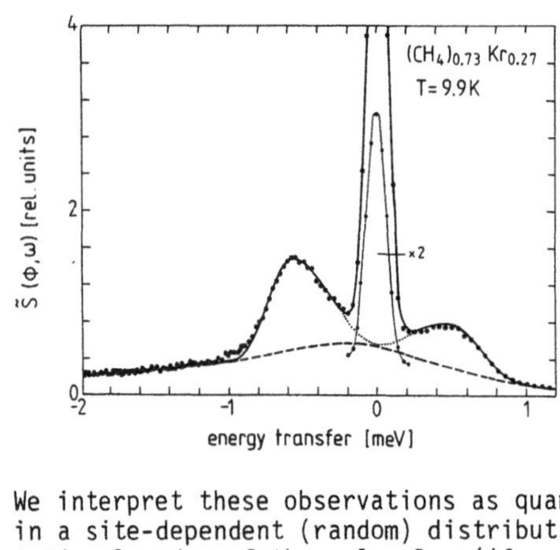

Fig.4 Spectra for x = 0.27 (λ_i = 4.8 Å). Thick solid lines result from our model (see below, solid = total scattering function, dashed = quasielastic contribution).

We interpret these observations as quantum mechanical excitations of the molecules in a site-dependent (random) distribution of potentials without long-range orientational order of the molecules ('frozen-in orientational disorder'). The energies of the surprisingly well-defined peaks are intermediate between the limits of rotational tunnelling and nearly free rotations, in agreement with NMR results /8/. The peak energies show that the cubic, site-independent 'crystalline field' still plays an important role for the rotational states.

The peak shift with increasing temperature is believed to be due to the growing thermal population of excited molecular states with an enhanced effective octopole moment (enhanced anisotropy), which reduces the energy splitting at neighbouring sites (in analogy to the tunnelling lines in CH_4II /4/). Conversion obviously is fast in region 2 of the phase diagram (compare /9/).

Our theoretical model for region 2 of the phase diagram /1/ approximates the measured spectra with a sum of a 'quasielastic' contribution increasing with rising temperature (see region 3) and an 'inelastic' part which is a superposition of 13 different lines for m = 0,...,12 nearest CH_4 neighbours excited to the first states above their groundstates (from symmetry A to symmetry T). The assumed molecular field is

$$V_M(m) = V_A + m \cdot V_B \tag{3}$$

with adjustable parameters V_A and V_B. V_A formally describes the contribution due to the disorder induced by dilution with Kr. The potential distribution of each subline has the same shape as adjusted empirically to the spectrum at T = 1.6 K where only the contribution with m = 0 is important. The molecular fields were converted to rotational energies with the assumed dependence

$$|\hbar\omega(V_M)| = 1.078 \cdot \exp\ (-V_M^2/2\delta^2) \quad \text{meV} \tag{4}$$

with δ = 18.4B, which reproduces the nearly free rotor energies for V_M = 0 (B = rotational constant = 655 µeV /6/). The different sublines were weighted with the appropriate combined binomial probabilities for the number of Kr/CH_4 and A-CH_4/T-CH_4 nearest neighbours (summed up for a fixed number m).

The concentration of excited molecules is $p_T = 1 - p_A$ with

$$p_A = 5/Z \tag{5}$$

and

$$Z = 5 + 9 \cdot \exp\ (-\Delta/k_BT) \tag{6}$$

(weight factors /9/, Δ= energy splitting).

As Fig.4 shows, the peak shift at low T could be qualitatively reproduced with the applied model (V_A = 13.6B, V_B = 1.3B); only the width at T = 4.8 K could not be accounted for completely. Thus the basic idea of our model seems to be a quite good approximation to the changes with temperature in the complex real system.

3. High-temperature rotations (T $\gtrsim$ 20...30 K)

In the high-temperature limit spectra at all Kr concentrations have a mainly quasielastic appearance, indicating dynamical disorder of the orientations in agreement with /3,10/, probably best described as continuous rotational diffusion which is disturbed by only weak hindering potentials. An example is shown in Fig.5.

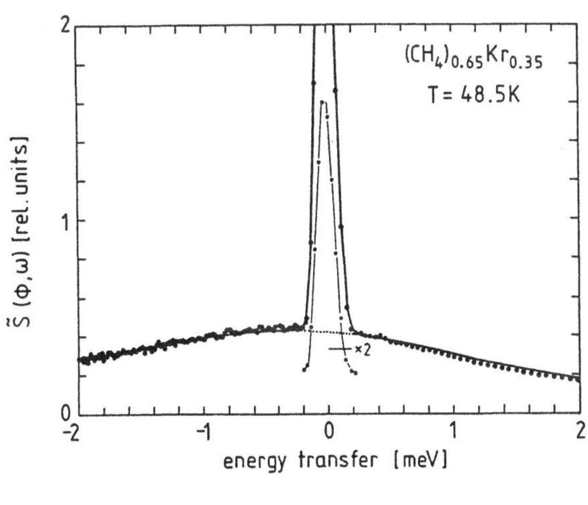

Fig.5 Spectrum for x = 0.35, T = 48.5 K (λ_i = 4.8 Å). The thick solid line results from a least squares fit with a scattering law for continuous rotational diffusion without hindering potentials (see /2,6/).

Using the model for region 2 it was possible to reproduce the observed continuous transition from region 2 to 3 of the phase diagram because of the assumption of a quasielastic scattering contribution (single Lorentzian function).

4. Conclusion

Solid CH_4-Kr mixtures have turned out to be a very interesting system for the study of dilution-induced orientational disorder in molecular crystals. The partial orientational order of phase II exists up to a Kr concentration x = 0.2. A simple statistical model accounts for the shape of the tunnelling lines at low Kr concentration. Frozen-in orientational disorder has been found at $x \gtrsim 0.2$ and low temperatures. Here the temperature dependence of the spectra could be qualitatively reproduced with the used model. The question of a possible orientational glass' (in analogy to spin glasses), which has been discussed for other molecular systems like $(o\text{-}H_2)_{1-x}(p\text{-}H_2)_x$, $(N_2)_{1-x}Ar_x$ and $(KCN)_{1-x}(KBr)_x$ (for references see /1/), cannot conclusively be answered by our data /1,2/.

Literature:

1. S. Grondey, M. Prager, W. Press and A. Heidemann, J.Chem.Phys. 85, 2204 (1986) and to be published
2. S. Grondey, Report of the KFA Jülich JÜL-2083, in press
3. W. Press and A. Kollmar, Sol.State Comm. 17, 405 (1975)
4. A. Heidemann, W. Press, K.J. Lushington and J.A. Morrison, J.Chem.Phys. 75, 4003 (1981)
5. W. Press, in: Single Particle Rotations in Molecular Crystals (Springer - Verlag, Berlin 1981)
6. A. Hüller and J. Raich, J.Chem.Phys. 71, 3851 (1979)
7. T. Yamamoto, Y. Kataoka and K. Okada, J.Chem.Phys. 66, 2701 (1977)
8. P. Calvani and H. Glättli, Sol.State Comm. 50, 169 (1984) and J.Chem.Phys. 83, 1822 (1985)
9. F. de Luca and B. Maraviglia, Phys.Lett. 96A, 212 (1983)
10. H. Kapulla and W. Gläser, in: Inelastic Scattering of Neutrons in Solids and Liquids (IAEA, Vienna 1973), p.841

Quantum Rotation of Ammonia Molecules in Disordered $Ca(NH_3)_6$

F. Leclercq[1], *P. Damay*[1], *A.J. Dianoux*[2], *and W. Press*[3]

[1]LESI, UA 253 du CNRS, HEI 13 rue de Toul, F-59046 Lille Cedex, France
[2]Institut Laue-Langevin, 156X, F-38042 Grenoble Cedex, France
[3]Institut für Experimentalphysik der Universität Kiel, Olshausenstr. 40–60, D-2300 Kiel 1, Fed. Rep. of Germany

Calcium hexaammine is a low electron density, metallic compound. In this expanded metal, the ammonia molecules show important rotational and reorientational motions down to 20 K as deduced from the study of the quasielastic neutron scattering (LECLERCQ et al./1/). A complete understanding of the inelastic scattering needs a simultaneous determination of the structure in the whole temperature range. Since the calcium compound crystallizes in a high symmetry group (bcc) and since no phase transition occurs as the temperature is decreased, the ammonia molecules appear to be in a frozen-in orientationally disordered state at low temperature. For temperature lower than 10 K, the Inelastic Neutron Scattering spectrum shows a very asymmetric line shape with a maximum near the energy of uniaxial rotation of the NH_3 / ND_3 groups. This is interpreted by quantum rotation in a (random) distribution of low hindering potentials. It is shown that these properties are very sensitive to non-stoichiometry effects.

Compounds formed between ammonia and metals (Li, Ca, Ba, Sr, Yb, Eu...) are a particular class of materials. The $Ca(NH_3)_6$ compound shows the following characteristics :

-It is a metallic low electron density compound ($n = 5.10^{21}\ e^-/cm^3$)
-The compound forms with a large volume expansion ($\Delta V/V = 45\%$).
-The system crystallizes in a high symmetry group (bcc) as seen by X-Ray and neutron diffraction at 77 K (MAMMANO /2/, VON DREELE et al. /3/). Above T = 20 K a diffusive motion is observed by neutron scattering. Since there is no phase transition found down to 5 K, ammonia molecules freeze in a disordered state at low temperature (molecular glass).
-Due to the large volume expansion at the formation and the metallic character of the compound, the ammonia molecules are in a unique environment. Indeed, long interatomic distances and strong electronic screening let expect that all local interactions (except the Ca-N and N-H bonds) are weak.

1 Structural Aspects

The experiments have been performed on the spectrometer D1A at the ILL. The deuterated sample (fine grained powder) was rotated throughout the experiment in order to eliminate effects due to preferential orientations.

The temperature-dependent measurements give already one important information: at all observed temperatures (6...140 K) the structure is body centered cubic. This indicates that the $-ND_3$ groups freeze in an orientationally disordered state at low temperature since the three-fold symmetry of the molecule is not compatible with the four-fold symmetry of the sites (00x) in the cubic lattice. A considerable $-ND_3$ delocalisation is found at the highest temperature (140 K). A better fit is obtained when N and D atoms are not restricted to well-localized positions (00x), but have a rather large delocalisation on a sphere around the calcium atom. A R-factor of 5.4 % is obtained with a form factor describing a modulated distribution:

$$F_{rot}(\vec{Q}) = b_{Ca} + \sum_{j=1}^{2} b_j n_j \sum_{l=0}^{4,6} a_l^{(j)} i^l j_l(Q\rho_j) K_{1l}(\Omega_Q)$$

where j = 1 for N atoms and j = 2 for D atoms.

j_l is the spherical Bessel function and K_{1l} the cubic harmonics.

These results are consistent with the high-temperature EISF obtained from Inelastic Neutron Scattering, which is not far from that of a sphere of radius r = Ca..H at 150 K (LECLERCQ et al./1/).

The structure at low temperatures is not yet completely understood. The fit is very poor if $-ND_3$ groups are placed on the expected (x00) positions even considering a disordered arrangement for the D atoms (D on a circle). Fourier analysis shows that there is a probability of about 16 % to find the $-ND_3$ groups with their centre of mass on positions (xxx). The inclusion of this additional site yields an improvement to R = 6.0 %. Reasonable Debye Waller factors are found for the rigid ND_3 molecules with the regular pyramidal geometry, but not for the Ca atoms (possibly because of their low weight in the expression for the structure factor).

The cell parameter was found to depend upon stoichiometry $Ca(NH_3)_{6-x}$. At 5 K the cell parameter a is equal to 8.99 Å for a "saturated sample" for which excess ammonia is found with x slightly negative, a = 8.92 Å for x around 0.1 and a = 8.82 Å for x = 0.5 . There was no excess calcium found is this last sample, indicating that the depletion of ammonia could possibly be continued up to rather high x-values before actually destroying the structure.

2 Inelastic Neutron Scattering

The experiments have been performed on the spectrometers IN6 and IN5 at the ILL using both hydrogenated and deuterated samples.

The spectra obtained for a $Ca(NH_3)_{5.9}$ compound are reported on Fig.1 at three temperatures. At the lowest temperature a broad peak is observed with a maximum at 0.67 meV. This maximum shifts to lower energy as the temperature increases and merges into the quasielastic central peak at 20 K.

The results for a deuterated sample at T = 1.7 K are given on Fig.2 for two different spectrometer resolutions. The maximum of the inelastic peak is then found at 0.32 meV.

For both samples, the peaks show a marked asymmetry; the maximum, or rather a cut-off on the high-energy side, is found very near the B value of uniaxial rotation of the ammonia molecule (B = 0.69 meV for NH_3 and B = 0.34 meV for ND_3

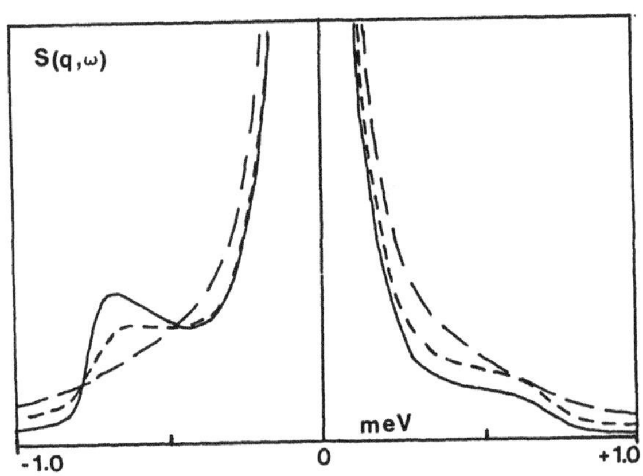

Fig. 1 - Inelastic spectra of the hydrogenated compound $Ca(NH_3)_{6-x}$, x= 0.10, obtained at three temperatures (——,5K; ----,10K; — —,20K) with the IN6 TOF spectrometer (λ=5.1Å, resolution : 120 μeV); the inelastic peak shows a maximum around 0.67 meV, close to the B value for uniaxial rotation of the $-NH_3$ group.

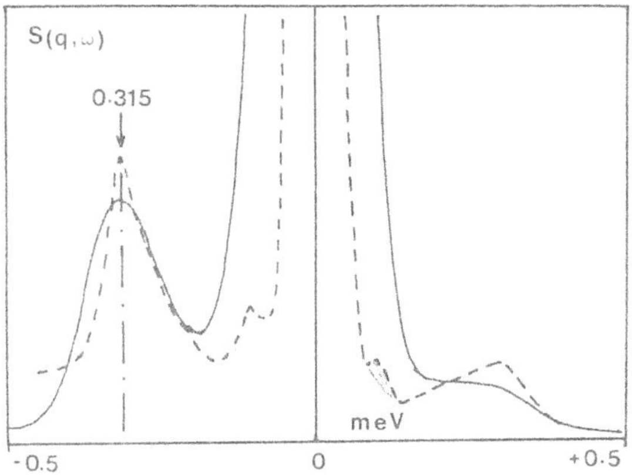

Fig. 2 - Inelastic spectra of the deuterated compound $Ca(ND_3)_{6-x}$, x=0.07 obtained for different spectrometer resolutions at T = 1.7 K (—— ,IN6, λ =5.1Å,FWMH=120μeV ----,IN5, λ =7.3Å,FWMH=40μeV, marked small side peaks on the IN5 spectrum are due to the sample environment). The observed inelastic peak shows a maximum at 0.32 meV and an asymmetric shape; on the high-energy side the width is given by the machine resolution.

using the N..D distance observed by HEWAT /4/ for pure NH_3 at T = 4K). The half-width on the high-energy side is given by the spectrometer resolution, while the shape on the low-energy side is much broader and does not depend on the instrumental resolution (see Fig.2); this may be interpreted as quantum rotational excitations in a distribution of potentials.

This result is not surprising if one considers the fact that the ammonia molecules are frozen in a disordered state. In positions (00x) the local potential could then be written as:

$$V(\varphi) = V_3 \cos 3\varphi + V_6^c \cos 6\varphi + V_6^s \sin 6\varphi + \ldots$$

but these local potentials average to zero. For the whole crystal the average potential may then be written as :

$$\langle V(\varphi)\rangle = \langle V_{12} \rangle \cos 12\varphi + \ldots \qquad \{ \langle V_3 \rangle = \langle V_6 \rangle = 0 \}$$

It must be noted that this low-temperature spectrum was totally insensitive to the cooling rate (10 minutes to 7 hours for cooling from 60 to 5 K).

The distribution of potentials gives rise to intensity close to the value corresponding to free rotation. This indicates that, contrary to metal hexaammine halides, the randomly disordered potential is always weak; this may be understood by a strong screening of the "crystal field" by the itinerant electrons.

A complete understanding of the low-temperature structure is necessary before modeling the distribution of V_3, V_6 local potentials, or eventually a V_{12} average potential. The main question is to determine if there is one or two sites for the ammonia molecules (only (00x) or possibly also (xxx)).

As the temperature increases the peak of the distribution is shifted towards lower energy splittings and simultaneously the spectrum becomes broader. This effect is usually attributed to dynamical disorder induced by fluctuating potential at higher temperature (HULLER et al. /5/).

It must be finally noticed that the quantum rotation is very sensitive to departures from stoichiometry. For the $Ca(NH_3)_{6-x}$ compound, the intensity of the inelastic peak decreases and its shape is considerably modified with increasing values of x; at the same time a broad line (width = 500 μeV) appears near 1.2 meV (see Fig.3). Higher order broad peaks at 2.3, 3.7 and 4.9 meV have also been detected for x = 0.40 (LECLERCQ et al. /1/). Presently we assume that this is induced by a structural change.

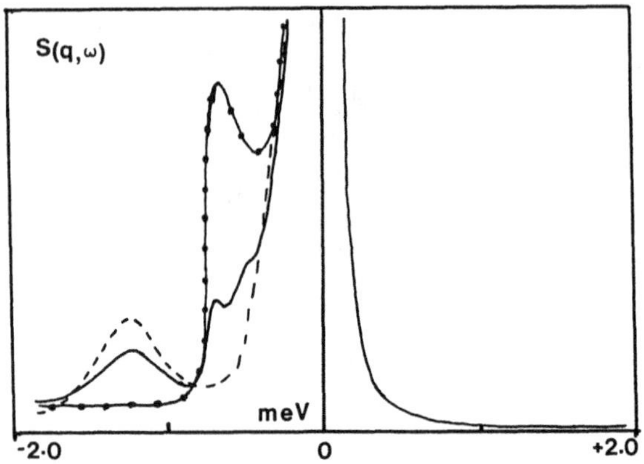

Fig. 3 - Influence of the departure from stoichiometry $Ca(NH_3)_{6-x}$ on the inelastic spectrum obtained with the IN6 spectrometer (λ = 5.1Å) at 1.5K (----, x = 0.40; ——, x = 0.15; -•-•-, x = 0.10).

Future efforts aim towards a better understanding of the effect of stoichiometry and of the ionic radius of the central metal ion, since similar behavior has already been detected for the other metal-ammonia compounds (Me = Li, Yb, ...).

3 Concluding Remarks

The calcium hexaammine compound is one of the few systems to display a quasi free rotation of a molecular group at low temperature.

As a metal, it may be compared to $Nb(OH)_x$ (MAGERL et al./6/). As an hexaammine, it is worthwile to mention the study of ammoniates salts carried out by BLANK et al./7/ in which the influence of different counter ions on the hindering potential is considered; almost-free rotation of the NH_3 group is observed in case of large, polarizable anions as, for example, in $Co(NH_3)_6(PF_6)_2$.

The study of rotational excitations in disordered systems is a rather recent topic, as pointed out in the review work of PRESS /8/. May be, except for the CH_4-Kr system (GRONDEY et al. /9/), a detailed description of the distribution of hindering potentials in these systems has still to be made, probably because a precise determination of the structure is difficult to obtain.

References.

1. F. LECLERCQ, P. DAMAY, P. CHIEUX : J. Phys. Chem., 88, 3886 (1984)
2. N. MAMMANO : In Metal Ammonia Solutions, Colloque Weyl II, ed. by J.J. LAGOWSKI and M.J. SIENKO, (Butterworths, London, 1970), p.367
3. R.B. VON DREELE, W.S. GLAUNSINGER, A.L. BOWMAN, J.L. YARNELL : J. Phys. Chem., 79, 2992 (1975)
4. A.W. HEWAT, C. RIEKEL : Acta Cristallographica, A35, 569 (1979)
5. A. HULLER, W. PRESS : In Neutron Inelastic Scattering, 1977, I.A.E.A.-SM-219/129, p.231, (Vienna, 1978)
6. A. MAGERL, A.J. DIANOUX, H. WIPF, K. NEUMAIER, I.S. ANDERSON : Phys. Rev. Letters, 56, 2, 159 (1986)
7. H. BLANK, J.K. COCKCROFT, G.J. KEARLEY : these proceedings
8. W. PRESS : In the Proceedings of the Meeting "Dynamics of Molecular Crystals" (Grenoble, june 1986) to be published in J. Chim. Phys.
9. S. GRONDEY, M. PRAGER, W. PRESS, A. HEIDEMANN : J. Chem. Phys., 85, 4, 2204 (1986); see also these proceedings

The Methyl Rotational Potentials of Various Isotopic Methyl Iodides and Their Coupling to Phonons

M. Prager[1], *J. Stanislawski*[1], *and W. Häusler*[2]

[1]Institut für Festkörperforschung der Kernforschungsanlage Jülich, Postfach 1913, D-5170 Jülich, Fed. Rep. of Germany

[2]Institut für Theoretische Physik der Universität Erlangen-Nürnberg, Glückstr. 6, D-8520 Erlangen, Fed. Rep. of Germany

The rotational potentials of methyl groups in protonated and deuterated methyl iodide were determined from the observed tunnel splittings and the librational energies. The potential becomes stronger in the deuterated matrix by about 10 %. For pure CD_3I the isotope effect on the librational modes shows a considerable further increase. No rotor-rotor coupling is found. The coupling to low-frequency phonons can explain the unusual increase of the tunnel splitting with temperature observed for T < 15 K.

1. Introduction

The methyl halides are the simplest molecules containing CH_3 groups and thus among the most simple molecules of all. They represent a symmetric prolate top, with the small momentum of inertia being equal to that of the CH_3 group. Methyl iodide has only one solid phase: Pnma with Z=4 molecules at c-sites /2/. The molecules are arranged in the a-c-plane as chains along the c-direction with alternating orientation of the molecules.

Various reasons make the methyl iodide interesting from the point of view of methyl group rotation: i) because of the molecular symmetry there is no intramolecular hindering potential. The methyl rotation is identical with the overall rotation of the whole molecule around its small axis of inertia. The rotational potential fully originates from the interaction with neighbouring molecules. ii) The temperature behaviour of the tunnelling modes, i.e. their coupling to to phonons /3,4/, might be especially simple because the libration of the methyl group is an external mode. In force models available for molecular crystals /5,6/ it can be found out by symmetry analysis with what phonons at all the methyl rotation can interact. Thus not the full density of states /4/ but individual phonon branches might be of importance. This could stimulate new theoretical efforts. iii) The lattice dynamics of molecular crystals consisting of polar molecules is an actually very active field /6/. Due to the simplicity of the molecules and the structure of the methyl halides CH_3I might be a candidate for a model system in this field.

---

1) This work is a part of a more comprehensive study of the methyl halides /1/

While methyl iodide is already intensively investigated with the Raman and FIR technique /7,8/ there is only one incoherent neutron scattering experiment known to us /9/ which was mainly concentrating on multiphonon scattering and not on the methyl group rotation. It is the scope of the present experiment, covering the energy range from 0.4 μeV to 60 meV and temperatures from 4K to 100K for various isotopic species, to determine the rotational potential of the methyl groups, the effect of isotopic substitution on this potential and the coupling of the quantum rotor to phonons.

2. Experiments and Results

The high-resolution experiments were performed at the backscattering spectrometer IN10 at the ILL, Grenoble, using the best obtainable energy resolution δE_{el} = 0.4 μeV. The energy range 2 < E [meV] < 60 was investigated with the thermal time-of-flight spectrometer SV22 at the FRJ2 in Jülich using various combinations of incoming neutron energy and energy resolution.

The high-resolution spectrum of CH_3I shows a tunnelling peak of resolution width at an energy transfer $\hbar\omega_t$ = 2.44 μeV (fig 1). With temperature this peak behaves quite unusually (fig. 2): Its frequency increases up to a temperature T ∿ 20 K. Thereafter it decreases as usually observed. The broadening, described by a Lorentzian, follows in the whole temperature range very well an Arrhenius behaviour with an activation energy E_a^{Γ} = 14 meV. Tunnelling spectra were also recorded for 2 % and 50 % CH_3I in CD_3I. A linear decrease with concentration is found yielding for infinitely low concentration $\hbar\omega_t$ = 2.02 μeV.

The prominent feature of the spectra in the meV region is a strong doublet line. Fig. 3 shows spectra of CH_3I, CH_3I in CD_3I and CD_3I. With increasing temperature

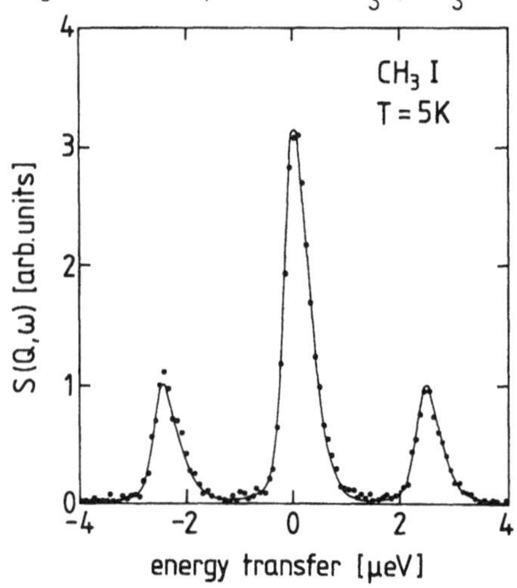

Fig. 1 Tunnelling spectrum of CH_3I. Energy resolution δE=0.4 μeV. Momentum transfer Q=1.88 $Å^{-1}$.

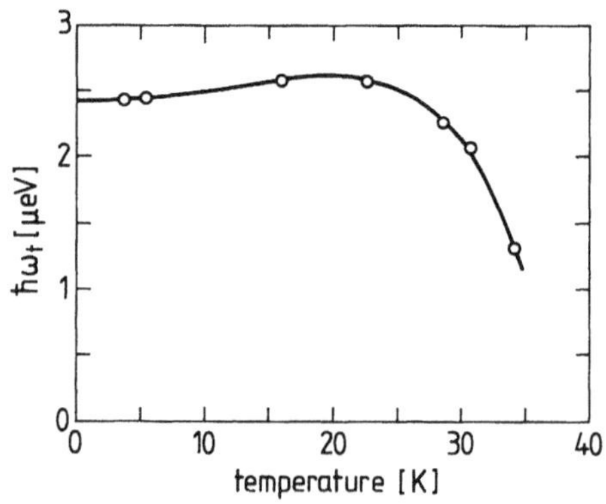

Fig. 2 Temperature dependence of the tunnel splitting in CH_3I. The solid line is a fit with the model described in the text.

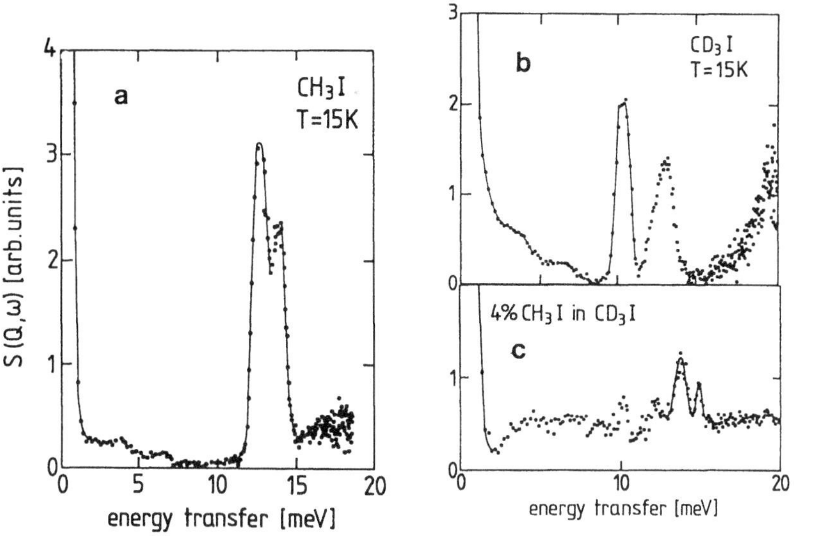

Fig. 3 Spectra of a) CH_3I, b) CD_3I and c) CH_3I in CD_3I (the spectrum of CD_3I is subtracted to show only the transition energies of the protonated molecules) in the meV region. The low-energy peak of each doublet represents the libration of the methyl groups. See the text.

all transition energies shift to lower values and the line intensities diminish. The lines at 13.27 meV in CH_3I, 13.8 meV in CH_3I diluted in CD_3I and 10.6 meV in CD_3I are faster damped with temperature than each second strong peak.

3. Discussion

As a consequence of the structure /1/ all methyl groups in solid methyl iodide are equivalent. The relative intensities of the tunnelling and elastic lines (fig. 1) are in agreement with the theoretically predicted values. No effect of coupling, suggestable from the head-to-head arrangement of the molecules and as found in lithium acetate /10/, is seen in the tunnelling spectrum. We first also believed that the doublet line in the meV region is a consequence of coupling. However, because of the different temperature dependences of the two lines and because of the different isotope effects (see below), they must be of different origin.

Since no effects of rotor-rotor coupling could be found we describe the methyl rotational energies by a single particle rotational potential:

$$V(\varphi) = \frac{V_3}{2} \cos 3\varphi + \frac{V_6}{2} \cos 6\varphi \ . \qquad (1)$$

From the eigenvalues $\hbar\omega_t$ = 2.44 µeV and E_{01} = 13.27 meV we obtain

$$VS = V_3 + V_6 = 24.9 \text{ meV}, \qquad \delta = \frac{V_3}{VS} = 0.92 \ . \qquad (2)$$

A second solution with V_3, $V_6 < 0$ can be excluded from the measured isotope effect (see below). Thus, in agreement with the general experience /10/, the deviation from a purely threefold cosine potential is small. The potential also describes rather well the second excited librational state found at 24.2 meV (calculated energy 24.6 meV). The second strong peak at 14.54 meV is assigned to a libration R_x, R_y of the molecule perpendicular to its symmetry axis.

Beside the temperature dependence, the isotope effect is useful to identify the character of the modes related with the observed transition energies. Generally the absolute strength of the potentials changes not or very little by deuteration, and the eigenvalues of the deuterated compound are obtained by simply scaling the masses. For librational modes the effect is roughly of the order $\sqrt{2}$. In CH_3I we expect exactly 1.35 by scaling the potential (2). However, in CD_3I we find only a shift by a factor 1.25 to 10.6 meV. The mode at 10.6 meV was identified as libration via its temperature dependence. This means that the rotational potential must have considerably increased by deuteration. Keeping fixed the shape of the potential (2), characterized by δ, we find VS = 29.5 meV or an increase by 25 %, probably caused by a relaxation of the CD_3I molecule into a new equilibrium orientation in the a-c-plane. Because of this strong isotope effect, earlier Raman and FIR lines were wrongly assigned /6/. The R_x, R_y libration has little decreased to 13.1 meV as expected.

An independent determination of this large isotope effect was achieved by using CH_3I as a probe in the matrix of CD_3I. Now two independent transition energies, $\hbar\omega_t$ = 2.02 µeV and E_{01}=13.8 meV are available to determine the rotational potential. The shape is found to be unchanged, while the strength has increased by 10 % to VS = 27.5 meV compared to pure CH_3I. It is reasonable that the potential is weaker than for pure CD_3I since the properties of both, the surrounding and the 'Auf'-molecule, enter in the interaction (e.g. octopole-octopole interaction in methane).

The tunnel splitting in methyl iodide shows a very unusual increase with temperature before it decreases in the well-known way /11/. A similar behaviour was recently found in some other materials also, e.g. /12/ and /13/. The large isotope effect has already shown that the molecular orientation in methyl iodide is sensitive to shallow changes of the interaction. Thus we expect that it also might be influenced by the temperature and change the rotational potential. Such a temperature behaviour can be described in a model which couples the quantum rotor to low-energy phonons via an interaction Hamiltonian

$$H^c_{int} = g^c \sqrt{2m\tilde{\omega}/\hbar}\; x \cos 3\varphi \qquad (3)$$

not considered so far /3,4/. Here g^c is the coupling constant, x the oscillator coordinate and φ the rotational angle. The Hamiltonian of the whole system is

$$H = H_{rot} + H_{osc} + H^c_{int} \qquad (4)$$

with $$H_{rot} = \frac{L_\varphi^2}{2\theta} + V_3 \cos 3\varphi ,$$

$$H_{osc} = \frac{p^2}{2m} + \frac{m}{2} \tilde{\omega}^2 x^2 .$$

Using the formalism developed in ref. 14 the tunnelling frequency corresponding to the excitation of n and n + 1 oscillator levels is found to change as

$$\hbar\omega_t(n+1) \,/\, \hbar\omega_t(n) = 1 + A^C, \quad \text{with } A^C > 0$$

for moderate coupling g^C. If the decrease of the tunnelling frequency at higher temperatures is taken into account phenomenologically, we get the following expression for the temperature dependence of the tunnelling frequency

$$\hbar\omega_t(T) = \hbar\omega_t(o) \left\{ \frac{1-\exp(-E^C/k_BT)}{1-(1+A^C)\exp(-E^C/k_BT)} - A^S \exp(-E^S/k_BT) \right\} . \tag{5}$$

The solid curve in fig. 2 is a fit using this formula with parameters $A^C = 0.22$, $A^S = 49.06$, $E^C = 2.15$ meV and $E^S = 12.94$ meV. While E^S is close to E_{01} as usually found, E^C represents the phonon energy. Indeed the density of states shows increased intensity in this energy range (fig. 3). The coupling parameter derived from these values is $g^C = 2.9$ meV, the strength of the rotational potential $V_3 = 18.2$ meV.

4. Conclusions

Rotational tunnelling and librations of methyl iodide could be well explained as a single particle rotation. No methyl rotor-rotor coupling, suggested from the structure, can be observed: the two transitions around 14 meV belong to two different molecular rotations (around and perpendicular to the symmetry axis) and are not a consequence of coupling. With deuteration the rotational potential of the methyl group increases by 25 %, probably due to a change of the equilibrium orientation. This large isotope effect beside the simple mass effect is unambiguously confirmed by using CH_3I as a probe in CD_3I. The very unusual increase of the tunnel splitting with temperature for $T < 15$ K could be explained by a coupling to low-frequency phonons via an interaction Hamiltonian $H^C_{int} \sim x \cos 3\varphi$ not considered so far.

It would be very useful for a more thorough interpretation of our data to investigate the lattice dynamics of methyl iodide.

References

1. M. Prager, J. Stanislawski and W. Häusler submitted to J. Chem. Phys.
2. T. Kawaguchi, M. Hijikigawa, Y. Hayafiyi, M. Ikeda, R. Fukishima and Y. Tomie, Bull. Chem. Soc. Jpn. 46, 53 (1973)

3. A. Hüller, Z. Phys. B36, 215 (1980)
4. A.C. Hewson, J. Phys. C15, 3841 and 3855 (1982)
5. P. Brüesch, Phonons: Theory and Experiments I Springer Series in Solid State Sciences Vol. 34 (Springer, Berlin, Heidelberg, New York 1982)
6. K.H. Link, H. Grimm, B. Dorner, H. Zimmermann, H. Stiller and P. Bleckmann, J. Phys. Chem. Solids 46, 135 (1985)
7. H. Takeuchi, J.L. Bribes, I. Hirada and T. Shimanouchi, J. Raman Spectrosc. 4, 235 (1976)
8. O.S. Binbrek, A. Anderson, B. Andrews and B.H. Torrie, J. Raman Spectrosc. 15, 406 (1984)
9. J.A. Janik, A. Bajorek, J.M. Janik, I. Natkaniec, K. Parlinski and M. Sudnik, Acta Physica Polonica XXXIII, 419 (1968)
10. S. Clough, A. Heidemann, A.H. Horsewill and M.N.J. Paley Z. Phys. B55, 1 (1984)
11. W. Press, Springer Tracts in Modern Physics, Vol. 92, (Springer, Berlin, Heidelberg, New York 1981)
12. M. Prager and W. Müller-Warmuth, Z. Naturforschung 39a, 1187 (1984)
13. L.P. Ingman, M. Punkkinen, A. H. Vuorimäki and E.E. Ylinen, J.Phys. C18, 5033 (1985)
14. W. Häusler and A. Hüller, Z. Phys. B59, 177 (1985)

Partially Deuterated Methyl Groups in Molecular Solids: The System 4-CH_2D-C_5H_4N with an Original Sixfold Symmetric Barrier

H.B. Brom[1], *H. den Adel*[1], *and B. Alefeld*[2]

[1]Kamerlingh Onnes Laboratorium der Rijksuniversiteit Leiden, Nieuwsteeg 18, 2311 SB Leiden, The Netherlands

[2]Institut für Festkörperforschung der Kernforschungsanlage Jülich, Postfach 1913, D-5170 Jülich 1, Fed. Rep. of Germany

Although by partial deuteration of a methyl group the correlation between nuclear spin states and rotational levels is broken, it is shown that at liquid helium temperatures for a sixfold symmetric potential, the rotational system is divided into non-converting symmetry species. This is in contrast to the case of a threefold symmetric potential where partial deuteration enables a fast establishment of thermal equilibrium between all rotational energy states. For the particular example of 4-methyl-pyridine (4MP), i.e. 4-CH_3-C_5H_4N, we have performed a unique doping experiment. By using undeuterated 4MP as a neutron probe in a sample of partially deuterated 4MP - because of localization the partially deuterated species is not seen in inelastic neutron scattering (INS) - the sixfold barrier height, V_6, is determined from the INS-data. Specific heat results on the other hand, - the undeuterated species does not contribute to the heat capacity - together with the INS-value for V_6, allow a further specification of the fourfold and twofold terms in the potential barrier.

1. Introduction

The influence of the asymmetry in the dynamics of a CH_2- or CHD_2-group on its rotational levels can be taken into account by an effective barrier with a lower symmetry than that of the original CH_3-potential. In the case of a threefold symmetric barrier like in nitromethane /1/, the barrier symmetry left after partial deuteration is trivial, breaking the original coupling between nuclear spin states and rotational levels. As a consequence there is a fast establishment of thermal equilibrium between the rotational levels. In the case of an original sixfold potential, there remains twofold symmetry in the system, the consequences of which for the properties of a CH_2D group are the main subject of this paper. For this study we selected 4-methyl pyridine (4MP) as the most favourable material. It has the largest tunnel splitting seen in a molecular solid, making quantum and symmetry effects optimally observable.

In undeuterated 4MP the energy levels for the torsional ground state have been determined by INS /2/. The splittings found can be reproduced by a purely sixfold symmetric barrier $V=\frac{1}{2}V_6(1+\cos 6\phi)$ with V_6/k = 180 K. At liquid helium temperature the tunnel-splitting contribution to the specific heat is only visible if one extends the duration of one data point to hours /3,4/, as expected for the required symmetry change. Also, a direct consequence of this symmetry is the huge nuclear dipolar signal which arises after a temperature jump of the lattice - the "Haupt" effect /5/. A first study on $4CH_2DC_5H_4N$ - or 4MP1D - mainly contains specific heat data and a few preliminary neutron results /6/. The aim of the new neutron experiments described in this paper was to obtain precise information about the energy level splitting between the tunneling states and about the phase transition at $T_p \approx 5.4$ K deduced from earlier heat capacity /6/ and NMR data /7/. The actual measurements were performed on a 4MP1D-sample 12% doped with undeuterated 4MP. Because on the one hand a contribution to the heat capacity of 4MP at low temperatures only shows up after hours, while on the other hand partially deuterated methyl groups are not expected to be seen by INS /8/, doping of the sample gives a unique possibility to use both INS and calorimetric experiments. The INS spectrum will reveal the undisturbed sixfold barrier of the system, while in the calorimetric measurements the typical symmetry-lowering effect of the partially deuterated groups will be visible.

2. Experimental Details and Results

The incoherent inelastic neutron scattering INS spectra of 4MP1D were measured at the institute von Laue-Langevin (IN5). The design and capability of this machine is discussed in detail elsewhere /9/. IN5 is a medium resolution, multichopper, time-of-flight spectrometer with banks of ^3He detectors that can be positioned over a wide range of scattering angles. For the present experiments an incident wave length of 6.9 Å was used and the detectors were grouped in mainly two blocks at 60° and 130°. At the large angles the maximum scattering amplitude was expected to occur. The elastic energy resolution was 46 μeV (FWHM). The results were examined for any q-dependence in the excitation energies and, in each case, none was found. The lack of q-dependence permitted the data from each bank of detectors to be summed; thus allowing a reduction in the time required to obtain a high quality spectrum. The experimental configuration of the instruments was kept the same for the experiments performed, and so only the cell scan at high temperature (50 K) was required for the background and "empty" cell corrections. The sample was filled as a liquid into a thin-walled, cylindrically shaped, Al-cell. The cell was part of a unit, which contained the necessary thermometry and which could easily be inserted into a precooled top-loading liquid helium dewar. The sample temperature was controlled to 0.05 K during data acquisition periods.

All the INS results shown have been corrected for scattering from the empty cell and further background. The occurrence of the phase transition at about 5.4 K led to an examination of the spectrum at as low a temperature, T, as possible (T = 1.5 K). The data obtained at this temperature are shown in Fig. 1a. The existence of a well-defined absorption (and emission) peak in the spectrum suggested the feasibility of a study of the temperature dependence. The appropriate data are shown in Figs. 1b, 1c and 1d. Note that the last two spectra correspond to the high-temperature phase. In Fig. 1e the temperature dependence of the energy splitting observed (energy gain and energy loss gave the

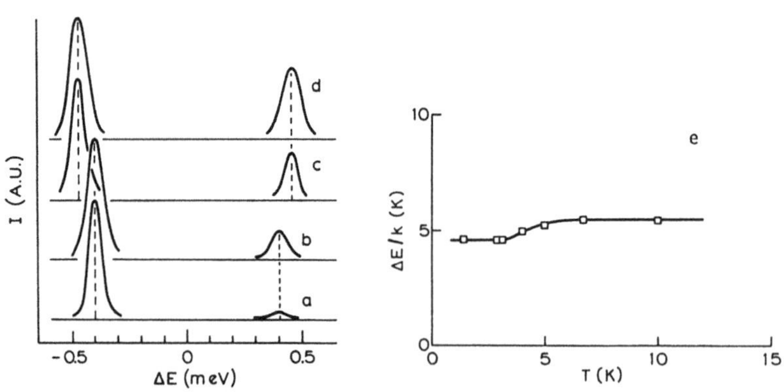

Fig. 1 The intensity of the inelastic scattered neutron beam in arbitrary units is shown as a function of the energy difference, ΔE, with the elastic peak, both for energy loss and energy gain at 4 different temperatures 1.4 K (a), 3.1 K (b), 6.7 K (c) and 10 K (d) resp.. Empty cell and background corrections are made. The shift in ΔE between 3.1 K and 6.7 K is clearly visible (e).

same result) is plotted versus T. A comparison of the sample temperature based on the ratio of the peak intensities, T', by assuming equal degeneracies of both levels to the sample temperature, T, as indicated by the germanium thermometer gave the following results for T'/T: 1.4 K/1.6 K, 2.8 K/2.8 K, 3.1 K/3.2 K, 4.0 K/3.4 K, 5.0 K/4.6 K, 6.7 K/5.8 K and 10.0 K/10.3 K.

As far as NMR is concerned, as in undeuterated 4MP, after a jump in lattice temperature (5 → 50 K) a dipolar signal appeared, which evolved into a final Zeeman equilibrium signal /5/. No relaxation study was pursued. For the specific heat data one is referred to DEN ADEL et al. /6/.

The 4-CH_2D-pyridine sample used in the experiment was commercially obtained from Merck, Sharp and Dohme (Canada). The ratio of 4-CH_2D-pyridine to 4-CH_3-pyridine of 88/12 was determined by high-resolution NMR at 300 MHz with a Bruker MSL300 using $CDCl_3$ as a solvent.

3. Discussion

The energy splitting of 5.3 K observed in the INS experiment - remember that the energy gain and loss figures pertain only to undeuterated 4MP - above the phase transition can be generated by a purely sixfold barrier of V_6/k = 230 K. This value is close to the V_6/k = 180 K value for undeuterated 4MP, because electronic orbitals are not affected by the isotope substitution. The energy splitting below the phase transition of $\Delta E/k$ = 4.7 K is the calculated splitting between the ground state and the first excited state in a purely sixfold barrier of V_6/k = 280 K. The analysis of the specific heat data will be limited to the region below 2.5 K in view of the additional contribution to C/R at higher temperatures due to the phase transition. The C/R ratio used in Fig. 2 is normalized to the amount of 4MP1D in the sample because the 12% 4MP dopant does not contribute to the heat capacity. For the analysis of these data it has to be realized that the 6 lowest levels of the V_6-potential will be shifted and the degeneracies will be lifted by the additional low-symmetry terms, expected for a partially deuterated system. An analysis of the width of the C/R distribution between 2.4 K and 1 K in terms of Schottky anomalies requires at least 3 of these 6 levels to contribute to the specific heat below 2 K. The height of the C/R data on the other hand for such a 3 level system is much too large if there is a normal Boltzmann distribution over these levels. A division of the system into two subsystems A and B, each consisting of two relevant levels with energy splittings $\Delta E(A)/k$ = 1.9 K and $\Delta E(B)/k$ = 4.4 K gives an optimal fit to the experimental data if the relative contributions are in the ratio 1:2 /10/. It is

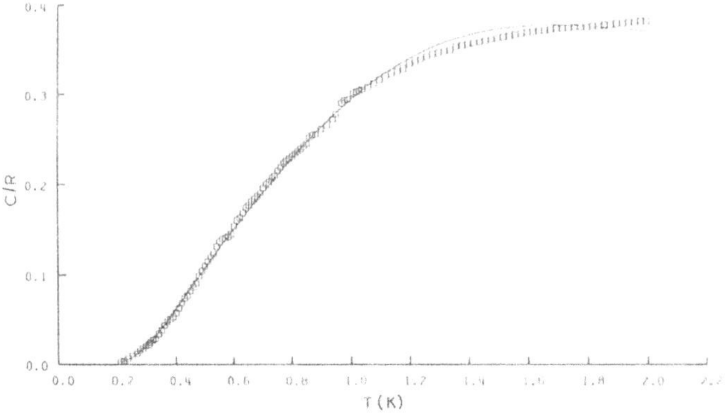

Fig. 2 The heat capacity per mol 4MP1D is compared to the calculated optimal result (drawn line) for a double two-level system with weighting factors 1 and 2.

suggestive to link this 1:2 ratio to the D-position with respect to the molecular frame. Both in the eclipsed and the staggered configuration there are 1 unique and 2 equivalent D-positions. The energy splittings given are then the allowed excitations for such a localized configuration. A less qualitative formulation of essentially the same point is that allowance of admixture of lower symmetry terms in the V_6-potential will produce an inequivalence of the rotational potential minima. The occupation of the levels, which in the case of V_4- and/or V_2-admixtures in the V_6-potential can be labelled by the irreducible representations of the C_{2v}-group A_1, A_2, B_1, B_2 /11/, will now show a non-trivial angular dependence: the A_1-level will have a preferential occupation of one minimum, while for the A_2-level the largest probability occurs in another minimum etc. If the V_4- and/or V_2-admixtures have a negative sign, the potential and occupation patterns, which will repeat themselves after π, show one unique and two equivalent minima. Phonon-mediated transitions are allowed between those levels having their occupation maximum in the same potential well: the system is divided into three disconnected subsystems of which there are two equivalent. In this way 4 of the 6 rotational levels of the ground vibrational state are taken into account; for 4MP1D the other 2 have to be at $\Delta E/k \geq 10$ K above the "ground" state of interest. As an example pertinent to 4MP1D, an admixture of a fourfold perturbation of 2% of the V_6-potential $V/k = \frac{1}{2}[280(1+\cos 6\phi) - 5(1+\cos 4\phi)]$ localizes the methyl group mainly in the 90° potential minimum for the lowest A_1 and B_2 levels with $\Delta E(A_1 - B_2)/k = 1.8$ K, while the A_2 and B_1 levels are clearly localized in the 30° minimum. The energy splitting between the lower A_2 and B_1 levels for this case is calculated to be 6.2 K. Further improvement to match the experimental level splitting might be possible by allowing admixtures of several lower order terms. The phase transition can be thought to be driven by the reduction of the total free energy of the lattice and rotational system together: the increase in lattice energy by the (slight) deformation being compensated by the lowering of the rotational free energy /12/.

Because the dipolar polarization phenomena are a property of the undeuterated methyl groups, a discussion of the polarization process in this diluted system is outside the scope of this article.

4. Conclusion

In 4MP1D, as in nitromethane, partial deuteration adds low symmetry terms to the effective rotational potential. Localization of the system into non-converting configurations is essential for the explanation of the specific heat data.

Acknowledgements

It is a pleasure to thank Dr. G. Kearly for his cooperation during the INS experiment at IN5 in Grenoble and Mr. C. Erkelens at the Gorlaeus laboratory for the high-resolution NMR analysis. One of us (HBB) gratefully acknowledges many valuable discussions with Prof. I. Svare and the assistance of Mr. A.F. Deutz in some of the computer calculations.

References

1. H.B. Brom, G. Mennenga, D. Cavagnat, J. Lascombe: J. Phys. C18 3407-3413 (1985)
2. B. Alefeld, A. Kollmar, B.A. Dasannacharya: J. Chem. Phys. 63, 4415 (1975)
3. P. Beckmann, S. Clough, J.W. Hennel, J.R. Hill: J. Phys. C10, 729 (1977)
4. H. den Adel, H.B. Brom, Z. Dokoupil, W.J. Huiskamp: Physica 111B, 160 (1981)
5. J. Haupt: Phys. Lett. 38A, 389 (1972)
6. H. den Adel, H.B. Brom, W.J. Huiskamp, B. Alefeld: Physica 112B, 309 (1982)
7. H. den Adel, H.B. Brom, R.A. Wind, D.J. Ligthelm: Bull. of Magn. Res. 2, 97 (1981)

8. K.J. Lushington, K. Maki, J.A. Morrison, A. Heidemann, W. Press: J. Chem. Phys. 75, 4010 (1981)
9. ILL (1981). Neutron scattering facilities at the HFR. Copies may be obtained from the Scientific Secretary at ILL, Grenoble, France
10. The fitting program "MINUIT", version 277, of the CERN was used.
11. M. Tinkham:: Group Theory and Quantum Mechanics (McGraw-Hill Book Company, New York 1964)
12. H. den Adel, H.B. Brom, D.J. Ligthelm and R.A. Wind, Physica 111B, 171 (1981)

Calorimetric and Dielectric Study of Partly Deuterated Ammonium Hexachlorostannate

J. Gilchrist, J. Odin, and J. Peyrard

Centre de Recherches sur les Très Basses Températures, C.N.R.S., 25, av. des Martyrs, B.P. 166X, F-38042 Grenoble Cedex, France

1. INTRODUCTION

Ammonium hexachlorostannate is a favorite for rotational tunnelling studies /1-3/. An unexpected dielectric relaxation was attributed, with some difficulty, to ammonium reorientations /4/. If correct, this would imply a specific heat anomaly somewhere below 20 K. Existing specific heat data /1/ only extend down to 20 K.

2. SPECIFIC HEAT OF NATURAL AND 50% d $(NH_4)_2SnCl_6$ FROM 1.3 TO 25 K

Crystals of approximately 500 mg were grown from aqueous solutions of 99.999% (metals basis) "Puratronic" product, and studied by adiabatic calorimetry so as to allow easy detection of any relaxation phenomenon with time constant 10 - 1000 s. The carbon thermometers were recalibrated at the end of each run. Table 1 summarises the results. The natural sample reached equilibrium rapidly at each temperature but below 3 K the 50% d sample scarcely reached equilibrium in 15 mns, and below about 2 K the tabulated C_p values, calculated from the temperature rise 15 mns after application of the heat pulse, are definitely underestimates. A second anomalous zone occurred around 17.5 K. The tabulated C_p value is the mean of four runs. After prolonged standing below 10 K, we found C_p = 19.78 J/mole K with a very slow approach to equilibrium (two runs), otherwise C_p = 19.47 J/mole K with fast approach to equilibrium. If the Arrhenius law of the dielectric relaxation /4/ is extrapolated, the relaxation time at 17.5 K is ca. 100 s, so the calorimetric anomaly could correspond to this relaxation. However, its weakness in the one sample and absence in the other suggests that the relaxation was caused by a few strongly dipolar excitations rather than many weak ones, probably not more than 0.01 per formula unit.

2.1 Natural $(NH_4)_2SnCl_6$

The heat capacity never obeys a T^3 law. Between 2.8 and 11 K it rises faster than T^3, elsewhere more slowly. The value at 2.8 K implies a Debye temperature not less than 104 K, so that below 2 K and above 4 K there is appreciable excess over the Debye specific heat. Below 2 K this may be due to the modification of the vibration spectrum as suggested by SVARE et al. /3/. Above 4 K the soft modes involving anion libration /5,6/ will provide a significant contribution, but ammonium librational modes will not contribute significantly below 25 K /1,2/. From 20 to 25 K our data agree accurately with MORFEE's /1/.

2.2 50% d $(NH_4)_2SnCl_6$

The heat capacity is enhanced at all temperatures by the part deuteration. The following considerations are likely to be significant: (a) the ammonium librational band will be reduced from ca. 3.2 THz towards 2.5 THz /5/; (b) the asymmetric ammonium ions will be more strongly phonon coupled, resulting in increased lifetime broadening of the librational spectrum /3/; (c) the anion librational modes may be coupled to the ammonium librations and be modified; (d) when the symmetry of the crystal field is lowered by defects, the asymmetric ammonium ions will become par-

Table 1 Interpolated values of the molar heat capacity C_p, /mJ/mole K/ at regular intervals of temperature, T /K/

T	C_p, 0% d	C_p, 50% d
1.4	7	769
1.6	9	670
1.8	12	591
2.0	15	548
2.2	20	505
2.4	25	448
2.6	31	422
2.8	38	392
3.0	47	366
3.2	58	344
3.4	70	334
3.6	85	332
3.8	102	331
4.0	122	334
5.0	271	453
6.0	558	757
7.0	1013	1272
8.0	1679	2061
9.0	2576	3057
10.0	3695	4325
11.0	5007	5814
12.0	6481	7490
13.0	8117	9364
14.0	9879	11389
15.0	11752	13568
17.5	16786	19625
20.0	22331	26052
22.5	28353	33143
25.0	34497	39175

tially localised in inequivalent pocket states. Tunnelling transitions between these states will be fairly strongly phonon-coupled and uninhibited by spin-symmetry requirements. (a) and (b) together could account for a significant enhancement at the beginning and end of the temperature range but (c) is necessary to account for the enhancement in the middle and (d) is necessary to account for the behaviour below 4 K since (a) and (b) alone would merely lead to an enhanced linear ($C_v \propto T$) term.

3. DIELECTRIC RELAXATIONS

Samples in powder form (as received, dried precipitate or crushed crystals) were pressed into pills and mounted between indium electrodes. Dielectric susceptibility and loss were measured between 10 Hz and 10 kHz with a transformer bridge. The main findings are summarised here. It is intended to publish more details later.

3.1 Relaxation of Fimland and Svare

We observed this relaxation in the natural compound between 23 and 30.5 K, where it continued to obey the Arrhenius law. After recrystallisation from aqueous solution the relaxation became 3 to 4 times stronger, but not after recrystallisation from 15% HCl solution. A weaker relaxation was obtained by precipitating the double salt by mixing NH_4Cl and $SnCl_4$ solutions, both in 20% HCl, and drying it in a dessicator with P_2O_5. Fully deuterated ammonium hexachlorostannate relaxed 20 - 30 times more slowly than the natural compound, partly-deuterated samples at intermediate rates. Rb_2SnCl_6 and Cs_2SnCl_6, precipitated from aqueous solutions, exhibited relaxations of comparable strength (weaker if pptd from strong acid) but at lower temperatures. The hexachloroplatinates of ammonium, potassium, rubidium and cesium all exhibited relaxations of similar strength. $(NH_4)_2PtCl_6$ relaxed more slowly than $(NH_4)_2SnCl_6$, with activation energy ca. 1.4 kcal/mole (cf. 1.2 kcal/mole /4/). No relaxations were observed, however, when the hexachloroplatinates were recrystallised or precipitated from 10% HCl.

We suggest the relaxations are associated with MCl_5OH^{2-} or $MCl_5OH_2^-$ impurity anions arising from the reactions in solution /7/:

$$MCl_6^{2-} + OH^- \rightleftharpoons MCl_5OH^{2-} + Cl^-,$$

$$MCl_6^{2-} + H_2O \rightleftharpoons MCl_5OH_2^- + Cl^-.$$

In the case of MCl_5OH^{2-} the relaxation would arise from the rotation of the hydroxyls about the tetrad axis (fig. 1). 2% of MCl_5OH^{2-} could account for the relaxations. The chlorine deficit reported by MORFEE /1/ corresponds to 2% of $SnCl_5OH^{2-}$. Two properties of ammonium hexachlorostannate helped to obscure the origin of the phenomenon: (a) the activation energy being equal - either by chance or because of a synchronisation effect - to the activation energy for ammonium reorientation as revealed by NMR /4/ and (b) the difficulty of pushing the equilibrium reaction completely to the left. Ammonium hexachloroplatinate has neither of these properties, the NMR activation energy being 0.68 kcal/mole /8/.

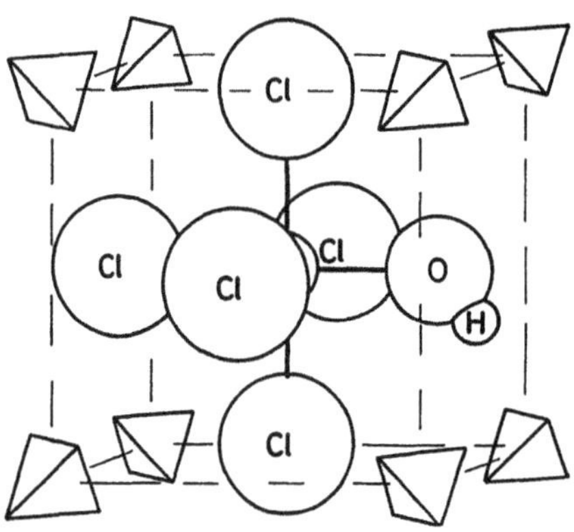

Figure 1 Octant of a unit cell bounded by eight NH_4^+ ions and containing a $SnCl_5OH^{2-}$ ion

3.2 Other relaxation of partly deuterated ammonium hexachlorostannate

Below 15 K the title compound exhibited another relaxation, ca. 200 times weaker than the preceding one, but absent in the natural or fully deuterated species (fig. 2). From 5 to 10 K, rate $\propto T^9$ and strength $\propto T^{-1}$ corresponding to r.m.s. dipole moment ca. 0.015 D per ammonium ion (0.005 D enhanced by Lorentz local field factor). Study of various samples from 5 to 85% d did not resolve any distinctive contributions of the different partly deuterated species, but mean relaxation rate tended to diminish with increasing % d. This relaxation apparently corresponds to the specific heat anomaly of the 50% d sample below 4 K. The dipole moment does not necessarily imply a distortion of the ammonium ion. The centre of charge may still coincide with the nitrogen nucleus, the latter shifted from the lattice site so as to minimise the zero-point energy of the ion. 0.005 D corresponds to a unit charge displaced by 10^{-13} m.

We have studied four other ammonium salts and found analogous relaxations with $(NH_4)_2PtCl_6$, $(NH_4)_2GeF_6$ and $(NH_4)_2Cr_2O_7$. We observed relaxations of $(NH_4)_2PtCl_6$ in

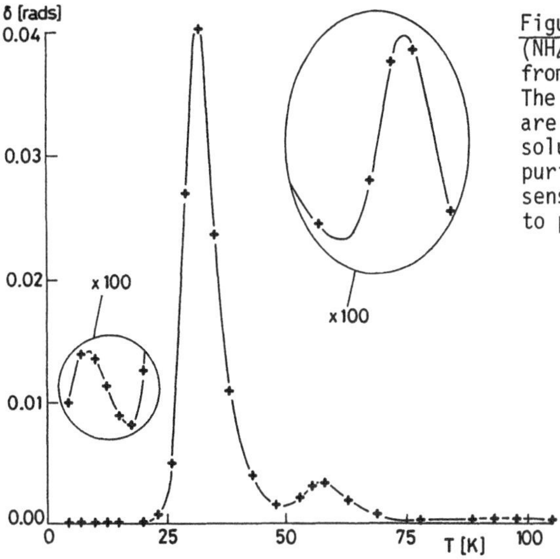

Figure 2 Dielectric loss angle of $(NH_4)_2SnCl_6$ - 85% d recrystallised from aqueous solution, at 1.2 kHz. The peaks at 32 K, 57 K and 96 K are sensitive to the acidity of the solution and are attributed to impurities. The peak at 8.5 K is insensitive to acidity but sensitive to percentage deuteration

the one-phonon tunnelling regime below 4 K, where relaxation rate depended very markedly on % d, and relaxations of $(NH_4)_2Cr_2O_7$ in the Arrhenius regime, between 25 and 40 K. For NH_3D^+ in ammonium hexachloroplatinate, the ratio of relaxation rate to Δ_0^2 (Δ_0 = tunnel splitting, estimated by interpolation between the values for NH_4^+ /9/ and ND_4^+ /10/ is comparable to that of the hydroxyl impurity in KCl /11,12/, indicating a comparable phonon coupling strength.

4. CONCLUSIONS

Hydroxyl is likely to be the major impurity in ammonium hexachlorostannate unless careful precautions are taken. It dominates the dielectric response in the relevant spectral range but does not hide the response due to partly deuterated ammonium ions. Its direct influence on heat capacity is very small, but it may exert an indirect influence in the case of the partly deuterated compound, by helping to reduce the crystal symmetry experienced by the ammonium ions. Either calorimetric or dielectric studies of partly deuterated low-barrier ammonium salts could prove to be a sensitive means of evaluating the lack of symmetry of the ammonium sites, due to impurities or to applied stress.

References

1. R.G.S. Morfee, L.A.K. Staveley, S.T. Walters, D.L. Wigley: J. Phys. Chem. Solids 13, 132 (1960)
2. M. Prager, W. Press, B. Alefeld, A. Huller: J. Chem. Phys. 67, 5126 (1977)
3. I. Svare, A.M. Raaen, B.O. Fimland: Physica 128B, 144 (1985)
4. B.O. Fimland, I. Svare: Physica Scripta 25, 694 (1982)
5. B.M. Powell, W. Press, G. Dolling: Phys. Rev. B 32, 3118 (1985)
6. D. Moses, M. Sutton, R.L. Armstrong, P.P.M. Meincke: J. Low Temp. Phys. 36, 587 (1979)
7. Gmelin Handbuch der Anorganischen Chemie, 8th ed., Vol. 46-C1, ed. by H. Bitterer (Verlag Chemie, Weinheim 1972)
8. I. Svare, A.M. Raaen, G. Thorkildsen: J. Phys. C 11, 4069 (1978)
9. M. Prager, A.M. Raaen, I. Svare: J. Phys. C 16, L181 (1983)
10. Z.T. Lalowicz, M. Punkkinen, E.E. Ylinen: J. Phys. C 12, 4051 (1979)
11. U. Bosshard, R.W. Dreyfus, W. Kanzig: Phys. Kondens. Materie 4, 254 (1965)
12. K. Knop, W. Kanzig: Helv. Phys. Acta 46, 889 (1974)

Tunneling to Reorientation Transition Effects in High Field ^{2}H-NMR Spectra of ND_4^+ Ions

L.P. Ingman, E. Koivula, Z.T. Lalowicz, M. Punkkinen, and E.E. Ylinen*
Wihuri Physical Laboratory, University of Turku, 20500 Turku, Finland

The high field ^{2}H-NMR spectra of the tunneling ND_4^+ ion are found to be separable into contributions of A, E and T symmetry species. Their temperature dependence as well as that of the tunneling frequency and spin-lattice relaxation are followed in an $(ND_4)_2SnCl_6$ single crystal. A common correlation time scale exists.

1. Introduction

The observables of NMR-methods, spectra and relaxation times, are sensitive to molecular motions in solids. Molecular reorientation may be characterized by a correlation time τ_c with a temperature dependence $\tau_c = \tau_c^o \exp(E_a/kT)$. This relation serves to obtain the activation energy E_a, which is a most universal parameter both in tunneling and reorientation ranges [1]. Recent studies by PRAGER et al [2] throw light on the origin of E_a and point out a perspective research field.

Narrowing of the spectrum having the "rigid" width δ is expected at such a temperature T_t, where the condition $\delta \cdot \tau_c \sim 1$ is fullfilled. The NMR absorption line narrows as spin interactions are being averaged down to an effective value. This is observed as a step in the temperature dependence of the second moment M_2. Additional observed features may help to specify the motion further.

Mainly three (CH_3, CD_3) and four (NH_4, ND_4, CH_4) spin systems have been studied in a variety of sites (for a review, see PRESS [3]). In such compounds some components of the NMR spectra are shifted by ν_t, where ν_t is the tunneling frequency. At higher temperatures these components should come back so close to the resonance frequency ν_o that they can be observed. In reality motional narrowing occurs almost always before ν_t decreases below the dipolar line width thus preventing the observation of a rigid-molecule spectrum. Only JOHNSON and MOTTLEY [4] observed an exception to this behavior in the NMR spectra

* Institut für Physikalische Chemie der Universität Münster, 4400-Münster, BRD and Institute of Nuclear Physics, 31-342 Krakow, Poland

of CH_3CD_2I. Very often motional narrowing starts before any significant decrease in ν_t. An example of such a behavior is provided by ammonium ion in NH_4I [5].

Because the quadrupolar interaction in CD_3 and ND_4 is roughly one order of magnitude stronger than the dipole-dipole interaction, it should be easier to find a ND_4-compound, where ν_t becomes, with rising temperature, smaller than the quadrupole frequency before the onset of motional narrowing. Also the one-particle character and related transformation properties lead to a better recognition of infrequent jumps between localized states. Our aim is therefore to search for motional effects in ^2H-NMR spectra of deuterated ammonium compounds.

2. Theoretical principles

2.1. The Spectrum

The solution of the eigenvalue problem for the general Hamiltonian

$$H = H_T + H_Z + H_Q \tag{1}$$

requires numerical calculations [6]. The tunneling (H_T) and Zeeman (H_Z) Hamiltonians supply zeroth order energies in space and spin coordinates. The quadrupole interaction Hamiltonian (H_Q), acting as a perturbation, contains terms in both coordinates. NMR acts on spin states, and senses in this way the time dependence imposed on H_Q by molecular motion.

The quadrupole Hamiltonian is a sum of single deuteron interactions and can be written in a symmetry-adapted form [6]

$$H_Q = \sum_{i=1}^{4} H_Q(i) = \sum_{\mu=-2}^{2} \sum_{x} B_x^{\mu} S_x^{-\mu} \tag{2}$$

where B_x^{μ} and S_x^{μ} are space and spin operators, respectively. x goes through the labels A, T', T" and T"', which refer to the irreducible representations of the tetrahedral point group. The terms $B_A^{\mu}S_A^{\mu}$ are totally symmetric relative to the motion and do not relax spins or the tunnel system. The other terms are time-dependent and induce transitions between the various states.

In the case of high magnetic field ($\nu_o > \nu_t >> \nu_q$) we are left with $B_x^o S_x^o$ terms, when calculating the spectra. The quadrupole interaction does not shift A or E levels in the first order. Because of the second order corrections, which are found to be propotional to $1/\nu_t$, the A doublets occur at $\pm\Delta\nu$, $\pm 3\Delta\nu$ relative to ν_o. The E doublets should appear at $\pm 2\Delta\nu$, $\pm 6\Delta\nu$. Here for example $\Delta\nu = 3.5$ kHz for $\nu_t = 0.5$ MHz.

Due to the common angular dependence these expressions are valid also for polycrystalline samples. The T symmetry components in a spectrum contain nummerous doublets with a strong angular dependence and reflect the tunnel splitting of the T levels. The T components are spread over the frequency range of ±140 kHz around ν_o [7].

2.2.Motional Narrowing

The above-sketched time-independent theory was applied succesfully on the analysis of low-temperature spectra of $(ND_4)_2SnCl_6$ [7]. The model accounts for the increase of the separation between the components of the A doublets, thus providing a direct measure of the tunneling frequency.

The varying tunneling splitting explains the observed features up to about 40 K, i.e. up to the point where motional narrowing of the central doublets starts to be significant. But as a matter of fact the temperature dependence of ν_t is already near 10 K readily related to torsional excitations, which destroy the coherence of the tunneling rotation. Motional effects can be observed for T components too. All these motional effects can be analysed in terms of the spin exchange theory described below.

The FID signal of a spectrum of ω_K frequencies can be written down as

$$G(t) = \overline{W}\,[\exp(i\underset{\sim}{\omega} + \underset{\sim}{\pi})t]\overline{1} \qquad (3)$$

where $\underset{\sim}{\pi}$ is the exchange rate matrix between ω_K and $\overline{W}$ is a vector of statistical weights of n states at equilibrium. Formal calculations of the NMR spectrum

$$I(\omega) = \mathrm{Re}\ {}_0\!\int^{\infty} G(t)\exp(-i\omega t)dt \qquad (4)$$

were not yet undertaken for a reorientating tetrahedral arrangement of four deuterons. We will use the standard example of exchange effects in a doublet spectrum as a guidance in our quantitative interpretation of ND_4^+ spectra.

The model considers a system characterized by the spectral frequencies ±d and by jumps performed from one to the other state with frequency Ω. This is a most essential simplification with respect to our case, where an adequate $\underset{\sim}{\pi}$ matrix may contain various frequencies between states represented by spectral components different not only (or necessarily) in sign, but also in the absolute value. The temperature evolution of the spectrum can be divided into four ranges, characterized by the rate of the involved motions relative to the non-averaged spectral frequency $d \simeq \pm 50$ kHz:

1. *ultraslow exchange limit*, $\tau_c \gtrsim 10^{-4}$s, the natural linewidth is observed;
2. *slow exchange*, $\tau_c \gtrsim 10^{-5}$ s, Lorentzian broadening of separate peaks is noticed, no shift observed;
3. *intermediate exchange*, $\tau_c \sim 10^{-6}$ s, the spectrum is spread out over a wide range;
4. *fast exchange limit*, $\tau_c \lesssim 10^{-7}$ s, Lorentzian line at the mean frequency.

3. Discussion of Experimental Results

We analyse the spectra presented in Fig. 1. The central part is seen clearly on the spectrum taken at 33 K (Fig. 1d). Two A doublets and the second E doublet are clearly recognisable. The first E doublet fills the gap between A components. The resolution in Fig. 1 is not sufficient to show the doublet structure at lower temperatures. The parameters ν_t and the width $\sqrt{M_2}$ for the Gaussian broadening function are sufficient to fit the central lines using the time-independent theory. The width of A and E components increases with temperature, actually fits to experimental spectra yield $M_2/10^{-8}\ T^2 = 0.07$ at 10 K, 0.12 at 35 K and 0.55 at 40 K.

The T components are seen in the spectrum at 4.2 K. A fit requires a $\pm 4^{\circ}$ distribution of ND-vectors. Torsional oscillations, but also crystal imperfections, may be responsible. The T components start broadening already at 8 K and are shifted somewhat towards the center at 15 K. The majority of the components are spread out at 33 K and we observe a flat background for the central spectrum. Rising of temperature brings more and more spectral components into the fast exchange limit. A central Lorentzian line builds up. The change of the shape is evident, when we consider the ratio $R = M_4 \cdot M_2^{-2}$, which for a Gaussian line equals 3. The value of R increases from 3 at 4.2 K to about 5 at 50 K.

The observed temperature dependence of T components may be compared with correlation times obtained from the preliminary measurements of the spin-lattice relaxation time T_1. At 20 K $\tau_c = 10^{-4}$ s as expected for slow exchange. At 33 K $\tau_c = 2 \cdot 10^{-5}$ s and most of the components are in the intermediate exchange range. At 44 K $\tau_c = 10^{-6}$ s corresponding to the fast exchange. Correlation times at 50 K and 70 K are equal to 10^{-7} s and $5 \cdot 10^{-9}$ s, respectively. The apparent activation energy E_a, at T < 40 K is smaller than obtained from the T_1 minimum.

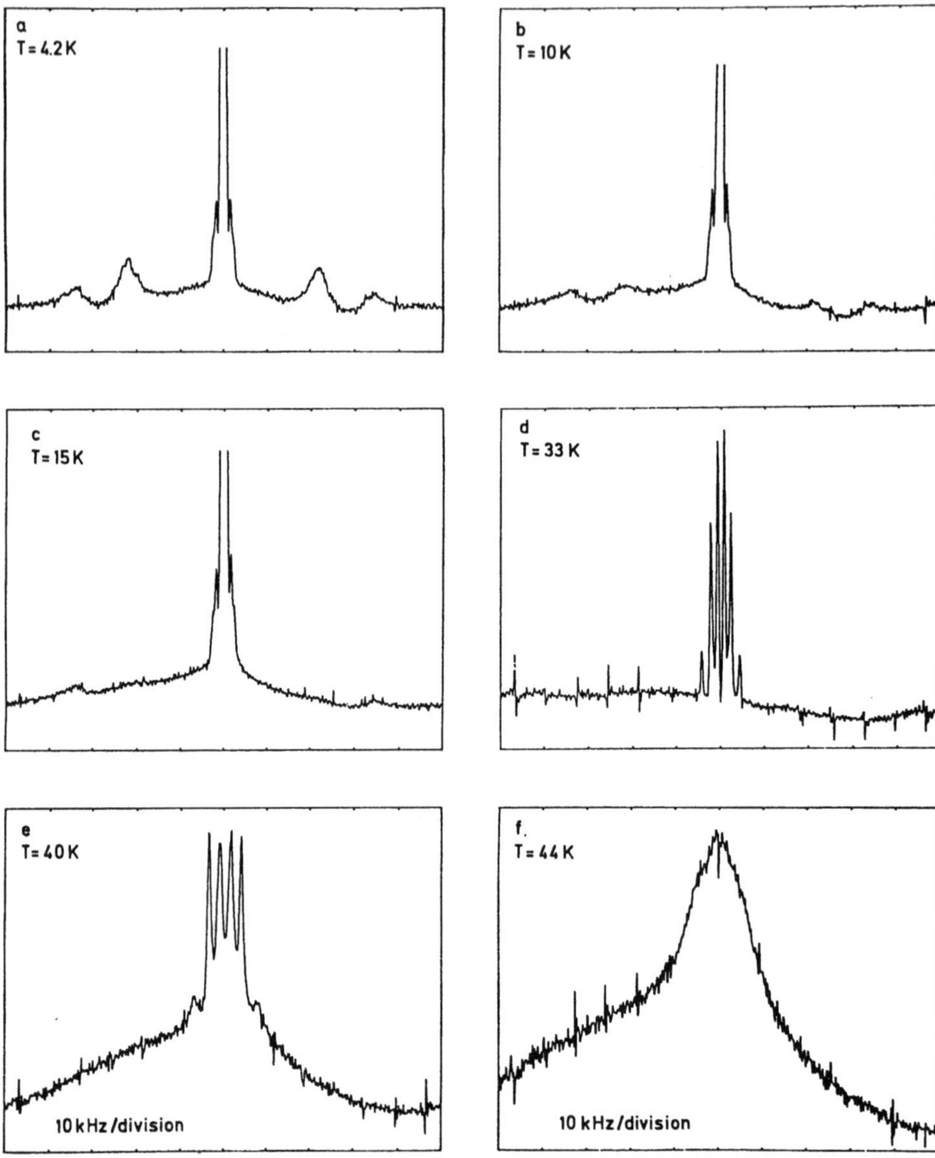

Fig.1. ^{2}H-NMR spectra of $(ND_4)_2SnCl_6$ single crystal measured at different temperatures at B_0 || [011] orientation.

The correlation times derived from the motional narrowing of T components and T_1 temperature dependence seem to agree. On the other hand, broadening of A and E components at a temperature as high as 40 K suggests lifetimes of those states of the order of a few ms. Symmetry species A seem to be, as expected, insensitive to molecular reorientations up to 40 K, but also symmetry conversion may be very slow.

Acknowledgement

ZTL is grateful to the Alexander von Humboldt Foundation for a fellowship.

References

1. S. Clough, A. Heidemann, A.J. Horsewill, J.D. Lewis, M.N.J. Paley: J. Phys. C: Solid State Phys. 15, 2495 (1982)
2. M. Prager, W. Press, A. Heidemann, C. Vettier: J. Chem. Phys. 80, 2777 (1984)
3. W. Press: Single Particle Rotation in Molecular Crystals, Springer Tracts Mod. Phys. 92 (Springer, Berlin, Heidelberg 1981)
4. C.S. Johnson, Jr., C. Mottley: Chem. Phys. Lett. 22, 430 (1973)
5. Z.T. Lalowicz, unpublished
6. Z.T. Lalowicz, W. Sobol: J. Phys. C: Solid State Phys. 16, 2351 (1983)
7. L.P. Ingman, E. Koivula, Z.T. Lalowicz, M. Punkkinen, E.E. Ylinen: (to be published)

Part IV

Translational Tunneling

Influence of Phonons and Electrons on Low-Temperature Translational Tunneling

H. Grabert [(a)]

Service de Physique du Solide et de Résonance Magnétique, Centre d'Etudes Nucléaires de Saclay, F-91191 Gif sur Yvette Cedex, France

The model of a particle moving in a multistable potential is encountered in various fields of physics. At low temperatures transitions between adjacent wells involve quantum tunneling which is strongly influenced by the coupling of the particle to its environment. Recent studies have led to some progress in the theoretical description of this problem. For interstitials in metallic crystals the tunneling rate is affected by the coupling to conduction electrons and phonons. At very low temperatures the defect tunnels coherently. This band-like transport is destroyed near a typical crossover temperature above which subsequent tunneling transitions become statistically independent. The incoherent tunneling rate first decreases with increasing temperature as an effect of the electron cloud surrounding the defect. The rate then goes through a minimum above which phonon assisted tunneling dominates leading to an increase of the rate. The basic ideas underlying the theoretical treatment of defects in metals and some recent theoretical predictions are presented.

1. INTRODUCTION

Diffusion of defects in metals has been an interesting subject of solid state physics since many years and has attracted a great deal of renewed interest lately. While earlier work has focussed mainly on the effects arising from the defect-phonon interaction [1-3], it has become clear more recently that the interaction with conduction electrons profoundly affects the defect motion at low temperatures where quantum tunneling prevails. Here, I will consider the diffusion of particles like protons or muons that are heavy compared with the electrons. Hence, it seems natural to assume that the electrons follow the defect adiabatically which would cause only a small renormalization of the effective defect mass. However, electron-hole pair excitations of the fermionic bath can have arbitrarily small energies leading to a nonadiabatic response of the screening cloud even to slow defect motions. This problem was recognized by Kondo [4] who showed that for low temperatures the electronic overlap integral becomes proportional to a power of temperature which should be reflected in a power law dependence of the diffusion coefficient on temperature [5].

The effects of environmental degrees of freedom on tunneling transitions were also the subject of recent theoretical work [6-8] showing that the problem can be treated by an extended WKB-approach employing functional integral techniques. In these theories the action functional of a particle in a rigid host crystal is supplemented by a term, the so-called influence functional [9], describing the effects of crystal excitations on the defect motion. In other words, the modes of the host crystal are eliminated and their influence is described in terms of a "damping" of the interstitial. The coupling to conduction electrons causes a damping coefficient which becomes frequency independent at low frequencies [10] while the effective damping arising from the phonon heat bath is strongly frequency dependent and vanishes at zero frequency [11]. As a consequence, the

(a) on leave from: Institut für Theoretische Physik, Universität Stuttgart, D-7000 Stuttgart 80, Germany

phonons lead only to a constant parameter renormalization at low temperatures and the temperature dependence of the diffusion rate is dominated by the coupling to the electronic bath.

In the following section I will first consider a defect tunneling between two interstitial sites. The influence of conduction electrons is shown to lead to a crossover between coherent clock-like tunneling at low temperatures and incoherent hopping-like tunneling at higher temperatures. Accordingly, the structure factor for neutron scattering from the tunneling center changes from inelastic to quasi-elastic scattering [12,13]. Such a behavior was observed recently for hydrogen trapped by oxygen in niobium [14]. As temperature is raised, the widths of the inelastic peaks broaden and they merge into one quasi-elastic peak. The width of the quasi-elastic peak shows a characteristic power-law dependence on temperature leading to a narrowing of the peak with increasing temperature.

The results on incoherent tunneling in a double well immediately extend to the situation of a defect moving in a periodic potential [8]. The behavior of the width of the quasi-elastic peak is now reflected in a corresponding temperature dependence of the diffusion coefficient. A decrease of the mobility with increasing temperature was observed in several recent experiments on muons in metals [15] and explained as an effect of the interaction with conduction electrons by Kondo [5]. When the temperature is increased further, the influence of the host phonons can no longer be described by a constant renormalization of the tunneling matrix. Section 3 gives a discussion of the temperature dependence of the incoherent tunneling rate of a defect coupled both to electrons and phonons. Since phonon-assisted tunneling leads to an increase of the rate with increasing temperature, the mobility has a minimum indicating a transition between electron and phonon dominated behavior. The last section contains some concluding remarks.

2. INFLUENCE OF ELECTRONS ON COHERENT TUNNELING

Let us consider a defect which tunnels between two trap sites in a metal. At low temperatures one can disregard excited defect states and the interstitial behaves effectively like a two-state system. We are interested then in the dynamics of transitions between the states 1 and 2 corresponding to the ground states at either site. Because of the small density of low-frequency phonons, the interaction with the lattice vibrations does not contribute to the dominant temperature dependence of the tunneling process for sufficiently low temperatures. Hence, the influence of phonons can be described in terms of a renormalized tunneling matrix Δ_p which is reduced by the phonon overlap integral. The interesting physics arises from the interaction with the conduction electrons.

The defect motion is described conveniently in terms of the transition probabilities $P_{ij}(t)$, $i,j=1,2$ which give the probability to find the interstitial at time t in state i when it was initially at time 0 in state j. For simplicity I will assume here that the two trap sites are not shifted against each other. Then, for reasons of symmetry and conservation of probability, we have $P_{11}(t)=P_{22}(t)$ and $P_{12}(t)=P_{21}(t)=1-P_{11}(t)$. Hence, it is sufficient to investigate, say, $P_{11}(t)$. Since all events with an even number of tunneling transitions within the period t contribute to $P_{11}(t)$ it may be written as

$$P_{11}(t) = 1 + \frac{1}{2} \sum_{n=1}^{\infty} (-1)^n \Delta_p^{2n} \int_0^t dt_{2n} \int_0^{t_{2n}} dt_{2n-1} \dots \int_0^{t_2} dt_1 \, Y_{2n}(t_1 \dots t_{2n}) \tag{1}$$

where Y_{2n} describes the influence of the electronic bath. In the absence of the defect-electron coupling one has $Y_{2n}=1$ and (1) gives $P_{11}(t)=[1+\cos(\Delta_p t)]/2$ describing the coherent oscillations of an undamped system. Besides the renormalization of the tunneling matrix the defect-phonon coupling also leads to a slow decay of the oscillation. This effect, however, is negligible against the much larger effects of the fermionic bath discussed below.

The influence of conduction electrons on a tunneling system was discussed recently by several authors [5,12,13,16-21]. At low temperatures the dominant effects come from the production and absorption of electron-hole pairs. These low-energy excitations of the Fermi liquid lead to an effective interaction between tunneling transitions described by the function

$$S(t)= 2K\ \ell n[\frac{\hbar\beta D}{\pi}\sinh(\frac{\pi t}{\hbar\beta})] \tag{2}$$

where $\hbar D$ is an effective bandwidth of the electronic bath and K is a dimensionless defect-electron coupling strength. In terms of the interaction S(t) the electronic influence function Y_{2n} takes the form

$$Y_{2n} = \cos^n(\pi K)\ 2^{-n} \sum_{\zeta_j=\pm1} \exp(-\sum_{j=1}^{n} S_j + \sum_{\substack{j,k=1\\ j>k}}^{n} \Lambda_{jk}\zeta_j\zeta_k) \tag{3}$$

where

$$S_j = S(t_{2j}-t_{2j-1}) \quad \text{and} \tag{4a}$$

$$\Lambda_{jk} = S(t_{2k}-t_{2j-1}) + S(t_{2k-1}-t_{2j}) - S(t_{2k}-t_{2j}) - S(t_{2k-1}-t_{2j-1}). \tag{4b}$$

Note that the transition probability (1) depends on Δ_p and D only through the quantity $\Delta_p D^{-K}$. Hence, the interaction (2) includes a renormalization of the tunneling matrix to which all excitations of the Fermi liquid contribute. The size of the effective bandwidth is chosen such that this renormalization is described properly. The high-energy excitations also screen the defect potential. As a consequence, the low-energy quasi-particle excitations of the electronic bath which cause the long-range component of the interaction (2) are coupled to the defect by a short-range pseudo-potential. The dimensionless coupling constant K characterizing the strength of the long-range interaction is related to the scattering phase shift produced by this pseudo-potential and is always less than 1/2 [17-20].

The formulas (1-4) describing the effect of conduction electrons on defect tunneling have many features in common with the corresponding results for a defect coupled to a phonon bath. The main difference arises from the fact that the density of electron-hole pairs becomes constant at low frequencies while the density of phonons vanishes as ω^2. The high density of low-energy excitations of the Fermi liquid leads to the long-range interaction (2) which is of the same form as the interaction caused by bosons with a corresponding density of states [22].

The differential cross section for neutron scattering from the defect is related to the structure factor [23]

$$S(\underline{k},\omega) = (2\pi)^{-1} \int_{-\infty}^{\infty} dt\ e^{-i\omega t} \langle e^{-i\underline{kq}}\ e^{i\underline{kq}(t)} \rangle . \tag{5}$$

For a defect tunneling between the sites $\underline{q}_1=(d/2)\hat{\underline{z}}$ and $\underline{q}_2=(-d/2)\hat{\underline{z}}$, where $\hat{\underline{z}}$ is a unit vector, one has

$$\langle e^{-i\underline{kq}}\ e^{i\underline{kq}(t)} \rangle = \cos^2(\frac{kd}{2}) + \sin^2(\frac{kd}{2})\langle\sigma_z\sigma_z(t)\rangle \tag{6}$$

where $\underline{k}=k\hat{\underline{z}}$, and where the effective two-state system of the defect is represented as a pseudo-spin. The symmetrized correlation function $\frac{1}{2}\langle\sigma_z(t)\sigma_z+\sigma_z\sigma_z(t)\rangle$ is related to the transition probabilities $P_{ij}(t)$ by

$$\langle\sigma_z(t)\sigma_z + \sigma_z\sigma_z(t)\rangle = P_{11}(t) + P_{22}(t) - P_{21}(t) - P_{12}(t) = 2P(t) \tag{7}$$

where $P(t)=2P_{11}(t)-1$. Now, using the well-known relation between the spectral functions of $\langle\sigma_z(t)\sigma_z+\sigma_z\sigma_z(t)\rangle$ and $\langle\sigma_z\sigma_z(t)\rangle$ one finds

$$S(\underline{k},\omega) = \cos^2(\frac{kd}{2})\delta(\omega) + \sin^2(\frac{kd}{2})\ \frac{\hat{P}(i\omega)+\hat{P}(-i\omega)}{1 + \exp(\beta\hbar\omega)} \tag{8}$$

where

$$\hat{P}(\lambda) = \int_0^\infty dt\, e^{-\lambda t}\, P(t) \tag{9}$$

is the Laplace transform of P(t). Since the first term in (8) gives rise to elastic neutron scattering, the cross section for inelastic neutron scattering is associated with the scattering function

$$J(\omega) = \frac{\hat{P}(i\omega)+\hat{P}(-i\omega)}{1 + \exp(\beta\hbar\omega)} \tag{10}$$

The formula (1) for the probability $P_{11}(t)$ can be evaluated in the so-called dilute bounce gas approximation (DBGA), where one neglects the interaction between pairs of tunneling events. A pair describes a forward and backward move of the tunneling particle and is frequently called a bounce. The pair interaction is described by the functions Λ_{jk} defined in (4b) which involve differences of interactions S(t) only and therefore decay for large pair separations. The DBGA can be shown to be very accurate for temperatures above $\hbar\Delta/k_B$ for all times and also for low temperatures for times which are not large compared with Δ^{-1} [24], where Δ is the effective tunnel splitting defined below. We are interested in the inelastic structure factor for frequencies of the order of the tunneling frequency Δ, therefore this approximation is very reasonable for the present problem. Within the DBGA the Laplace transform of $P(t) = 2P_{11}(t)-1$ emerges as [25,26]

$$\hat{P}(\lambda) = [\lambda + \Delta(\frac{\hbar\beta\Delta}{2\pi})^{1-2K} \frac{\Gamma(K+\hbar\beta\lambda/2\pi)}{\Gamma(1-K+\hbar\beta\lambda/2\pi)}]^{-1} \tag{11}$$

where

$$\Delta = [\cos(\pi K)\Gamma(1-2K)]^{1/(2-2K)}\, \Delta_p(\Delta_p/D)^{\frac{K}{1-K}} \tag{12}$$

is the effective tunnel splitting [21], while $\Gamma(z)$ denotes Euler's gamma function. It is now convenient to introduce the dimensionless quantities $\nu=\omega/\Delta$, $\tau=k_B T/\hbar\Delta$, and $j(\nu)=J(\nu\Delta)\Delta$. Then (10) and (11) combine to yield [13]

$$j(\nu)= \frac{2}{1+\exp(\nu/\tau)}\, \mathrm{Re}[\frac{1}{f(\nu)+i\nu}] \tag{13}$$

where

$$f(\nu)= (2\pi\tau)^{2K-1} \frac{\Gamma(K+i\nu/2\pi\tau)}{\Gamma(1-K+i\nu/2\pi\tau)}\,. \tag{14}$$

The relations (13) and (14) determine the cross section for inelastic neutron scattering as a function of the dimensionless temperature τ and the dimensionless coupling parameter K.

Let us first discuss two limiting cases. At T=0 we obtain $f(\nu)=(i\nu)^{2K-1}$ and the spectral function becomes

$$j(\nu)= \Theta(-\nu) \frac{2|\nu|^{1-2K}\sin(\pi K)}{[|\nu|^{2(1-K)} - \cos(\pi K)]^2 + \sin^2(\pi K)} \tag{15}$$

displaying an inelastic peak which arises from <u>coherent</u> clock-like tunneling of the defect. The interaction with the conduction electrons destroys the coherence on a time scale of order $[\sin(\pi K)\Delta]^{-1}$ and leads to a complicated non-Lorentzian lineshape. The peak has its maximum at

$$\omega_m = -\Delta\, [\cos(\pi K)(1+\sqrt{F(K)})/(3-2K)]^{\frac{1}{2(1-K)}} \tag{16}$$

where $F(K)=1+(1-2K)(3-2K)/\cos^2(\pi K)$. Compared with an undamped tunneling system where $\omega_m=-\Delta$, the peak is shifted towards the origin.

On the other hand, for high temperatures, $T \gg \hbar\Delta/k_B$ i.e. $\tau \gg 1$, we may neglect the ν-dependence of $f(\nu)$. We then have

$$j(\nu) = \frac{1}{1+\exp(\nu/\tau)} \frac{2\gamma}{\nu^2+\gamma^2} \tag{17}$$

which is seen to be a quasi-elastic peak of Lorentzian form with a dimensionless linewidth

$$\gamma = f(0) = \frac{\Gamma(K)}{\Gamma(1-K)}(2\pi\tau)^{2K-1} \ . \tag{18}$$

This particular form of the scattering function arises from incoherent tunneling between the two sites with a tunneling rate $\Gamma = \gamma\Delta/2$. Note that γ is proportional to T^{2K-1} and hence decreases with increasing temperature.

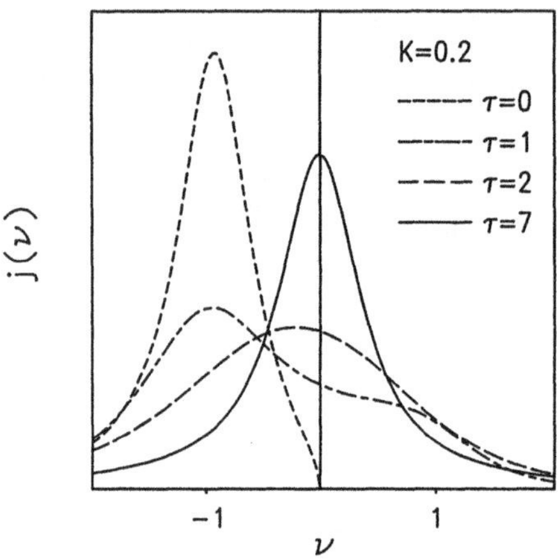

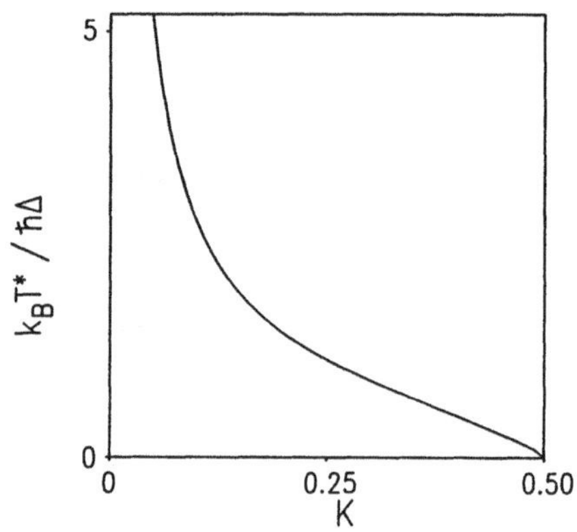

Figure 1: For a system with dimensionless coupling constant K=0.2 the dimensionless structure factor $j(\nu)$ is shown for various values of the dimensionless temperature τ.

Figure 2: The crossover temperature τ^* is shown as a function of K.

In Fig.1 the spectral function $j(\nu)$ is depicted for K=0.2 and various temperatures between T=0 and $T=7\hbar\Delta/k_B$. Clearly, the spectrum shows a crossover between inelastic neutron scattering at low temperatures and quasi-elastic scattering at high temperatures. A crossover temperature T^* may be defined by the condition that the symmetrized spectrum $\frac{1}{2}[j(\nu)+j(-\nu)]$ has a point of inflection at $\nu=0$, so that the spectrum is doubly peaked below T^* and has a single quasi-elastic peak above T^*. In dimensionless units one finds [13]

$$\tau^* = \frac{1}{2\pi}\left[\frac{\Gamma(K)}{\Gamma(1-K)}(\pi\cot(\pi K) + \sqrt{\psi'(K)-\pi^2/2})\right]^{\frac{1}{2(1-K)}} . \tag{19}$$

Here $\psi'(z)$ denotes the tri-gamma function. Fig.2 shows the dimensionless crossover temperature as a function of the coupling parameter K. We see that T^* is proportional to 1/K for small K. T^* is the temperature where roughly the smooth transition between coherent tunneling and incoherent tunneling occurs.

In the limit K<<1 of weak electron-defect coupling the spectral function (13) approaches a sum of two Lorentzian lines. Taking only terms of lowest order in K into account one obtains from (13) and (14)

$$j(\nu) = \frac{1}{1+\exp(\nu/\tau)} \frac{4\gamma}{(\nu^2-\nu_0^2)^2+4\gamma^2} \tag{20}$$

where $\nu_0=1+K[\mathrm{Re}\Psi(i/2\pi\tau)-\ell n(1/2\pi\tau)]$ and $\gamma=(\pi K/2)\coth(1/2\tau)$. Hence, for weak damping of the tunneling system the linewidth is given by

$$2\gamma\Delta= \pi K\Delta \coth(\hbar\Delta/2k_BT). \tag{21}$$

This result arises from a Korringa-type relaxation of the two-state system and can easily be obtained from lowest-order perturbation theory in the defect-electron coupling [27]. The perturbative treatment is valid for $\gamma<<1$ so that the full expression (13) must be used when $\pi K/2$ or $\pi K\tau$ is not small compared with 1.

3. TEMPERATURE DEPENDENCE OF INCOHERENT TUNNELING

Above the crossover temperature T^* introduced in the previous section the defect hops randomly from one interstitial site to the other and there is no correlation between subsequent tunneling transitions. As a consequence the rate for transitions between two sites is the same whether the defect moves in a double well or in a periodic crystal potential. In the latter case the tunneling rate Γ determines the quantum diffusion coefficient D_q. For crystals with cubic symmetry one has

$$D_q= \frac{1}{6}zd^2\Gamma \tag{22}$$

where z is the number of neighboring interstitial sites and d the distance between two sites.

Due to the fact that effects of quantum coherence are negligible, the rate for incoherent transitions between two sites can be calculated by a number of methods yielding equivalent results [5,8,24,25,28-30]. When the two sites are degenerate in energy the rate can be written as

$$\Gamma = (\Delta_0/2)^2 \int_{-\infty}^{+\infty} dt \exp[-Q(t)] \tag{23}$$

where Δ_0 is the bare tunneling matrix and where the function Q(t) describes the influence of crystal excitations. Frequently, the response of the crystal to the motion of the defect can be treated in a linear approximation since each single mode of the metal is excited only weakly by the tunneling particle. Naturally, the combined effect of these modes on the defect motion is not necessarily weak. Within the linear approximation for the response of the metal its influence on the defect is described in terms of the function

$$\chi(t) = \frac{i}{\hbar} <[F(t),F(0)]> \tag{24}$$

where the average is over the equilibrium state of the metal and where

$$F = d^{-1}(H_1-H_2) \tag{25}$$

is a "force" operator in which H_j (j=1,2) is the Hamiltionian of the host crystal with the defect held fixed in the initial and final state, respectively. The function Q(t) is then given by

$$Q(t) = -\frac{d^2}{\pi\hbar}\int_0^\infty d\omega \frac{\chi''(\omega)}{\omega^2} [\frac{\cos(\omega t)}{\sinh(\beta\hbar\omega/2)} - \coth(\beta\hbar\omega/2)]. \tag{26}$$

in which $\chi''(\omega)$ is the imaginary part of the half-sided Fourier transform of $\chi(t)$. The susceptibility $\chi''(\omega)$ of the environment consists of an electronic and a phonon part

$$\chi''(\omega) = \chi''_{el}(\omega) + \chi''_{ph}(\omega). \tag{27}$$

In normal metals, the low-frequency electronic susceptibility can be shown to behave as

$$\chi''_{el}(\omega) = \frac{2\pi\hbar}{d^2} K\omega . \tag{28}$$

Accordingly, the electronic part of the function (26) reads

$$Q_{el}(t) = 2K\, \ell n[\frac{\hbar\beta D_0}{\pi}\cosh(\pi t/\hbar\beta)] + 2W(K) \tag{29}$$

where D_0 is a cutoff frequency which is large compared with Δ_0 but small enough in order that the two-state approximation for the defect and the quasi-particle picture for the excitations of the Fermi liquid are valid. The term W(K) depends on high-frequency properties but is independent of t in the limit $\hbar\beta D_0 >> 1$ so that for low temperatures it can be absorbed into an effective bandwidth $D = D_0 \exp[W(K)/K]$. The function (29) is then related to the interaction (2) introduced earlier by $Q_{el}(t) = \mathrm{Re}\, S(t - i\hbar\beta/2)$.

When the defect-lattice interaction is linear in the displacements of the host atoms, the low-frequency part of the phonon-susceptibility $\chi''_{ph}(\omega)$ reads

$$\chi''_{ph}(\omega) = \frac{\pi\hbar}{d^2} c_j \omega^{2j+1} \qquad (j=1,2),\ \omega << \omega_D, \tag{30}$$

where ω_D is the Debye frequency. The case j=1 applies to defect tunneling between tetrahedral sites in bcc materials like Nb and Fe, while the case j=2 applies to tunneling between octahedral sites in fcc crystals like Cu and Al. The coupling constant c_j can be calculated from the long-range displacement field caused by the defect. Now (26) and (30) combine to yield [30]

$$Q_{ph}(t) = V_j(t) + U_j + 2W_j , \tag{31a}$$

$$V_j(t) = (-1)^j c_j \frac{\pi}{\hbar\beta}(\frac{\partial}{\partial t})^{2j-1} \tanh(\frac{\pi t}{\hbar\beta}) \ [1+O(1/\hbar\beta\omega_D)] , \tag{31b}$$

$$U_j = 2c_j \Gamma(2j)\zeta(2j)(\hbar\beta)^{-2j} \ [1+O(1/\hbar\beta\omega_D)] \tag{31c}$$

where $\zeta(n)$ is a Riemann number. Here, $2W_j$ is the time-independent contribution at T=0. This quantity which depends on the high-frequency behavior of $\chi''_{ph}(\omega)$ causes the phonon dressing of the tunneling matrix according to

$$\Delta_p = \Delta_0 \exp(-W_j). \tag{32}$$

Combining these results the rate formula (23) takes the form

$$\Gamma = (\Delta_p/2)^2 \ (\pi/\hbar\beta D)^{2K} \ \exp(-U_j) \int_{-\infty}^{\infty} dt \ \frac{\exp[-V_j(t)]}{\cosh^{2K}(\pi t/\hbar\beta)} \ . \tag{33}$$

Now, since U_j and $V_j(t)$ vanish for T=0, the phonon influence does not contribute to the leading temperature dependence of the rate at low temperatures. Setting $U_j = V_j(t) = 0$ one obtains from (33)

$$\Gamma_0 = \frac{\sqrt{\pi}}{4} \frac{\Delta_p^2}{D} \frac{\Gamma(K)}{\Gamma(K+1/2)} (\frac{\pi k_B T}{\hbar D})^{2K-1} \tag{34}$$

which in terms of the effective tunnel splitting Δ introduced in (12) may be written as $\Gamma_0 = \gamma\Delta/2$ where γ is given by (18). Hence, we recover our previous result for the tunneling rate in the incoherent region above T^*. The power-law behavior, $\Gamma_0 \propto T^{2K-1}$, of the tunneling rate arises from the frequency-independent damping of the defect by the screening cloud [5,8]. Since K<1/2, the rate (34) decreases with increasing temperature. This behavior should be observable for temperatures in the range $T^* << T << T_{ph}$ where T_{ph} is defined below.

For higher temperatures the rate formula (33) must be evaluated more carefully. Here I will restrict myself to the case j=1 and to temperatures well below the Debye temperature $\hbar\omega_D/k_B$ where the explicit expressions (31b) and (31c) are valid. The temperature-dependent tunneling rate then takes the form

$$\Gamma = \Gamma_0 \exp(-T^2/3T_{ph}^2)\ M(K,K+\tfrac{1}{2},T^2/T_{ph}^2) \tag{35}$$

where Γ_0 is given by (34) and where

$$T_{ph} = \hbar/\pi k_B\sqrt{c_1} \tag{36}$$

is a typical temperature above which the influence of phonons dominates the temperature dependence of the defect motion. Further, $M(K,K+\frac{1}{2},z)$ is a Kummer function [31]. When the result (35) is expanded in terms of T/T_{ph} one finds [30]

$$\Gamma = \Gamma_0\{1 + \frac{4K-1}{3(1+2K)}(T/T_{ph})^2 + \frac{1}{18}\,\frac{3+8K+16K^2}{(1+2K)(3+2K)}(T/T_{ph})^4 + \ldots\}. \tag{37}$$

Since the last and higher order terms always lead to an increase of the rate with increasing temperature, the rate goes through a minimum [5]. For temperatures near and above T_{ph} phonon-assisted tunneling enhances the rate. Fig.3 shows the temperature-dependent rate (35) for various values of K. We see that for small K<1/4 the minimum is more pronounced since the first correction to Γ_0 in (37) has a negative coefficient. A similar analysis can also be carried out for the case j=2 where $T_{ph}=\hbar/\pi k_B c_2^{1/4}$ [30].

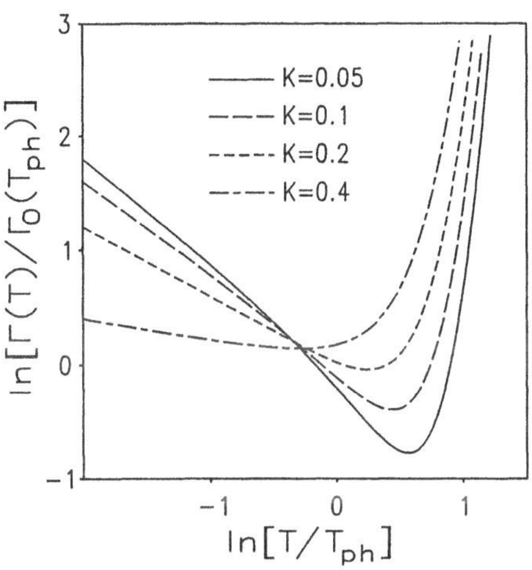

Figure 3: The temperature-dependent tunneling rate $\Gamma(T)$ normalized to $\Gamma_0(T_{ph})$ is shown as a function of T/T_{ph}.

The explicit analytical result (35) is based on the assumptions of a linear coupling to Debye phonons and $k_BT_{ph} \ll \hbar\omega_D$. While nonlinear interactions with the displacement of the host atoms always contribute to the temperature-independent phonon-dressing factor (32) they lead only to corrections of higher order in T to the interactions (31b) and (31c) which determine the temperature dependence of the rate. However, in particular in the case of weak defect-phonon coupling where T_{ph} is not much smaller than the Debye temperature, these nonlinear phonon couplings may modify the behavior of the rate already near the region of the mobility minimum. Naturally, the treatment of nonlinear interstitial-lattice couplings [32-34] does not allow for an analytical evaluation of the rate formula but is important for a quantitative comparison between theory and experimental data in the entire range of temperatures.

4. CONCLUSIONS

In this section I will shortly outline some limitations and extensions of the theory presented above. So far I have completely disregarded inhomogeneous broadening arising from lattice imperfections and the finite concentration of defects. Disorder will start to dominate the behavior of the tunneling rate when k_BT is of the order or below the typical energy difference between the ground states at neighboring sites. The effect of such a bias energy can easily be incorporated into the calculation of incoherent tunneling rates [8,25,34,35]. Since disorder tends to localize the defect, the quantum diffusion coefficient will be lowered by inhomogeneous broadening.

A bias energy also tends to weaken quantum mechanical coherence between neighboring wells. As a consequence, coherent defect tunneling will only be observable when the typical energy shift between adjacent sites is small compared with k_BT^*. Here, I have only discussed coherent tunneling for trapped defects moving between two sites. This theory [13] cannot straightforwardly be extended to particles in a periodic crystal field. In the latter case coherence between many sites will lead to the formation of Bloch states at very low temperatures. Then, the quantum diffusion coefficient is expected to vary as 1/T with temperature [36]. A detailed theory of the crossover between coherent band-motion and incoherent hopping for defects coupled to conduction electrons is presently not available.

The theory presented here has disregarded so-called electron-assisted processes. Although the coupling constant for these events is proportional to the tunneling matrix and therefore small [4] they may become relevant at very low temperatures [37]. Furthermore, I have restricted myself to normal metals. In the superconducting state the excitations of the fermionic bath have a low-frequency cutoff leading to pronounced differences regarding the influence on tunneling systems [38-39]. Finally, I have not discussed the experimental situation which is reviewed elsewhere in this volume [14,15].

ACKNOWLEDGEMENTS

I would like to thank D.Richter and H.Wipf for stimulating my interest in defect tunneling in metals. A large part of the work reported here results from collaborations with S.Dattagupta, S.Linkwitz, H.R.Schober, and U.Weiss. It is also a pleasure to thank G.-L.Ingold and P.Schramm for some help in preparing this manuscript.

REFERENCES

[1] C.P.Flynn and A.M.Stoneham, Phys.Rev.B1,3966(1970)
[2] Yu.Kagan and M.I.Klinger, J.Phys.C7,2791(1974)
[3] H.Teichler and A.Seeger, Phys.Lett.82A,91(1981)
[4] J.Kondo, Physica 84B,40(1976)
[5] J.Kondo, Physica 125B,279(1984)
[6] A.O.Caldeira and A.J.Leggett, Ann.Phys.(NY)149,374(1983)
[7] H.Grabert, U.Weiss and P.Hanggi, Phys.Rev.Lett.52,2193(1984)
[8] U.Weiss and H.Grabert, Phys.Lett.108A,63(1985)
[9] R.P.Feynman and F.L.Vernon, Ann.Phys.(NY)24,118(1963)
R.P.Feynman, Statistical Mechanics (Benjamin, New York, 1972);
[10] F.Guinea, Phys.Rev.Lett.53,1268(1984)
[11] J.P.Sethna, Phys.Rev.B24,698(1981); B25,5050(1982)
[12] J.Kondo, Physica 141B,305(1986)
[13] H.Grabert, S.Linkwitz, S.Dattagupta, and U.Weiss, Europhys.Lett.2,631(1986)
[14] See article by H. Wipf, D. Steinbinder, K. Neumaier, P. Gutsmiedl, A. Magerl, and A.J. Danoux (this volume) and references therein
[15] For a review see: D.Richter (this volume)
[16] C.C.Yu and P.W.Anderson, Phys.Rev.B29,6165(1985)
[17] L.D.Chang and S.Chakravarty, Phys.Rev.B31,154(1985)
[18] F.Guinea, V.Hakim, and A.Muramatsu, Phys.Rev.B32,4410(1985)
[19] K.Yamada, A.Sakurai, and S.Miyazima, Progr.Theor.Phys.73,1342(1985)
[20] P.Hedegård and A.O.Caldeira (to be published)
[21] Yu.Kagan and N.V.Prokof'ev, ZhETF (Sov.Phys.JETP) 90,2176(1986)

[22] S.Chakravarty and A.J.Leggett, Phys.Rev.Lett.52,5(1984)
[23] See e.g.: D.Richter, in: Springer Tracts in Modern Physics, Vol.101 (Springer, Berlin 1983)
[24] A.J.Leggett, S.Chakravarty, A.T.Dorsey, M.P.A.Fisher, A.Garg, and W.Zwerger (to be published)
[25] H.Grabert and U.Weiss, Phys.Rev.Lett.54,1605(1985)
[26] A.Garg, Phys.Rev.B32,4746(1985)
[27] B.Golding, J.E.Graebner, A.B.Kane, and J.L.Black, Phys.Rev.Lett.41,1487(1978)
[28] R.Silbey and R.A.Harris, J.Chem.Phys.80,2615(1984)
[29] C.Aslangul, N.Pottier, and D.Saint-James, J.Physique 46,2031(1985)
[30] H.Grabert, U.Weiss, and H.R.Schober, Hyperfine Interact. (to appear)
[31] Handbook of Mathematical Functions (M.Abramowitz and I.A.Stegun, eds., Dover, N.Y.,1965) p.504
[32] H.R.Schober and A.M.Stoneham, Phys.Rev.B26,1819(1982)
[33] A.Klamt and H.Teichler, Phys.Stat.Sol.(b)134,533(1986)
[34] J.Kondo, Hyperfine Interact. (to appear)
[35] M.P.A.Fisher and A.T.Dorsey, Phys.Rev.Lett.54,1609(1985)
[36] J.Jäckle and K.W.Kehr, J.Phys.F13,753(1983)
[37] K.Vladár, G.T.Zimányi, and A.Zawadowski, Phys.Rev.Lett.56,286(1986)
[38] J.L.Black and P.Fulde, Phys.Rev.Lett.43,453(1979)
[39] C.C.Yu and A.V.Granato, Phys.Rev.B32,4793(1985)

Quantum Diffusion of Light Interstitials in Metals

D. Richter

Institut Laue-Langevin, 156X, F-38042 Grenoble Cedex, France

Recent theoretical considerations suggest that at low temperatures the transport properties of light interstitials in metals are dominated by the interaction with the conduction electrons. After briefly surveying the state of the theory, we examine experimental results for muon diffusion in Aℓ and Cu. In Aℓ muon diffusion is observed via diffusion-limited trapping, and local properties of the impurity as well as long-range diffusion are playing a role. Using a large variety of impurities it was possible to evaluate the diffusion coefficient over nearly 4 orders of magnitude in temperature. Other than in Aℓ, in Cu muon diffusion is observed directly through motional narrowing due to elementary jumps. Qualitatively, the observed jump rates show a very similar behaviour as in Aℓ. Quantitatively, they are 3 orders of magnitude slower. In both cases a low T power law behaviour ($D \sim T^{-\alpha}$) is found. After passing through a minimum the diffusion coefficient first increases linearly, thereafter exponentially with temperature. Using Kondo's theory /1/ together with recent developments in the small polaron theory, the experimental data are discussed.: In the low T limit a good explanation of the observed diffusion behaviour is possible. The intermediate range ($D \sim T$) is indicative for a one-phonon process while at high T the usual multiphonon processes are dominant.

1. INTRODUCTION

The hydrogen isotopes including the positive muon cover nearly a factor of thirty in mass. Dissolved in a metal they exhibit transport phenomena spanning the range between thermal activation and tunneling motion. Theoretically these diffusion processes have been described in terms of the small polaron concept considering the elastic interactions to the host lattice. Recently Kondo extended the theory and included explicitly also the interactions to the conduction electrons /1/. Their influence results in two effects : (i) The screening electron cloud which accompanies the light particle reduces the tunneling matrix element by a factor proportional to a power law in temperature. (ii) Electron scattering causes a damping of the particle motion which is proportional to T. The latter dominates and with decreasing temperature a weakly increasing diffusion coefficient is predicted. Finally below a certain cross over temperature T* the hopping motion changes to coherent propagation /2/.

The idea of an explicit electronic contribution to the motional properties of light interstitials in metals has been stimulated mainly by the outcome of muon diffusion experiments in metals, which disagreed strongly with the low T predictions of conventional small polaron theory /3-8/. Later on, also hydrogen atoms trapped at oxygen impurities in Nb turned out to be excellent candidates for an experimental access to this new motional mechanism /9/. In this short review we concentrate on experimental results on the motion of muons in metals taking Aℓ and Cu as examples. Neutron scattering experiments on H-tunneling at O-impurities in Nb are reported elsewhere at this conference /10/.

We commence with a short overview on the present state of theory, then we outline briefly the muon spin rotation technique and its application to diffusion problems. Thereafter we turn to the experimental results. While in Cu the muonic motion occurs on a time scale such that single muon jumps can be observed directly through motional narrowing, in Aℓ muon diffusion is very rapid and can only be accessed via diffusion-limited trapping. In both materials at low T a power low behavior of the diffusion coefficient is observed with $D \sim T^{-\alpha}$. (α = 0.7 for Aℓ; α = 0.6 for Cu). A minimum value of D is reached for Aℓ around 5K and for Cu in the order of 50 K. Towards higher T in Aℓ the diffusion coefficient increases first linearly and then exponentially with temperature while in Cu such a linear regime has not yet been identified with certainty. Using Kondo's theory, together with recent developments in the small polaron theory, the experimental data are discussed. In the low T limit a good explanation of the observed diffusion behaviour is possible. The intermediate temperature range ($D \sim T$) seems to be indicative for a one-phonon process while at high T the usual multiphonon processes are dominant.

2. THE ROLE OF CONDUCTION ELECTRONS IN SMALL PARTICLE TRANSFER

Conventional small polaron diffusion theories /11,12/ assume implicitly that there is no other effect of the conduction electrons than their influence on the potential energy at the interstitial positions. Then at low temperatures the tunneling matrix element is given by /11/

$$J_p = J_o\, e^{-S(T)} \quad . \tag{1}$$

Here J_o represents the transfer integral for tunneling of the bare particle between sites R_1 and R_2 and $e^{-S(T)}$ is a reduction factor which takes into account the fact that the lattice distortion has to be translated. In a "golden rule" ansatz the transition probability between two sites R_1 and R_2 can be written as :

$$W = 2\pi\, J_p^2\, \rho_{eff} \tag{2}$$

where ρ_{eff} is the final density of states. Kagan and Klinger /13/ proposed that ρ_{eff} is distributed due to indirect phonon scattering processes and arrived at $W \sim T^{-9}$ which has, however, never been observed. Recently Kondo /1/ and Yamada /14/ realized that interstitial diffusion of light particles in a metal cannot be described by a lattice dynamical approach alone, but that the influence of conduction electrons is crucial. Thus, in addition to the phonon coupling, the Hamiltonian for the diffusing particle has to include particle electron interaction :

$$H_i = H_o + \sum_{\underline{k}\sigma} \varepsilon_k a^+_{\underline{k}\sigma} a_{k\sigma} + \sum_{\underline{k}\,\underline{k}'\sigma} V_o \quad e^{i(\underline{k}-\underline{k}')\underline{R}_i}\, a^+_{\underline{k}\sigma}\, a'_{\underline{k}\sigma} \quad . \tag{3}$$

There i describes the position of the H-like particle, $a^+_{\underline{k}\sigma}$ and $a_{\underline{k}\sigma}$ are electron creation and annihilation operators, V_o is the H-electron interaction parameter and H_o is the usual phonon term. Kondo calculated the electronic part of the transition probability using a perturbation expansion of Eq.(3). This expansion contains divergent terms of the type $e^{i(\underline{k}-\underline{k}')(\underline{R}_1-\underline{R}_2)/(\varepsilon_k-\varepsilon_{k'})}$ which however, according to Kondo, can be summed up to a finite value. Grabert et al. treated Eq. (3) using path integral methods /2/. In the hopping regime apart from numerical factors both treatments lead to identical results, while in the tunneling regime significant differences appear. A comparison of both results shows that Kondo's solution is the weak coupling limit of Grabert's treatment.

First we consider the hopping regime, where no phase relations of the particle wave functions at the initial and final sites exist. Then, the calculations result in a renormalisation of the tunneling matrix element to

$$J_{eff} = J_o\, e^{-S} \left(\frac{kT}{\varepsilon_F}\right)^K \tag{4}$$

where ε_F is the Fermi energy. The parameter K describes the electron-H coupling and is related to the phase shifts δ_ℓ of the electrons which scatter at the potential set up by the interstitial impurity /15/

$$K = 2(2\ell+1)\left[\frac{1}{\pi}\tan^{-1}\frac{\sqrt{1-J_0^2(k_Fa)}\ \tan\delta_\ell}{\sqrt{1+j_0^2(k_Fa)\ \tan\ \delta_\ell}}\right]^2 \tag{5}$$

where $j_0(k_Fa)$ is the zero order spherical Bessel function $a = (R_1-R_2)$ and K_F is the Fermi wave vector. With Eqs. (2) and (4) the low T hopping rate can be calculated, if we consider that the final electron density of state is reduced by the smearing of the Fermi surface ($\rho_{eff} \sim T^{-1}$) /1/

$$W = \frac{J_p^2}{h}\ \frac{\cos\ \pi K}{\sqrt{\pi}}\ \Gamma(K)\ \Gamma(\frac{1}{2}-K)\ (\frac{1}{\varepsilon_F})^{2K}\ (\pi kT)^{2K-1} \tag{6}$$

where $\Gamma(X)$ is the Γ-function.

For temperatures $kT \lesssim J_{eff}$ a transition to coherent propagation is expected. For a double well system Grabert /2/ calculated the cross over temperature T^* as a function of coupling strength K and found kT^*/J_{eff} to be in the order of 1. At higher temperatures electronic and lattice contributions to the transition probability have to be combined in calculating the transition rate. Using Holstein's expression for the conventional lattice-induced small polaron transfer and assuming further that electronic and lattice contributions can be factorized, the transition rate W can be obtained numerically for all temperatures /1/.

3. THE μSR TECHNIQUE

Muon spin rotation (μSR) experiments (for an introduction see Ref. 16) exploit the parity violation in the muon decay which causes a preferred positron emission in the muon spin direction. The quantity measured in a μSR study is the time evolution of the polarization of muons stopped in matter. For this purpose knowledge of the muon residence time in a sample is required. In the most common approach this is achieved by measuring the individual lifetime of each muon : the incoming muon starts a clock and the decay positron stops it, when it hits a detector telescope. During an experiment typically 10^6-10^7 such events are recorded and stored in a time histogram. Figure 1 shows schematically the basic μSR setup with transverse

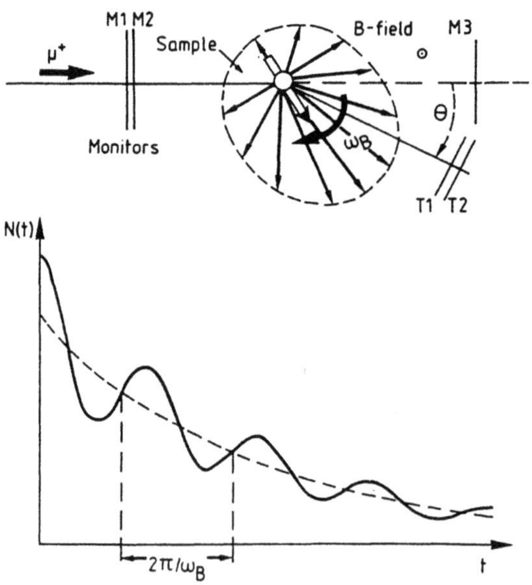

Fig. 1 . Upper part : Schematic presentation of transverse μSR setup. Counters M1, M2 in coincidence, and M3 in anticoincidence register the incoming muon, telescopes T1 and T2 register the emitted positron, ω_B is the muon Larmor frequency in a perpendicular B field. The length of the arrows symbolizes the spin-dependent positron emission probability.
Lower part : Time histogram of a μSR spectrum. The dashed line gives the muon radioactive decay function. The solid line represents the actual oscillating signal in an external field.

field geometry, which is the most common technique for diffusion studies. The incoming muon which retains its polarization during the fast thermalization process in a metal starts to precess with the Larmor frequency $\omega_B/2\pi = \gamma_\mu B = 13.55$ kHz/G and the resulting time histogram is also shown in Fig. 1 : The muon decay function is superimposed by an oscillatory component which has a maximum whenever the muon spin points in the direction of the counter. For the time-dependent counting rate we have

$$N(t) = N_0 \exp(-t/T_\mu)\,[1 + aP(t)\cos(\omega_B t + \phi)], \tag{7}$$

where a is the experimentally achieved asymmetry, ϕ is a constant phase angle, $T_\mu = 2.20$ µs is the muon life time, and P(t) is the damping of the spin polarization. It is caused by random fluctuations of the local field originating, e.g. from the nuclear dipole moments in nonmagnetic solids. Muon stopped at different sites in a metal experience slightly different Larmor frequencies and lose their phase coherence; this results in a damping of the measured polarization. The damping corresponds to a linewidth and a characteristic line shape in the frequency spectrum (after Fourier transformation). The phenomenology of µSR analysis is identical to that of NMR. For an immobile muon the line shape is Gaussian,

$$P(t) = \exp(-\sigma^2 t^2), \tag{8}$$

where $\sigma^2 = 1/2\,\gamma^2 B_{dip}^2$ is the second moment of the frequency distribution due to internal fields. For a mobile muon the field fluctuations are averaged and motional narrowing occurs. Then, the depolarization has the form

$$P(t) = \exp(-2\sigma^2 \tau_\mu t), \tag{9}$$

where τ_μ is the average time during which correlations between the frequencies exist and is proportional to the residence time at a certain site. Under the assumption of a Gaussian Markovian process for the precession frequency with a correlation time τ_μ the line shape profile can be evaluated for the whole range of mobilities /17/.

Besides transverse experiments, also longitudinal and zero yield µSR is applied. The stochastic theory for this case has been worked out by Kubo and Toyabe /18/ and applied to µSR by Yamazaki /19/. Experimentally the longitudinal relaxation function $P_z(t)$ is obtained in forward backward geometry where the ratio

$$\frac{N_f - N_b}{N_f + N_b} = A\,P_z(t) \tag{10}$$

is measured. Apart from experimental difficulties, longitudinal experiments have several advantages : (i) in the static limit in zero field for a given σ the damping is larger than in a transverse set up. In general this results in a higher statistical accuracy. (ii) In zero field for an immobile muon after an initial decay, $P_z(t)$ increases asymptotically for long times to 1/3 of its initial value, since precession due to z-components of B_{dip} should not change P_z on the average. Slow motional processes suppress this secondary increase of $P_z(t)$ and can thus be detected with great sensitivity. (iii) Applying weak longitudinal fields, in the motional narrowing limit τ_μ and σ and can be determined separately, which according to Equ. (9) is impossible in a transverse experiment.

4. MUON DIFFUSION EXPERIMENTS IN Al

In pure Aℓ muons are very mobile and as a consequence a damping of the transverse muon depolarization function cannot be observed at any temperature /4/. Therefore, up to now muon diffusion experiments in Al were only possible indirectly via diffusion-limited trapping. Pure Al is doped with well-known small amounts of substitutional impurities like Mn, Mg, Li, Ag /3/ or radiation defects like vacancies after electron irradiation /20/. Then the temperature dependence of the muon depolarization rate is measured. It contains information on the muon trapping rate at the defect $1/\tau_1$ as well as on the escape rate $1/\tau_0$. For diffusion-limited trapping the diffusion coefficient D and $1/\tau_1$ are related by /16/.

$$1/\tau_1 = 4\pi\, r_t\, c_t\, D \tag{11}$$

where c_t is the trap concentration and r_t is the trapping radius. Both r_t as well as the escape rate $1/\tau_0$ are characteristic for each impurity and contain information on the muon potential near the trap. The trapping and escape rates can be evaluated from the data using a two-state trapping model /21, 22/ which considers repeated capture and release processes at the traps: A muon diffusing in the undisturbed lattice is caught by a trap after an average term τ_1; a muon in a trap escapes on the average after a time τ_0. The polarization decay in the free state is taken as the depolarization function in the homogeneous lattice; the depolarization at the trap in the simplest approximation is given by a Gaussian decay (see Equ.(8)) which assumes an immobile muon in the trapped state. This model leads to a set of integral equations which can be well approximated by a system of differential equations for the polarization decays from muons disintegrating in the trapped state $Q_0(t)$ or in the free state $Q_1(t)$, respectively /22/:

$$\begin{aligned} \frac{dQ_1}{dt} &= 2\sigma^2\, \tau_\mu(\exp(-t/\tau_\mu) - 1) - 1/\tau_1 Q_1 + 1/\tau_0\, Q_0\,, \\ \frac{dQ_0}{dt} &= 2\sigma_t^2\, \tau_0(\exp(-t/\tau_0) - 1) - 1/\tau_0 Q_0 + 1/\tau_1\, Q_1\,. \end{aligned} \tag{12}$$

The experimentally observed depolarization function P(t) then is the sum of Q_1 and Q_0. σ_t^2 is the second moment of the field distribution in the trapped state.

In order to get reliable data for the diffusion coefficient it is important to study trapping at different impurities and thus to separate local properties of a trap from the overall diffusion behaviour. Fig. 2 presents an overview of experimental results for the damping parameter σ (the data were fitted with Equ.(8)) from Aℓ samples which were doped with Mn, Li, Ag, Ge, Ga or Si. Fig. 3 shows the concentration dependence of the depolarization rate for AℓMn with different Mn concentrations. From Fig. 2 we observe that the general shape of $\sigma(T)$ is independent from the type of impurity. This already indicates that the measured temperature dependencies are not due to peculiar local processes at the impurity but reflect muon diffusion behaviour in Aℓ. The measured concentration dependencies confirm this interpretation.

In the range between 10-100 K the depolarization rates are due to trapping at the impurities, which is a process possible if the diffusion of the muons in the Aℓ - lattice has increased enough that the perturbed regions are reached with high probability during the muonic life-time. Above a characteristic temperature the trapped muons are released by thermal activation and diffuse very fast in the metal giving negligible σ-values. Applying the two-state model (Equ.(12)) a fit of the AℓMn data in the range between 5 and 50 K yields a trapping rate depending linearly on C_{Mn} as required by Equ.(11). For the temperature dependence a power law behaviour $1/\tau_1 \sim T^\alpha$ with $\alpha \simeq 1$ is found /3/. The positions of the observed peaks in σ are related to the binding energies at the trap. It is interesting to note that the binding energies at the traps which are evaluated from the escape rates $1/\tau_0$ do not correlate with the elastic distortions created by the impurity atoms but can be understood from the produced electronic disturbances /23/.

The increase of the depolarization rates at the lowest temperatures is common for all doped samples, but is practically absent in very pure aluminium /4/. This feature has been interpreted as evidence for another trapping process at low temperatures, where the diffusion rate increases with decreasing temperature. Again the

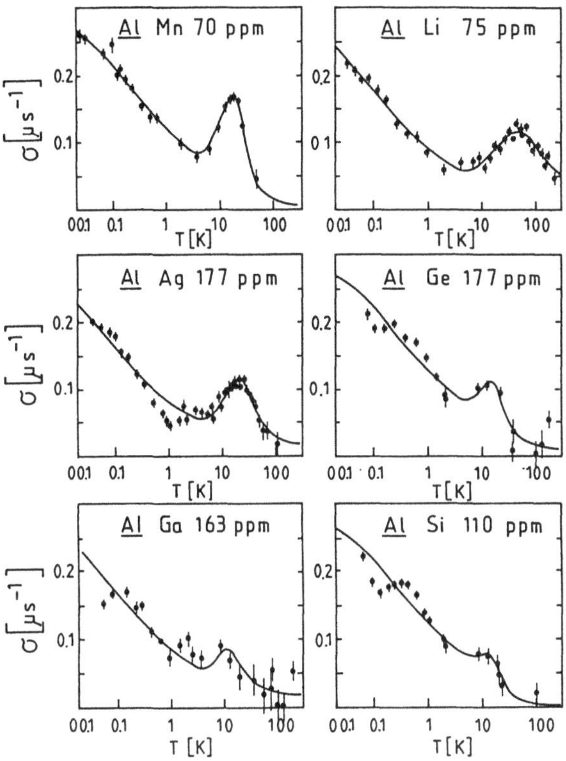

Fig. 2. Gaussian damping parameter σ (see Equ.(8)) for Aℓ damped with different impurities. The solid lines are the result of a combined fit of the two-state diffusion and trapping model (Equ.(12)) using Equ.(13) for the diffusion coefficient.

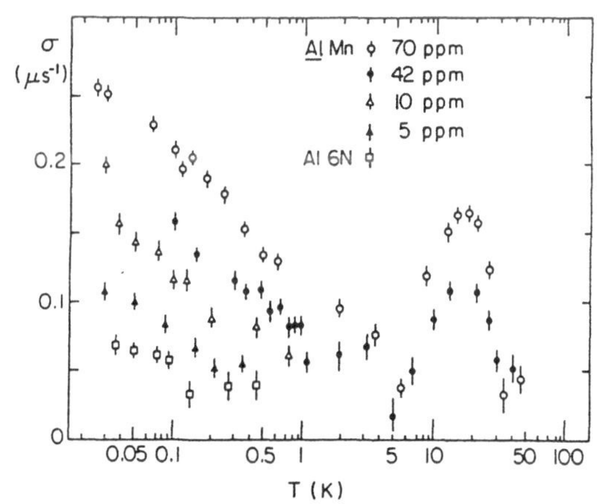

Fig. 3. Gaussian damping parameter σ for AℓMn samples with different Mn contents. In the range between 5 and 50 K the results for only two concentrations are displayed.

concentration dependence, as studied for Al Mn, forms strong evidence for such an interpretation. The slope of the σ vs T curves in the 0.03 - 1 K region seems to be largely independent of impurity type, which indicates that it is governed by a process characteristic for the host metal. Below 0.2 K in the Si-and Ge-doped (and to a lesser extent also in the Ga-doped) samples the increase of σ(T) is not monotonous. There is a shallow minimum around 0.1 K which was not observed in the Li, Mn, and Ag-doped samples. This 'anomaly' has to be related to local processes near the trap and is discussed in ref. 23.

In order to get an accurate result for the muon diffusion coefficient in Aℓ the data from all the doped Aℓ-samples have been fitted over the whole temperature range (0.02 K < T < 300 K) using a common temperature dependence for D over the full range. The fit has been performed applying the differential equation representation of the two-state trapping model (Equ.(12)) /24/. Since the impurity trapping data bear no information on D above 50 K, diffusion coefficients derived from trapping at irradiation-induced vacancies /20/ were included in the fit. There, for the diffusion coefficient, an activated behaviour with $E_a \approx 370$ K was derived. Though these data are subject to large uncertainties in their absolute scale due to problems in the estimation of the actual vacancy concentration, their temperature dependence should be correct. In order to represent the temperature variation of the diffusion coefficient over the largest possible range we have fitted a function which includes the above-mentioned activated process with $E_a = 370$ K. The overall fit to a function of the type

$$W = A\, T^{-0.7} + B \,.\, \exp(-20/T) + CT^{-1/2} \exp(-370/T) \tag{13}$$

is shown by the solid lines in Figure 2. The first exponential (exp(-20/T)) has been used for technical reasons and can be replaced by $\tilde{B}\, T^{1.3}$ without changing the results.

In order to obtain absolute numbers for the muon jump rate, it is necessary to assume a trapping radius for one of the impurities. Based on the fact that Ag and Li cause only minor elastic distortions and that the electronic configuration for both impurities seems to allow an access to nearest neighbour sites /23/, we assume that the radius is a/2 for these impurities corresponding to a nearest neighbour octahedral site. Then the muon hopping rate in Aℓ is described by the coefficients $A = 3.4 \cdot 10^8$ [s^{-1}], $B = 2.2 \cdot 10^9$ [s^{-1}] ($\tilde{B} = 1.23 \cdot 10^7$ [s^{-1}]) and $C = 5.5 \cdot 10^{13}$. The thus obtained D(T) is presented in Figure (4).

5. MUON DIFFUSION IN CU

Other than in Aℓ in Cu the characteristic muon jump rate falls into the time window of a μSR experiment and for a long time the muon diffusion in Cu appeared to be a "text book" case since it seemed to follow simple theoretical expectations. Fig. 5 presents early μSR data on Cu taken by Gurevich et al. in 1972 /25/. From

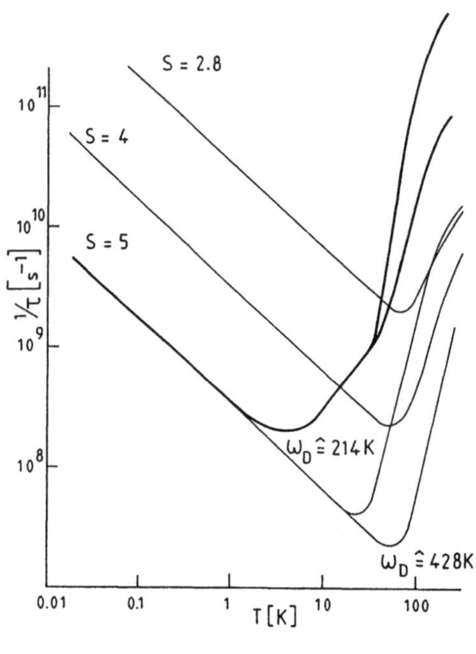

Fig. 4. Muon jump rate in Aℓ as a function of temperature. Thick solid line : Result from the evaluation of the experimental data. The two lines at high temperatures are calculated by changing the prefactor C in Eq. (13) by 10 in order to account for possible uncertainties in the vacancy concentration /20/. Thin solid lines : Theoretical curves for different S. For S = 5 curves for two different Debye temperatures are included.

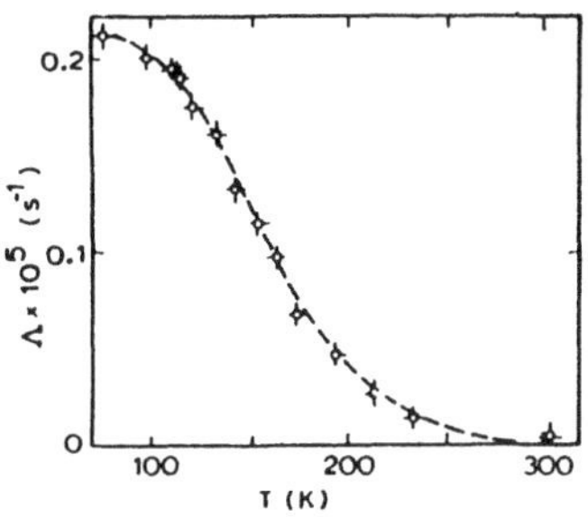

Fig. 5 - Muon depolarization data obtained from Cu as a function of temperature (ref. /25/).

a low T static linewidth the damping parameter starts to decrease around 80 K and falls off monotonically until it reaches complete motional narrowing around 300 K. In this temperature range the muon correlation time $1/\tau_\mu$ to a good approximation follows an Arrhenius relation /26/

$$1/\tau_\mu = 10^{7.61(4)} \exp(-48.4(1.5)[\mathrm{meV}]/ht). \tag{14}$$

From a comparison with H diffusion data (prefactor $(1/\tau)_0 = 10^{14}s^{-1}$; E_a=400 meV /27/) subbarrier motion for the muon is evident. It can account for the strongly reduced prefactor - no longer determined by an attempt frequency - and the drastically smaller activation energy - the muon does not climb the high barrier but tunnels through it. Finally, Teichler /28/ analysed the data in terms of the Flynn-Stoneham small polaron theory /12/ and arrived at E_a = 75 meV and J_0 = 18 μeV.

In 1980 experiments at CERN /4/ revealed that the muon depolarization rate decreases again below 5 K. Several explanations were put forward : (i) partial delocalization of the muon on metastable sites /29/, (ii) diffusion and trapping as in Aℓ implying that in sufficiently pure Cu, as in Aℓ muon diffusion would be extremely fast and /30/ (iii) finally a diffusion coefficient which increases with decreasing T /5/. Thereafter, careful lineshape studies under transverse geometry /6/ zero field studies, which can distinguish better between slow dynamic and static phenomena (see chapter III) /5,7/ and finally combined measurements applying weak and zero longitudinal fields /8/ established the diffusion picture. Fig. 6 displays a collection of the in author's opinion most reliable data for the muon jump rate in Cu which are available by now. The dashed line reproduces the high T data displayed in Fig. 5. As in Aℓ $1/\tau_\mu$ passes through a minimum, the temperature of which is translated to a higher value ($T_M \approx$ 50 K). Below this minimum the muon jump rate increases with a power law in T. The best fit gives

$$1/\tau_\mu = 4 \,.\, 10^5\, T^{-0.6}\, [s^{-1}] \quad . \tag{15}$$

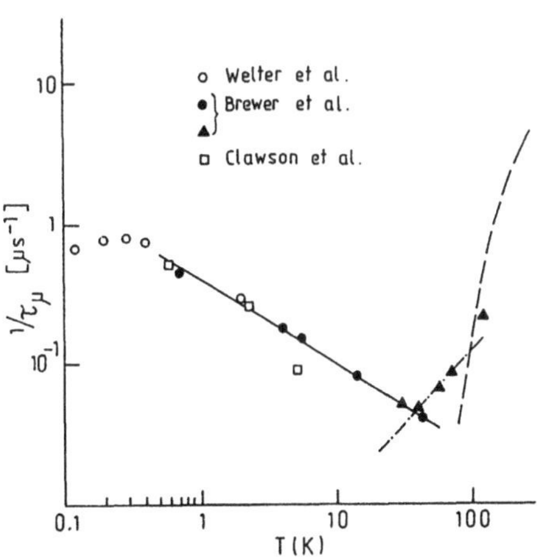

Fig. 6 - Muon jump rates in Cu. Solid line : Result of Equ.(15); dashed line : High T behaviour according to ref. /25,26/. Dashed dotted line : linear T dependence.

Compared to Aℓ we find a very similar power law. The magnitude of the muon jump rate, however, changes by 3 orders of magnitude. Above the minimum the data points appear to be above the low T extrapolation of the high T result shown as a dashed line. Their temperature dependence is compatible with a linear behaviour as found in Aℓ (dashed dotted line). However this T region is not yet enough explored to come to a final conclusion. Other than in Aℓ at low temperatures (below 0.5 K) the jump rate levels off and remains at a value of $\sim 7 \cdot 10^5\ s^{-1}$.

6. COMPARISON WITH THEORY

Qualitatively muons in Aℓ and Cu exhibit a similar diffusion behaviour characterized by a power law at low temperatures and an exponential growth at high T. In between the two limiting temperature dependencies for Aℓ and very recently most likely also for Cu /8/ an intermediate T-range has been identified where D grows approximately linear in T.

We commence with the low-temperature data. There, from the exponent in the temperature power law $1/\tau_\mu \sim T^{-\alpha}$ the muon electron coupling coefficient can be derived (Equ.(6)). With α = 1-2 K we get K = 0.15 and K = 0.2 for muons in Aℓ and Cu respectively.

Equ.(5) allows a theoretical estimation of K provided there is information on the electron phase shifts δ_ℓ. For a first approach we rely on calculations by Puska and Nieminen who reported Fermi level scattering phase shifts for atoms embedded in a homogeneous electron gas /31/. They solved the Kohn-Sham density functional equations by self-consistent iteration and published tables which provide δ_ℓ as a function of an electron gas density parameter and the nuclear charge Z. Approximating the electron density by a uniform distribution of the conduction electrons and interpolating the table we find a dominating contribution from the S-wave scattering for both metals. The scattering phases are $\delta_0^{A\ell}$ = 1.205 [rad] and δ_0^{Cu} = 1.289. With $K_F^{A\ell}$ = 1.75 Å^{-1} and K_F^{Cu} = 1.35 Å^{-1} and assuming jumps between O-sites /3,32/ Equ. (5) yields $K^{A\ell}$ = 0.27 and K^{Cu} = 0.33. As observed K^{Cu} comes out larger than $K^{A\ell}$. This result originates from the higher electron density in Aℓ than in Cu. The magnitudes of the interaction parameters, however, are too large. In the framework of the functional density calculations lower K-values would require a higher effective electron density. Obviously, more realistic models are necessary, in order to arrive at a better theoretical prediction.

From the low T approximation of Kondo's theory (Eq. (6) together with Eq. (1)) it is evident, that for a given Fermi energy ($\varepsilon_F^{A\ell}$ = 0.8 Ry, ε_F^{Cu} = 0.51 Ry the magnitude of the hopping rate is largely determined by the phonon-dressed tunneling matrix element $J_p = J_0\, e^{-S}$. In isotropic Debye approximation S(0) is given by $S = 5\, E_a/2\hbar\omega_D$ /12/ where E_a is the activation energy for diffusion in the high T limit of the small polaron theory. E_a can be taken (i) from an evaluation of high T diffusion data in terms of the small polaron theory; or (ii) can be estimated from the elastic properties of the metal and the volume expansion ΔV caused by the muon /16/ :

$$E_a = 0.258\left(\frac{\Delta V\ (C_{11} + 2C_{12})}{3a}\right)^2 \frac{1}{a\ C_{44}} \tag{16}$$

where C_{ij} are the elastic constants and a is the lattice constant. In order to compare with our low T-data we have to keep in mind that Eq. (6) refers to the probability for a muon transition to one of the adjacent sites. The total jump probability between O-sites in a fcc lattice is 12 times higher. Thus for a direct comparison the jump rates of Figs. 4 and 6 have to be divided by 12. Based on the different estimates for E_a, table I summarizes the results for J_0 and S. Unfortunately there exists no reliable ab 'initio' calculations to compare with. Such calculations would require a precise knowledge of the potential surface for the muon along its path through the lattice.

Table I : Experimental results for the bare tunneling matrix element J_0 based on the low T data using different estimates for E_a.

	E_a from high T slope			E_a from Equ. (16)*		
	E_a [meV]	S	J_0 [meV]	E_a [meV]	S	J_0 [meV]
Cu	75 /28/	6.3	0.070	70	5.9	0.046
Aℓ	50**	3.4	0.12	40	2.8	0.060

* Following the systematics of H in fcc metals /33/ we used $\Delta V = 2.9$ Å^3. As a consequence of the larger zero point motion a larger ΔV for the muon is conceivable.

** This estimate is based on scaling the experimental slope of 32 meV by the ratio of the small polaron result for Cu and the slope of the Cu-data in an Arrhenius representation.

In order to evaluate the theoretical predictions for the cross over temperature to coherent propagation, we need the tunneling matrix elements for T = 0. According to Grabert (Equ.(18) in ref. 2) they are calculated to $J_{eff}^{A\ell}(T = 0) = 3.2 \cdot 10^{-7}$ eV and $J_{eff}^{Cu}(T = 0) = 1.62 \cdot 10^{-9}$ eV. These very small tunneling matrix elements imply cross over temperatures of $T_{A\ell}^{*} \simeq 10$ mK and $T_{Cu}^{*} \simeq 30$ µK, in both cases out of the range of the present experiments. Finally, we comment on the low T plateau in the muon jump rate observed in Cu : Introducing finite energy differences ΔE between two sites, Kondo obtains a modified expression for the low T jump rate /1/. This electron-assisted hopping approaches a constant value at low T, the magnitude of which is determined by the width of the energy difference distribution function E_W. Physically, the levelling-off originates from the cut-off in the response of the electron system which for $kT < E_W$ is determined by the now fixed level distributions of the muons and not anymore by kT. In fitting the low-temperature Cu data Kondo obtained $<\Delta E^2>^{1/2} \approx 30$ µeV. Considering the isotope distribution and the impurity contents of the investigated Cu samples, this value is roughly in the order of what is expected, though the weak sensitivity of the experimental results /8/ to large changes in the isotope composition are not satisfactorily explained.

On the high T side beyond the linear range multiphonon contributions become important, and the complete theory including phonon-assisted transitions has to be applied. This involves numerical integration. For 3 selected values of S the result of such a calculation for Aℓ is included in Fig. 4 (thin lines). Thereby, the parameters S and ω_D determine the shape of the theoretical curves at higher temperatures. As expected from the relation between E_a and S larger values of S result in a steeper high T slope. At the same time a larger S also shifts the minimum towards lower temperatures. Both effects lead to an improved agreement with the experimental data. However, even for the largest S-value of 5.1 the discrepancies on the high T-side amount to more than one order of magnitude in the rate. Agreement at the low and high T side would require S = 8. Such a strong coupling to the phonons however is difficult to justify.

A change of the Debye frequency towards lower values influences the theoretical temperature dependencies strongly. This is indicated by a line calculated for $\omega_D = 0.5\,\omega_D^{A\ell}$. But again there is no justification for a strong local change of ω_D. There remains the possibility that the simple Holstein ansatz for the phonon part is an oversimplification and processes beyond the Condon approximation or a fluctuating tunneling matrix element have to be considered. Similar problems were encountered by Kondo when he tried to perform such a fit to the Cu data /1/. Again in order to reproduce the high temperature part he had to assume a considerably lowered Debye temperature $\omega_D = 111$ K /33/ instead of $\omega_D^{Cu} = 361$ K.

As displayed above both for Cu and Aℓ the data near the diffusion minimum indicate a roughly linear T dependence of the jump rate. This phenomenon is well established for Aℓ, where it is reported by 3 different laboratories /3, 34, 35/, but as can be seen from Fig. (6) appears also to be present in the Cu data.

Theoretically such a linear T-dependence corresponds to a jump process where only one phonon is exchanged with the lattice. For jumps between equivalent sites in ideal lattices such particle transfers violate the requirement of energy conservation and are forbidden. Therefore the prevalent low T process in conventional small polaron theory involves two phonons leading to high powers in T. Recently Teichler and Seeger /36/ investigated the occurrence of one-phonon processes in the small polaron framework within the Condon approximation. According to their calculations a one-phonon process may occur (i) either in disturbed systems or (ii) between crystallographically inequivalent sites. In isotropic Debye approximation the one-phonon jump rate between energetically distorted sites (energy difference ΔE) becomes

$$\Gamma^{1ph} = J^2_{eff} \frac{d^2 P^2 \Delta E^2}{12\pi\rho\hbar^6 c^7} kT \quad (kT >> \Delta E) \tag{17}$$

where d is the jump distance, ρ the mass density, c the sound velocity and P the double force tensor. Since ΔE depends strongly on the amount of strain creating impurities, trapping experiments should reveal strong deviations from the linear concentration dependence of the trapping rate (Eq. 11). Such a prediction is not supported by Mn-concentration-dependent experiments on AℓMn where $1/\tau_1 \sim C_t$ was found /3/. Between inequivalent sites the one-phonon rate again in isotropic Deby approximation is given by

$$\Gamma^{1ph} = \frac{J^2_{eff}}{2\pi\rho c^5\hbar^4} \left\{ \frac{d^2 Tr(p_i+p_f)^2}{216\ h^2 c^2} (\Delta\varepsilon)^2 + \frac{[Tr(p_i-p_f)]^2}{4} \right\} \frac{\Delta\varepsilon}{\exp(\Delta\varepsilon/kT) - 1} \tag{18}$$

where $\Delta\varepsilon$ is the energy difference between sites and p_i and p_f are the double force tensors in initial and final state. Both in Aℓ /3/ and in Cu /38/ there is evidence for O and T-site occupation and this jump sequences involving both types of sites are conceivable. However, in order to observe a linear T-dependence of the muon jump rate the energy differences between both sites have to be much smaller than the temperature of observation ($\Delta\varepsilon^{A\ell} << 10K$, $\Delta\varepsilon^{Cu} << 100$ K). Such small energy differences between O and T sites are very unlikely (for Cu e.g. a value of 100 meV was estimated theoretically /37/). Neglecting this problem we calculate Γ^{1ph} from Equ.(18) assuming $[Tr(p_i-p_f)/2]^2 \sim 1\ eV^2$ and find that under the condition of linearity in T only the second term contributes significantly to the rate. We obtain $\Gamma^{1ph}_{A\ell} \simeq 10^6$ T and $\Gamma^{1ph}_{Cu} \simeq 10^3$T. Both values are within about one order of magnitude of the experimental results.

7. CONCLUSION

In this paper muon diffusion results in Aℓ and Cu were presented. In both cases they cover more than 3 orders of magnitude in temperature and display clearly quantum diffusion behaviour. In particular this is evident from the low T results, where a diffusion coefficient increasing with decreasing temperature has been found. In this range the muon motion is governed by the interaction with the conduction electrons as proposed recently by Kondo /1/. In the framework of his theory the low T muon jump rates are described using reasonable parameters. As a consistency check, the transition temperatures to coherent transport are calculated and are found to be well below the experimental T-range. The experimentally observed power law behavior ($D \sim T^{-\alpha}$) is in accordance with theoretical predictions. At intermediate temperatures the linear T-dependence of the diffusion coefficient indicates the existence of one-phonon assisted hopping processes. In the high T range, where multiphonon process prevail, severe deviations from a simple small polaron description become evident and a combination of more improved versions of the small polaron theory with Kondo's electronic interaction theory are needed to proceed further.

ACKNOWLEDGEMENTS

This paper summarizes mainly the efforts of the Uppsala-Jülich CERN-μSR collaboration and many discussions with the different members are gratefully acknowledged.

REFERENCES

1. J. Kondo, Physica 126B, 377 (1984).
2. H. Grabert, S. Linkwitz, S. Dattagupta, U. Weiss, Europhysics Letters, in press, and this conference.
3. K.W. Kehr, D. Richter, J.M. Welter, O. Hartmann, E. Karlsson, L.O. Norlin, T.O. Niinikoski, A. Yaouanc, Phys. Rev. B26, 567 (1982).
4. O. Hartmann, E. Karlsson, L.O. Norlin, T.O. Niinikoski, K.W. Kehr, D. Richter, J.M. Welter, A. Yaouanc, J. Le Hericy, Phys. Rev. Lett. 44, 337 (1980).
5. C.W. Clawson, K.M. Crowe, S.E. Kohn, S.S. Rosenblum, C.Y. Yuang, J.L. Smith, J.H. Brewer, Physica 109 + 110B, 2164 (1982).
6. J.M. Welter, D. Richter, R. Hempelmann, O. Hartmann, E. Karlsson, L.O. Norlin, T.O. Niinikoski, D. Lenz, Z. Phys. B52, 303 (1983).
7. R. Kadono, J. Imizato, K. Nishiyama, K. Nagamine, T. Yamazaki, D. Richter, J.M. Welter, Physics Lett. 109A, 61 (1985).
8. J.H. Brewer, M. Celio, D.R. Harshman, R. Keitel, S.R. Kreitzmann, G.M. Luke, D.R. Noakes, R.E. Turner, E.J. Ansaldo, C.W. Clawson, K.M. Crowe, S. Kohn, S.R. Rosenblum, J. Smith, C.Y. Huang, Proceedings 4th conference on Muon Spin Rotations, June 23rd - 27th 1986, Uppsala. Hyperfine interactions, in press.
9. A. Magerl, A.J. Dianoux, H. Wipf, K. Neumayer, I.S. Anderson, Phys. Rev. Lett. 56, 159 (1986).
10. H. Wipf, D. Steinbender, K. Neumaier, P. Gutsmiedl, A. Magerl, A.J. Dianoux, this conference.
11. T. Holstein, Ann. Phys. N.Y. 8, 325-343 (1959).
12. C.P. Flynn, A.M. Stoneham, Phys. Rev. B1, 3966 (1970).
13. Yu. Kagan, H.J. Klinger, J. Phys. C7, 2791 (1974).
14. K. Yamada, Prog. Theor. Phys. 72, 195 (1984).
15. K. Yamada, A. Sakurai, S. Miyazima, Prog. Theor. Phys. 73, 1342 (1985).
16. For a review see E. Karlsson, Phys. Rep. 82, 271 (1982). D. Richter, Springer Tracts in Modern Physics 101 (1983).
17. See e.g. A. Abragam, Nuclear Magnetism (Oxford Press, Oxford 1961)
18. R. Kubo, T. Toyabe in Magnetic Resonance and Relaxation, ed. by R. Blinc, North-Holland Publ. Co., Amsterdam (1967).
19. T. Yamazaki, Hyperfine Int. 6, 115 (1979).
20. D. Herlach in Recent Developments in Condensed Matter Physics, Vol. 1, ed. by J.T. Devreese (Plenum Press, New York, 1981).
21. D. Richter, T. Springer, Phys. Rev. B 18, 126 (1978).
22. K.W. Kehr; G. Honig, D. Richter, Z. Phys. B32, 49 (1978).
23. O. Hartmann, E. Karlsson, B. Lindquist, E. Wäckelgard, R. Wäppling, D. Richter, R. Hempelmann, T.O. Niinokoski, in reference /8/.
24. D. Richter in ref. /8/.
25. I.I. Gurevich, E.A. Meleshko, I.A. Muratova, B.A. Kikolsky, V.S. Roganov, V.I. Selivanov, B.V. Sokolov, Phys. Lett. 40A, 143 (1972).
26. V.G. Grebinnik, I.I. Gurevich, V.A. Zhukov, A.P. Manych, E.A. Meleshko, I.A. Maratova, B.A. Nikolskii, V.A. Suetin, Sov. Phys. JETP 41, 777 (1976).
27. L. Katz, M. Guinan, R.J. Borg, Phys. Rev. B4, 330 (1971).
28. H. Teichler, Phys. Lett. 64A, 78 (1977).
29. A. Seeger, Phys. Lett. 83A, 33 (1982).

30. J. Chappert, O. Hartmann, E. Karlsson, T.O. Niinikoski, L.O. Norlin, A. Yaouanc, Solid State Commun. 44, 13 (1982).
31. M.J. Puska, R.M. Nieminen, Phys. Rev. B27, 6121 (1983).
32. M. Camani, F.N. Gygax, W. Ruegy, A. Schenck, H. Schilling; Phys. Rev. Lett., 39, 836 (1977).
33. J. Kondo in Ref. /8/.
34. E. Sato, T. Hatano, Y. Suzuki, M. Imafuku, M. Sunaga, M. Doyama, Y. Morozumi, T. Suzaki, K. Nagemine, Hyperfine Int. 17-19, 203 (1984).
35. N. Haas, K.P. Döring, K. Fürderer, M. Gladish, D. Herlach, W. Jacobs, H.-J. Mundinger, J. Rosenkranz, H.E. Schäfer,M. Schmolz, A. Seeger in ref. /8/.
36. H. Teichler, A. Seeger, Phys. Lett. 82A, 91 (1981).
37. H. Teichler, Z. Phys. Chemie N.F. 114, 155 (1979).
38. G. Flik, J. Bradbury, W. Cooke, M. Leon, M. Paciotti, M. Schillaci, K. Maier, H. Rempp and A. Seeger in ref. 8.

The Influence of Electrons on the Tunneling State of a Hydrogen Atom in a Metal

H. Wipf[1], *D. Steinbinder*[1], *K. Neumaier*[2], *P. Gutsmiedl*[2], *A. Magerl*[3], *and A.J. Dianoux*[3]

[1]Institut für Festkörperphysik, Technische Hochschule Darmstadt, Hochschulstr. 2, D-6100 Darmstadt, Fed. Rep. of Germany

[2]Walther-Meissner-Institut für Tieftemperaturforschung, Walther-Meissner-Str. 8, D-8046 Garching, Fed. Rep. of Germany

[3]Institut Laue-Langevin, 156X, F-38042 Grenoble Cedex, France

The thermal destruction of a tunneling state in a molecular crystal, which becomes observable in a neutron spectroscopic experiment say above 10 K, is generally attributed to interactions with phonons. If the tunnel system is in a metallic environment there is an additional relaxation path originating from the interaction with electrons. We have studied by inelastic neutron scattering the tunneling of a single proton in $Nb(OH)_x$ with x = 0.002 and x = 0.0002 from 0.1 to 10 K in both a superconducting and a normal conducting environment (applied magnetic field 0.7 T). These data show that the interaction of the H tunnel state with conduction electrons dominates the observed damping in this temperature range. In addition, different values for the tunnel splitting are found in the superconducting state (J^S = 226 μeV) and in the normal conducting state (J^N = 206 μeV). This renormalisation and the damping are described consistently by only one coupling parameter K = 0.053.

1. INTRODUCTION

Tunneling of atomic species is a commonly found phenomenon in solid state physics and chemistry. However, the features observed in a measurement depend very much on the nature of the tunneling entity and on its environment.

In molecular crystals rotational tunneling of CH_3, CD_3, NH_4 or similar units has been studied in many crystals and indeed a large fraction of this conference is devoted to this class of material [1]. Due to the crystalline structure all the potential wells (pocket states) connected with the tunnel system are equivalent. Modern research efforts concentrate e.g. on the transition from the low-temperature quantum state (T < 10 K) to the classical rotational hopping regime (T > 50 K). In the models proposed it is generally assumed that this transition is due to a coupling of the tunnel system with the phonons of the crystal.

A very different model is used to describe tunneling in amorphous materials invoking transitions within asymmetric double minima potentials. These two-level systems (TLS) are characterized by a broad distribution in both the matrix element and the asymmetry of the pocket states. This yields an excitation spectrum which is smeared out over a wide energy range. Although the tunneling entities have not been identified to date, the rationale for the model is given by the structural disorder of amorphous materials. This model for TLS has been used very successfully to describe the low-temperature anomalies for a very large variety of materials ranging from dielectric glasses to amorphous metals.

Defects in lattices with variable concentrations (e.g. CN^- or OH^- in alkali halides) constitute a third class of tunneling centers. The tunneling unit is known as it is for molecular crystals. At low defect concentration the tunnel spectrum shows well-defined excitations. With increasing concentration mutual interactions between the defects lead to asymmetric pocket states like for TLS and, consequently, the excitations become smeared out in energy.

H tunneling in $Nb(OH)_x$ falls into the same category and the above transition has been the subject of recent studies [2,3]. In addition, $Nb(OH)_x$ has further interesting properties which are not found or not accessible in other systems : the Nb host is a superconductor with a transition temperature of $T_c = 9.2$ K. This gives the unique possibility to measure a tunnel system below T_c in a superconducting or a normal conducting environment (applied magnetic field). Actually, $Nb(OH)_x$ can be looked upon as the model system to study the interaction of a tunnel state with conduction electrons, and this paper will concentrate on this particular aspect.

Sections 2 and 3 deal with the geometry of the (OH) defect in the Nb host and the scattering law. Then, experimental data will be presented (chapter 4) and discussed in chapter 5.

2. THE GEOMETRY OF THE (OH) DEFECT IN Nb

O defects in Nb occupy octahedral interstitial sites. At low temperature each O can trap one H. The H vibrational spectrum reveals that the H is trapped on a tetrahedral site of few Å away from the O [4]. This is supported by the diffuse scattering pattern of $Nb(OD)_x$ from which a O-D distance of 4.0 ± 0.5 Å has been derived [5]. A tunnel state for the H can only develop if the pocket states are equivalent sites in the distortion field of the trapping defect and if the distance between the pocket states is small. Figure 1 shows the cubic unit cell for the Nb host, the trapping O located at an octahedral site (big dot) and tetrahedral sites (various small symbols). Tetrahedral sites which are equivalent with respect to the O are marked by the same symbols. Only the sites shown by the full square symbols agree with all the above conditions. Note, that there are 8 pairs of these particular tetrahedral sites around an O. They are, however, connected via non-equivalent sites. This is the reason why H tunneling in $Nb(OH)_x$ is restricted to two pocket states.

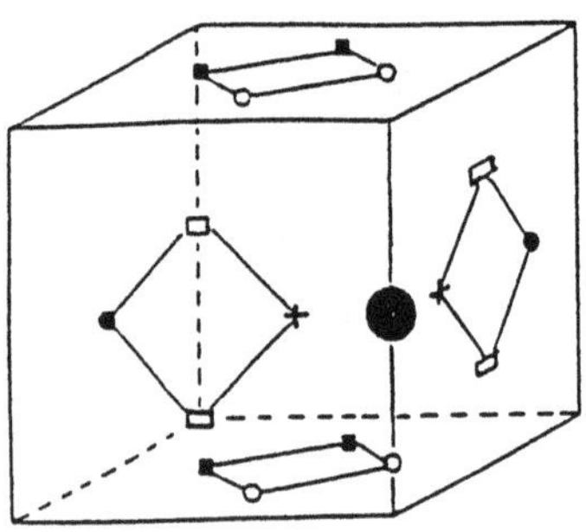

Fig. 1 Configuration of the trapping O defect with the delocalised H (see text for details).

3. THE SCATTERING LAW

Neutron scattering is a direct spectroscopic technique to study excitations on an atomic scale. The cross section for inelastic scattering from tunnel transitions in a TLS averaged for polycrystalline samples is given by [2,6]

$$\frac{d^2\sigma}{d\Omega d(\hbar\omega)} = \frac{\sigma^{inc}}{4\pi}\frac{\kappa_f}{\kappa_i}e^{-2W}\left(\frac{1}{2}-\frac{\sin Qd}{Qd}\right)\int_{-\infty}^{\infty}dJ\,Z(J)\int_{-\infty}^{\infty}d\Delta\,Z(\Delta)\,\frac{J^2}{J^2+\Delta^2}\,\frac{O(\hbar\omega)}{1+\exp(\hbar\omega/k_BT)} \tag{1a}$$

with

$$O(\hbar\omega) = \Gamma E^2/(((\hbar\omega)^2-E^2)^2+(\Gamma\hbar)^2\pi) ; \tag{1b}$$

σ^{inc} denotes the incoherent neutron-hydrogen cross section, e^{-2W} is the Debye-Waller factor, $\vec{Q} = \vec{k}_f - \vec{k}_i$ are the wave vectors before and after the scattering event, $\hbar\omega$ is the energy transfer of the neutron, d describes the distance in between the potential minima and k_BT is the temperature of the sample. Z(J) and Z(Δ) are distribution functions for the tunnel matrix element and for the asymmetry. Finite values for the distributions Z(J) and Z(Δ) are attributed to interactions between (OH) defects. For the low defect concentrations of our samples (x = 0.0002 and x = 0.002) these effects are small but still noticeable. This results in an inhomogeneous linebroadening which we will not be concerned with in this paper [6].

Here we will concentrate on the energy-dependent part of the scattering law. It is accounted for by the dynamical response function $O(\hbar\omega)$ which accounts for a damping due to an interaction of the tunnel state with its environment. We will treat $O(\hbar\omega)$ as a damped oscillator function (eq. 1b) characterized by the tunnel splitting J and by the damping Γ.

4. EXPERIMENTAL

The Nb host is a superconductor with a transition temperature of 9.2 K. Therefore, H tunneling in $Nb(OH)_x$ below this temperature can be measured in either a superconducting or a normal conducting environment (applied magnetic field 0.7 T). Fig. 2 shows two spectra taken at 0.1 K with and without a field. In addition to the elastic scattering there is a well-defined inelastic peak from the H tunnel state (the excitation on the neutron energy gain side is not visible at this low temperature). The two salient features observed are : i) the peak measured in the normal conducting state is reduced from 226 μeV to 206 μeV and ii) a dynamical width of about 45 μeV is found in the normal conducting state.

5. DISCUSSION

Recently the influence of electrons on tunnel systems has found an upsurge in theoretical interest. Problems considered deal not only with TLS in a conductor but they also include the low temperature quantum diffusion of light particles in metals and the transition from the quantum regime to the classical diffusion regime [7-11].

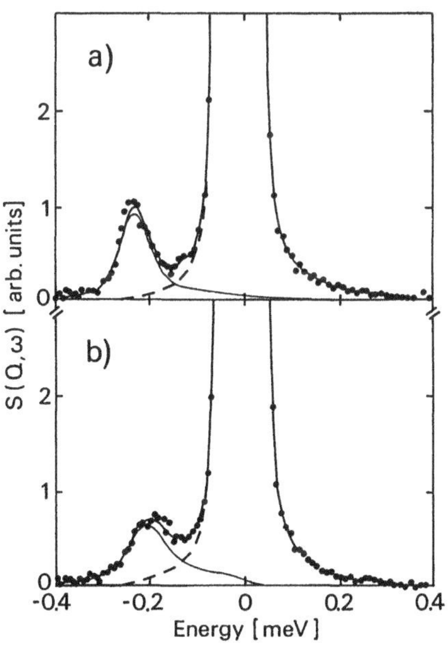

Fig. 2 Neutron spectra measured at 0.1 K in the superconducting (part a) and in the normalconducting state (part b). The dashed line gives the elastic energy resolution of the spectrometer and the thin solid line shows the fitted inelastic scattering. The thick solid line is the sum of both scattering contributions. The data were taken with the time focusing time-of-flight spectrometer IN6 at the ILL.

The coupling strength K between the tunnel state and the electrons is a dominant input parameter for the theories. Electrons with angular momentum ℓ and scattered at the potential from the defect with a phase shift α_ℓ contribute to the coupling K like [12]

$$K = 2(2\ell + 1)\left(\frac{1}{\pi} \tan^{-1} \frac{\sqrt{1 - j_0^2(k_f d)}\ \tan\alpha_\ell}{\sqrt{1 + j_0^2(k_f d)\ \tan^2\alpha_\ell}}\right)^2 . \tag{2}$$

Here j_0 and k_f are the spherical Bessel function and the Fermi wave vector. A calculation of K which takes into account the geometry of the H tunnel state and a realistic band structure for Nb has not yet been performed. Therefore K has to be determined from experimental data. Teichler has shown [7] that it can be extracted from the ratio of the tunnel splitting in the normal conducting and superconducting state at T = 0 K according to

$$\ell n \left(\frac{J^N}{J^S}\right) = K\left(\ell n \frac{J^N}{J^S + 2\Delta(0)} + 1.1 \left(1 - \frac{J^S}{2\Delta(0)}\right)\right). \tag{3}$$

Here $\Delta(T)$ is the electronic gap energy with $\Delta(0)$ = 1.53 meV.
With J^N = 0.206 and J^S = 0.226 we obtain K = 0.053 with an estimated error of about 15 %. This shows that H in $Nb(OH)_x$ is only weakly coupled to the conduction electrons.

Due to this weak coupling the damping is well described by a Korringa formalism. In the superconducting state the damping is given by [2,6,9,13]

$$\Gamma = 4\pi K \frac{k_B T}{1 + \exp\left(\Delta(T)/k_B T\right)} \tag{4a}$$

and in the normal conducting state by

$$\Gamma = \pi K J^N \coth\left(\frac{J^N}{2 k_B T}\right) . \tag{4b}$$

In the limit $T \to 0$, equs. 4a and 4b yield zero homogeneous damping for the superconducting state and a damping of 35 µeV for the normal conducting state if the above value of K = 0.053 is used. The agreement with the experiment is remarkable considering that there is no adjustable parameter.

6. SUMMARY

Our measurements of the tunnel splitting of the H vibrational ground state in $Nb(OH)_x$ in the normal conducting and superconducting states show a direct coupling with the electrons. For the normal conducting state we find a reduction of the matrix element (9 %) and a finite damping (Γ/J^N = 0.2) in the limit of low temperatures. All these features can be described quantitatively within the framework of recent theories by only one coupling parameter K = 0.053. This value implies that H in $Nb(OH)_x$ represents the case of small coupling between a tunnel state and conduction electrons (in contrast to e.g. μ^+ in Cu and Al).

REFERENCES

1. See also W. Press : "Single-Particle Rotations in Molecular Crystals" in Springer Tracts Mod. Phys. 92 (Springer, New York 1981)
2. A. Magerl, A.J. Dianoux, H. Wipf, K. Neumaier and I.S. Anderson : Phys. Rev. Lett. 56, 159 (1986)
3. P. Gutsmiedl, N. Schiekhofer, K. Neumaier and H. Wipf, contribution to this Workshop

4. A. Magerl, J.J. Rush, J.M. Rowe, D. Richter and H. Wipf, Phys. Rev. B 27, 927 (1983)
5. A. Magerl et al. : to be published
6. H. Wipf, D. Steinbinder, K. Neumaier, P. Gutsmiedl, A. Magerl and A.J. Dianoux : to be published
7. See contributions to this workshop from H. Grabert, H. Teichler, A. Weiss et al., and D. Richter
8. J. Kondo : Physica 126B, 377 (1984)
9. J.L. Black : "Glassy Metals 1" in Topics in Applied Physics 46, ed. H.J. Güntherrodt (Springer, Berlin 1981)
10. C.C. Yu and A.V. Granato, Phys. Rev. B 32, 4793 (1985)
11. H. Grabert, S. Linkwitz, S. Dattagupta and U. Weiss, Europhys. Lett. 2, 631 (1986)
12. K. Yamada, A. Sakurai, S. Miyazima, Prog. Theor. Phys. 73, 1342 (1985)
13. The coupling strength is also given as a product of the electronic density of states at the Fermi level ρ with an appropriate matrix element $V_\perp$ (see e.g. Refs. 2 and 9). Corresponding formula become identical for $2K = (\rho V_\perp)^2$.

H and D Tunneling in Nb: Dependence on Defect Concentration and Trapping Center

P. Gutsmiedl[1], *M. Schiekhofer*[1], *K. Neumaier*[1], *and H. Wipf*[2]

[1]Walther-Meissner-Institut für Tieftemperaturforschung, Walther-Meissner-Str. 8, D-8046 Garching, Fed. Rep. of Germany

[2]Institut für Festkörperphysik, Technische Hochschule Darmstadt, Hochschulstr. 2, D-6100 Darmstadt, Fed. Rep. of Germany

Low-temperature tunneling of H or D interstitials trapped by impurity atoms such as O or N in Nb has been investigated by various methods such as specific heat, thermal conductivity, ultrasonic (anelastic) and neutron spectroscopic measurements. The experiments show (i) that the O (or N) impurity atoms trap hydrogen interstitials at low temperatures, and (ii) that the trapped hydrogen is delocalized among (most likely) two nearest-neighbor tetrahedral interstitial sites (for a detailed discussion see /1,2/).

In this contribution, we report on an experimental study made by low-temperature specific heat measurements ($0.03 < T < 4$ K) on a series of NbO_xH_y, NbO_xD_y, NbN_xH_y and NbN_xD_y samples ($0.00005 < x,y < 0.02$) in which we investigated systematically H and D tunneling with respect to its dependence on (i) the trapping center and (ii) the defect concentration.

Our samples were prepared from polycrystalline Nb rods (70 mm long, 6 or 12 mm in diameter) received from Materials Research GmbH (MARZ grade 99.99 %). To reduce the content of interstitial C the rods were decarburized at 1700 C in an O_2 atmosphere ($\sim 10^{-4}$ mbar) /3/, and subsequently annealed under UHV conditions ($\sim$ 2200 C, $\sim 10^{-10}$ mbar), yielding residual resistivity ratios of about 1200, which corresponds to a total impurity of 40 ppm /4/. Thereupon, for O or N and H or D doping they were again annealed in an appropriate atmosphere first of O_2 or N_2 ($\sim$ 1800 C, $2\cdot10^{-6}$ up to $4\cdot10^{-5}$ mbar) /3/, and then of H_2 or D_2 ($\sim$ 1200 C, 1 up to 700 mbar) /5/. After the last annealing, the rods were quenched to room temperature in a He jet stream in order to prevent precipitation of O or N, respectively. The O and N concentrations were determined by residual resistivity measurements /4/ (before H or D doping), and the H and D concentration by vacuum extraction of small pieces cut from the individual rods (the accuracy of the quoted concentrations ranges between 15 and 30 %). Our specific heat data (experimental accuracy 3 %) were taken with a standard quasiadiabatic method in a dilution refrigerator (magnetic fields were screened to less than 2 μT).

Figure 1 shows the dependence of the specific heat on the trapping center and the H-isotope. The data are presented in a log-log plot versus temperature. The results for Nb doped either with O or H are included, as well as the specific heat (broken line) calculated for pure superconducting Nb. The main characteristic of Fig. 1 is the large excess specific heat of samples with trapped hydrogen isotopes. Two features are especially important, (i) the isotope dependence, and (ii) the dependence on the trapping impurity. The isotope dependence is already known /6/ and will be not discussed here. The influence of the trapping impurity is less pronounced than that of the H-isotope but also well defined. The specific heat anomalies of the H-isotopes trapped onto nitrogen are shifted to lower temperature compared with those due to the O-H(D) complexes, indicating smaller tunneling energies.

The concentration dependence is clearly illustrated in Fig. 2 by the four O-D doped samples for which the magnitude of the specific heat at 0.1 K differs only by a factor of ~ 2 although the concentration of O-D pairs varies by a factor of 70, indicating strong interaction effects between neighboring tunneling pairs. The

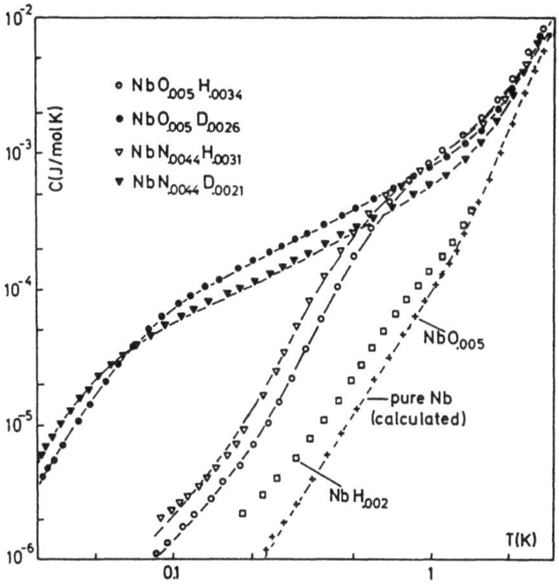

Fig. 1: Specific heat of (superconducting) Nb samples partially doped with O or N and H or D demonstrating the dependence on (i) the H-isotope and (ii) the trapping impurity.

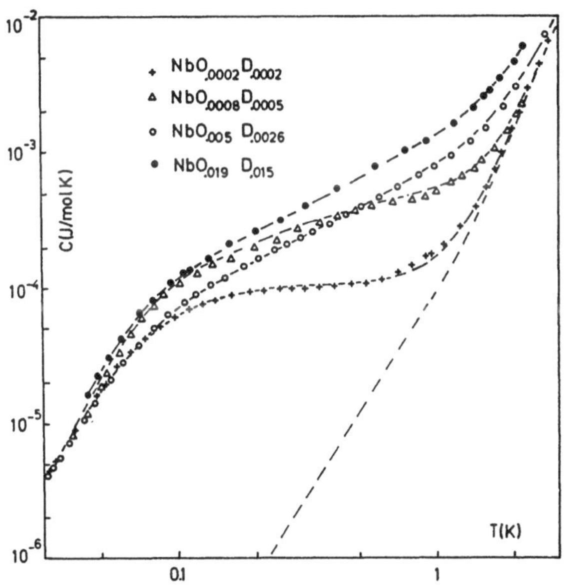

Fig. 2: Specific heat of NbO_xD_y samples whose concentration of tunneling complexes varies between 0.0002 and 0.014.

low-temperature cut-off of the specific heat anomaly is independent of O-D concentration, suggesting the same tunneling energy for all the O-D samples.

In Fig. 3 are represented the results obtained for 2 samples containing both nominally 150 ppm O and H. An impurity analysis for one of the samples indicated 130 ppm O, 40 ppm N, 55 ppm C and 150 ppm H. The specific heat anomaly below 1 K is drastically different for the 2 samples. Beneath the nearly doubled magnitude of the anomaly for one sample, a well pronounced low-temperature tail is observed.

These experimental results are quantitatively analysed within the framework of a model developed in Ref. 7 and 8. For the considered two-site tunneling system the energy difference E between the two tunneling eigenstates is given by $E=(J^2+\varepsilon^2)^{1/2}$

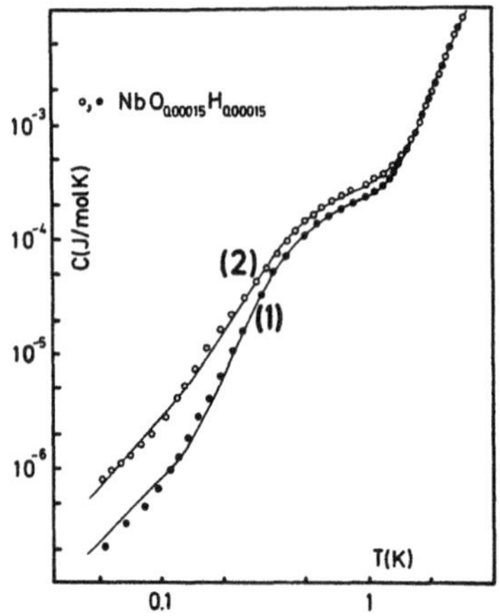

Fig. 3: Specific heat of two NbO_xH_y samples with nominal x,y = 0.00015. See text for details.

where J is the tunneling matrix element, and ε is the energy shift between the two sites due to stresses caused by neighboring complexes. For a statistical pair arrangement, the distribution function Z(ε) of the random elastic stresses ε is assumed to have a Lorentzian shape /9/,

$$Z(\varepsilon)=\pi^{-1}\varepsilon_0/(\varepsilon_0^2 + \varepsilon^2) \tag{1}$$

where ε_0 is the average value for the energy shift. Similarly, a distribution Z(J) with

$$Z(J)=\pi^{-1}\Delta J/(\Delta J^2 + (J-J_0)^2) \tag{2}$$

has been introduced. It is supposed that the widths of both distributions will increase with increasing defect concentration when interactions become more important.

The specific heat C(E) for a given tunneling system with an energy difference E is given by the Schottky expression

$$C(E)=k_B(E/k_BT)^2 \frac{\exp(-E/k_BT)}{(1+\exp(-E/k_BT))^2} \quad . \tag{3}$$

Starting from zero temperature, C(E) increases at first exponentially, passes through a maximum at T_{max} = .42 E and decreases finally proportional to T^{-2}.

In the present experiments, the measured excess specific heat C_{exc} is the sum of the specific heat contributions of all tunneling systems with their respective energy difference E. C_{exc} can be written in terms of the distribution functions Z(J) and Z(ε) as

$$C_{exc} = N \iint Z(J)\ Z(\varepsilon)\ C(E)\ d\varepsilon dJ \tag{4}$$

where N is the number of tunneling complexes.

The distribution in ε and J yield important effects on the specific heat. The specific heat of systems with a well-defined J_0 but with a distribution in ε ($\varepsilon_0 > J_0$) is reduced at T_{max} by a factor proportional to ε_0 and is extended with increasing ε_0 more and more to higher temperatures. In contrast, the distribution Z(J) renders the exponential increase of C_{exc} below T_{max} into a less steep one.

Using (4) we made fits to the measured excess specific heat of these (lines in Fig. 1,2 and 3) and other samples which are for the sake of clarity not shown here. For each sample we obtained three fit parameters, (i) the tunneling matrix element J_0, (ii) the width of the tunneling matrix element ΔJ, and (iii) the energy shift parameter ε_0.

As the main result of this study the following tunneling energies are found:

H trapped by O	$J_0(O\text{-}H)$ = 230 ± 10	μeV
D trapped by O	$J_0(O\text{-}D)$ = 20 ± 1	μeV
H trapped by N	$J_0(N\text{-}H)$ = 170 ± 10	μeV
D trapped by N	$J_0(N\text{-}D)$ = 14 ± 1	μeV

They are independent of the concentration of tunneling complexes, at least in the studied range up to 1 at %. The values for $J_0(O\text{-}H)$ and $J_0(N\text{-}H)$ are in perfect agreement with recent results of inelastic neutron scattering studies, partly on the same samples /10,11/. The tunneling energies for trapped D are known only from these and previous /6/ specific heat measurements.

Together with the displacement of near neighbors in Nb created by interstitially dissolved O and N recently determined by DOSCH et al. /12/ with coherent elastic-diffuse neutron scattering, this set of tunneling energies provides a solid experimental basis for a microscopic model of H tunneling in metals.

The width of the tunneling energy ΔJ increases with the concentration of tunneling complexes up to 15 % of J_0. But even for the smallest concentrations its value is 5 % of J_0, which is probably either due to the still active interaction between tunneling complexes, or to residual impurities such as C. The energy shift parameter ε_0, even for the smallest concentrations not negligible, increases roughly linearly with the concentration:

$$\varepsilon_0(\text{meV}) = (10 \pm 3) \cdot N(\text{at\%}) \qquad (5)$$

confirming the model of stress-induced energy shifts from neighboring complexes /7/. This proportionality of ε_0 with N is the reason for the specific heat at T_{max} being within a factor of 2 independent of the concentration of tunneling pairs.

The fit to the data of the two samples shown in Fig. 3 yielded puzzling results, namely the values for J_0 = 170 μeV and for the concentration 0.004 at% for sample (1) and 0.007 at% for sample (2). Taking into account our results for the tunneling energies and the impurity analysis for one of these samples, we have to conclude that only H trapped by N is tunneling and contributing to the specific heat anomaly. This implies that for the remaining hydrogen no trapping centers are available. It is possible that the final quench to room temperature during the sample preparation was too slow to avoid O precipitation, the quenching speed for N is less critical /13/. Another important feature is the broad low-temperature tail of the specific heat of sample (2) yielding a width of the tunneling energy ΔJ of 27 μeV much larger than ΔJ = 7 μeV of the other sample. The origin of the difference is unknown, presumably it is also due to the sample preparation.

In summary, with our specific heat measurements we have determined the tunneling energies of H and D trapped either by O or by N. These values provide a solid experimental basis for a microscopic model for H-isotopes trapped by interstitial O or N in Nb. But extreme careful sample preparation and analysis is necessary if quantitative conclusions from experimental results should be drawn.

References

1. A. Magerl, J.J. Rush, J.M. Rowe, D. Richter, H. Wipf Phys. Rev. B27, 927 (1983)
2. H. Wipf, K. Neumaier: In Electronic Structure and Properties of Hydrogen in Metals, eds. P. Jena, C.B. Satterthwaite, Plenum Publishing Corp., New York and London 1983, p. 485
3. E. Fromm, H. Jehn, Metall. Trans. 3, 1685 (1972)
4. H. Schultz, Mat. Sci. Eng. 3, 189 (1968/69)
5. E. Veleckis, R.K. Edwards, J. Phys. Chem. 73, 683 (1963)

6. H. Wipf, K. Neumaier, Phys. Rev. Lett. 52, 1308 (1984)

7. H. Wipf, A. Magerl, S.M. Shapiro, S.K. Satija, W. Thomlinson Phys. Rev. Lett. 46, 947 (1981)

8. H. Wipf, A. Magerl, S.M. Shapiro, S.K. Satija, W. Thomlinson Proc. Miami. Int. Symp. on Metal-Hydrogen Systems, Miami 1981

9. A.M. Stoneham, Rev. Mod. Phys. 41, 82 (1969)

10. A. Magerl, A.J. Dianoux, H. Wipf, K. Neumaier, I.S. Anderson Phys. Rev. Lett. 56, 159 (1986)

11. A. Magerl, private communication

12. H. Dosch, A. v. Schwerin, J. Peisl, Phys. Rev. B34, 1654 (1986)

13. E. Fromm, J. Less-Common Metals 14, 113 (1968)

Sound Velocity Measurements on Hydrogen-Doped Niobium in the Superconducting and Normal States

W. Morr[1], G. Weiss[1], and H. Wipf[2]

[1]Institut für Angewandte Physik II der Universität Heidelberg, Albert-Überle-Str. 3–5, D-6900 Heidelberg, Fed. Rep. of Germany

[2]Institut für Festkörperphysik, Technische Hochschule Darmstadt, Hochschulstr. 2, D-6100 Darmstadt, Fed. Rep. of Germany

Hydrogen interstitials in niobium, trapped by additional impurities such as nitrogen or oxygen, form tunneling systems with energy splittings $E \approx 2$ K. This has been demonstrated by low-temperature specific heat measurements /1,2,3/, by inelastic neutron scattering experiments /4,5/, and also by sound velocity measurements /6/. Additionally, hydrogen tunneling systems give rise to relaxation processes resulting in a peak of the ultrasonic attenuation at 3K if the niobium host is in the superconducting state /7,8,9/. By suppressing superconductivity with a magnetic field this attenuation peak vanishes, proving that the dynamics of the tunneling systems are governed by their interaction with conduction electrons /8,9/. Recently this important role of the conduction electrons has also been verified in inelastic neutron scattering experiments /5/.

In this contribution we present measurements of the sound velocity of H- and D- doped niobium between 0.1 K and 20 K in the superconducting and the normal state. Such experiments are especially well suited to determine tunneling matrix elements and energy spectrum as well as the relaxation times of the hydrogen tunneling systems.

Our sample was a niobium single crystal containing approximately 1500 ppm O and N. It was doped with either 3000 ppm H or, after extraction of H, with 2400 ppm D in the respective gas atmosphere and appropriate temperatures as described earlier /2,3/. Longitudinal sound was propagated along the [100]-direction and relative changes of the sound velocity were measured using a phase-sensitive detection method with resolution better than $5 \cdot 10^{-7}$. In all experiments the sound velocity decreases steeply above T=10 K. This effect of lattice anharmonicity can be taken into account when the data are compared to those of undoped samples. Our results, already corrected for this effect, are shown in Fig. 1. The remaining temperature dependence of the sound velocity is caused by the interaction of the sound wave with the tunneling systems. The main features are: (i) an overall increase of the sound velocity with temperature in all experiments, the effect being larger for the deuterized sample (triangles) than for the hydrogenated one (circles), (ii) the sound velocity is always higher in the superconducting (open symbols) than in the normal state (full symbols), merging for temperatures above 5 K, and (iii) in the superconducting state a steplike decrease is observed around 3 K for H-tunneling systems and around 4 K for D-tunneling systems.

According to the theory developed for describing the low-temperature acoustic behaviour of amorphous solids, two types of interaction between tunneling systems and the sound wave contribute to the temperature dependence of the sound velocity, the resonant and the relaxation process /10/. The contribution of the resonant process

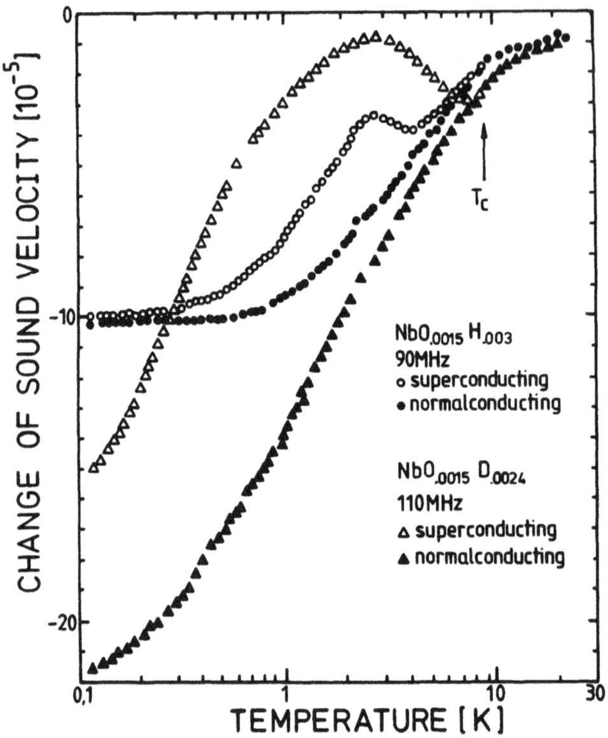

Fig. 1: Temperature dependence of the sound velocity of Nb:O doped with either H or D. The normal conducting state was achieved with a magnetic field applied parallel to the [100]- direction

can be expressed as /6/

$$\left.\frac{\Delta v}{v}\right|_{res} = -\frac{N\gamma^2}{\rho v^2}\,\frac{\Delta_o^2}{E^2}\,\frac{1}{E}\,\tanh\frac{E}{2kT} \qquad (1)$$

where ρ is the mass density and v the sound velocity. Δ_o is the tunneling energy and E the total energy splitting of the tunneling system. $E = (\Delta_o^2+\Delta^2)^{1/2}$ where Δ is the energy difference of the two wells forming the double well potential. γ is the deformation potential and N the density of tunneling systems with given Δ_o and E. The temperature dependence of (1) is due to the thermal occupation of the tunneling systems. The reduction of the sound velocity caused by the resonant process vanishes at high temperatures, when $kT >> E$. Already with a fixed value of E and $E = \Delta_o$ (1) is able to describe the experiment on the H-doped sample very well in the superconducting state below 2 K, especially the increase of $\Delta v/v$ at a well-defined temperature. A similar interpretation was given by /6/ resulting in a value of $E=\Delta_o=1.4$ K. Equation (1) also explains that in the case of deuterium where the tunnel splitting is expected to be smaller the increase of $\Delta v/v$ starts already at lower temperatures.

The steplike decrease of the sound velocity of the H-doped sample in the superconducting state around 3 K is accompanied by an attenuation peak and is caused by the relaxation process. This process originates from the fact that the sound wave modulates the energy splitting of the tunneling systems according to their deformation potentials and therefore disturbs the equilibrium population. The tunneling systems try to reestablish this equilibrium with their characteristic relaxation rates τ^{-1}. The contribution of the relaxation process to the sound velocity is given by /10/

$$\left.\frac{\Delta v}{v}\right|_{rel} = -\frac{N\gamma^2}{2\rho v^2}\frac{\Delta^2}{E^2}\frac{1}{kT}\,\mathrm{sech}^2\frac{E}{2kT}\frac{1}{1+\omega^2\tau^2} \tag{2}$$

where $\omega/2\pi$ is the frequency of the sound wave. (2) obviously reduces the sound velocity further, strongest for $E \approx kT$ as expressed by the factor $T^{-1}\mathrm{sech}^2(E/2kT)$ which reflects the change of the equilibrium population with energy. The temperature dependence of (2) is mainly given by the last factor, i.e. by the temperature dependence of the relaxation rate. In amorphous metals relaxation of the (intrinsic) tunneling systems is caused by their interaction with thermal phonons and with conduction electrons. The latter has been discussed as a Korringa-like process /11/ the relaxation rate of which is given in the normal conducting state by

$$\tau_n^{-1} = K\frac{\Delta_o^2}{E^2}\,E\coth\frac{E}{2kT} \tag{3}$$

where K includes an appropriate scattering potential and the density of states of the electrons per atom at the Fermi level. In the superconducting state the opening of the energy gap $2\Delta_s$ in the electron density of states has to be taken into account /12/. In the simplest case, $E \ll \Delta_s$, this results in

$$\tau_s^{-1} = 4K\frac{\Delta_o^2}{E^2}\frac{kT}{1+e^{\Delta_s/kT}} \quad . \tag{4}$$

Equation (2) in connection with this strongly temperature-dependent relaxation rate describes very well the step-like decrease of the sound velocity around 3 K, where $\omega\tau$ becomes unity. In agreement with shape and frequency dependence of the attenuation peak these contributions to the relaxation process are mainly coming from tunneling systems with $E \approx \Delta \approx 5$ K /8/.

Taking the values for the coupling strength K between tunneling systems and conduction electrons from this analysis, one finds that in the normal conducting state $\tau^{-1} \gg \omega$ even for $T \to 0$. Therefore the last factor in (2) becomes unity and the temperature dependence is determined only by the density of states of the tunneling systems. Comparison of the sound velocity data of the superconducting with those of the normal state obviously allows to distinguish between contributions from the resonant and from the relaxation process. In addition, the two processes depend differently on the ratio between tunnel splitting and total energy splitting, the resonant process being proportional to $(\Delta_o/E)^2$ and the relaxation process to $(\Delta/E)^2$.

For the density of states a model suggests itself which successfully explained recent specific heat measurements on Nb with different OH concentrations /3/. There a low-energy cut-off in the density of states, independent of defect concentration, was found. It was postulated that all tunneling systems have the same Δ_o identical with this low-energy cut-off. For the asymmetry energy Δ, however, a Lorentzian distribution centred around zero was assumed with a width $\bar{\Delta}$ increasing with the number of defects. Inserting such a distribution in (1) and (2) we find very good aggreement with our experiment on the H-doped sample, except for the fact that at low temperatures theory predicts the same sound velocity for the superconducting and the normal state. In our experiment, however, the sound velocity in the flat low-temperature part is unambiguously higher in the superconducting state.

As pointed out recently /13/ a reduced tunnel splitting in the normal conducting state could explain such an effect via the resonant process described by (1). It has been predicted already some time ago /14/ that the interaction between a tunneling particle and conduction electrons should slow down the movement and thus reduce the tunneling energy. From our measurements we estimate this reduction to be approximately 5% of the tunnel splitting Δ_0=1.6 K in the case of hydrogen. This is a much smaller effect than that claimed in a very recent paper /15/ where the experiments have been carried out only down to 0.5 K.

Finally, we want to discuss briefly the results of the deuterized sample. Keeping in mind that it is the same sample with respect to the O,N-concentration, the distribution of asymmetry energies should remain, since it is very likely to be caused by the strain fields of these defects. In fact, our experiment is almost perfectly described by assuming Δ_0(D)=0.32 K together with a Lorentzian distribution in Δ having the same width $\bar{\Delta}$ =3 K as in the case of hydrogen.

In conclusion we have demonstrated that sound velocity measurements are extremely valuable for studying the quantum behaviour of localized hydrogen atoms. In the case of the superconducting Nb-host not only the density of states but also the influence on the dynamics of the tunneling systems can be studied.

References

1. G.J. Sellers, A.C. Anderson, H.K. Birnbaum, Phys. Rev. B10, 2771 (1974)
2. C. Morkel, H. Wipf, K. Neumaier, Phys. Rev. Lett. 40, 947 (1978)
3. H. Wipf, K. Neumaier, Phys. Rev. Lett. 52, 1308 (1984)
4. H. Wipf, A. Magerl, S.M. Shapiro, S.K. Satija, W. Thomlinson, Phys. Rev. Lett. 46, 947 (1981)
5. A. Magerl, A.J. Dianoux, H. Wipf, K. Neumaier, I.S. Anderson, Phys. Rev. Lett. 56, 159 (1986)
6. G. Bellessa, J. de Physique 44, L-387 (1983)
7. D.B. Poker, G.G. Setzer, A.V. Granato, H.K. Birnbaum, Z. Phys. Chem. 116, 39 (1979) and Phys. Rev. B29, 622 (1984)
8. J.L. Wang, G. Weiss, H. Wipf, A. Magerl, in Phonon Scattering in Condensed Matter, eds. W. Eisenmenger, K. Lassmann, S. Döttinger (Springer, Berlin, Heidelberg 1984) p. 401 and 442
9. E. Drescher-Krasicka, A.V. Granato, J. de Physique 46, C10-73 (1985)
10. S. Hunklinger, W. Arnold, in Physical Acoustics Vol. XII, eds. W.P. Mason, R.N. Thurston (Academic, New York 1976) p. 155
11. B. Golding, J.E. Graebner, A.B. Kane, J.L. Black, Phys. Rev. Lett. 41, 1487 (1978)
12. J.L. Black, P. Fulde, Phys. Rev. Lett. 43, 453 (1979)
13. C.C. Yu, A.V. Granato, Phys. Rev. B32, 4793 (1985)
14. J. Kondo, Physica 84B, 40 (1976)
15. K.R. Maschhoff, E. Drescher-Krasicka, A.V. Granato, in Proceedings of the Fifth Intern. Conf. on Phonon Scattering in Condensed Matter, Urbana, Illinois (1986)

Quantitative Theory of the Tunnel Level Splitting for Hydrogen Trapped at Oxygen in Niobium

H. Teichler

Institut für Metallphysik der Universität Göttingen, and
Sonderforschungsbereich 126, D-3400 Göttingen, Fed. Rep. of Germany

Abstract: A brief review is given on the status of quantitative theoretical estimates for the low-temperature hydrogen tunnel level splitting in metals and its renormalization by environmental phonons and electrons. An approximation scheme is presented applicable to normal and to superconducting electrons. From that for H at O in Nb the value $K \approx 0.05$ is estimated for the nonadiabatic electron-hydrogen coupling constant by comparing the experimental data on the level splitting in normal and superconducting Nb. Together with our present theoretical results on the 'bare' hydrogen tunnel integrals and on the phonon renormalization this yields for H at O in Nb a level splitting of 0.04 meV (compared to 0.23 meV from experiments).

1. Introduction

The present contribution concerns the problem of quantitative theoretical estimates of hydrogen tunnel level splittings in metals and their renormalization due to interaction with environmental phononic and electronic degrees of freedom. This question arises in various areas of coherent quantum motion in solids. We shall consider as a particular example hydrogen trapped near an oxygen impurity in niobium, where according to current interpretations [1,2] hydrogen is capable to carry out tunnel transitions between two equivalent positions. This example has been selected since microscopic models [3-7] seem to describe well the quantum properties of H in Nb as far as lattice and phonon aspects are concerned, and since experimental data are available on the H-level splitting in superconducting [1,2] and normal [8] Nb which provide challenging information on the electronic renormalization effects.

Estimation of the tunnel level splitting means in detail determination of the 'bare' tunnel integrals, of the phonon, and of the electron renormalization factors. All three quantities shall be discussed here where we restrict to the low-temperature regime of coherent motion [9,10]. - For the phonon part of the problem we take advantage of the large ratio between the mass of host-lattice atoms and hydrogen interstitial and apply the adiabatic decoupling of host-lattice and interstitial dynamics well known from the 'small-polaron' model [10,11] for H in metals. This permits the introduction and numerical evaluation of the adiabatic H wave function and of the 'bare' tunnel integrals [4-6]. Moreover, it yields the renormalization of the level splitting due to the coupling to the phonons [12].

For the electronic part of the problem, instead of working with the real-time Green's function technique recently applied successfully [13-15] to the normal conducting environment, we introduce a simpler approximation scheme for the low-temperature range which is applicable to the normal as well as to the superconducting state. Our scheme adopts some ideas from KONDO's [16] variational approach to the non-adiabatic electron effects and takes into account the 'most divergent' terms [16]. The second order perturbation treatment of the superconducting environment by YU and GRANATO [17] emerges from our approach as its lowest approximation. By means of our scheme the coupling constant K [9] for non-adiabatic electron-hydrogen interaction below is estimated from the experimental data on the tunnel level splitting of H at O in normal [8] and superconducting Nb [1,2].

2. Characterization of the Model

Our discussion is based on a two-centre tunnel system consisting of a particle tunneling with 'bare' tunnel integral J between two centres r and l, phonons (frequencies $\omega(q)$, polarization $\underline{e}(q)$), and electrons (energies $\eta(\underline{k})$). The particle interacts with the electrons and phonons via

$$V = \Sigma_{j=r,l} a_j^+ a_j \{\Sigma_{\underline{k},\underline{k}',\sigma} V(\underline{k}-\underline{k}';j)\, c^+_{\underline{k}\sigma} c_{\underline{k}'\sigma} - \Sigma_q\, \gamma(q;j)\, (b_q + b^+_{-q})\}. \quad (1)$$

(Here the operators a,b,c refer to the 'particle', resp. phonons and electrons). As already mentioned, the adiabatic 'small-polaron' decoupling [10,11] is used for the particle-phonon correlations, where we assume that the particle follows adiabatically the motion of the host-lattice atoms. Then for each configuration of the lattice (described by a set of displacements $\{\underline{u}(m)\}$ of the atoms $\{m\}$) the particle wave funtion can be defined and the transfer integral becomes configuration dependent, i.e., $J = J(\underline{u}(m))$ [18] (where the $\underline{u}(m)$ are related to the b_q, b_q^+ in the usual way).

For this system the symmetric ground state and its antisymmetric counterpart shall be modelled by

$$|\Psi_{+/-}> = 2^{-1/2} \{\exp(-iW(r) - iQ(r))\cdot a_r^+ \pm \exp(-iW(l) - iQ(l))\, a_l^+\}\, |\emptyset>, \quad (2)$$

$$Q(j) = -i\Sigma_q \lambda(q;j)\, (b_q - b^+_{-q}) \;,\quad j=l,r \quad (3)$$

$$W(j) = \Sigma_{\underline{k},\underline{k}',\sigma} w(\underline{k},\underline{k}';j)\, c^+_{\underline{k}\sigma} c_{\underline{k}'\sigma} \;. \quad (4)$$

In (2) $|\emptyset>$ denotes the ground state of the system without tunneling particle and the operators Q(j), W(j) simulate the change in the phonon system and in the electron distribution due to the presence of that particle. Within this model the level splitting between even and odd states becomes.

$$\Delta E = 2\, \mathrm{Re} < \exp(iQ(r))\; J(\underline{u}(m))\; \exp(-iQ(l)) >_{ph}\; <\exp(iW(r))\, \exp(-iW(l))>_{el}. \quad (5)$$

Here the brackets mean the expectation values with regard to the free phonon resp., free electron ground state. In (5) advantage has been taken of our assumption that J is independent of the electronic coordinates, which permits factorization of ΔE into an electronic and a phonon part.

'Optimum' expressions for Q(j), W(j) can be determined via RITZ's principle by minimising the ground state energy $<\Psi_+|H+V|\Psi_+>$. For the genuine polaron problem such a variational approach has been studied already years ago, e.g., by EMIN [19] regarding the optical phonons. Here we propose to use it for the coupling to the electrons, too, and thereby to generalize KONDO's ansatz [16]. (However, application of the variational approach to the electrons needs additional approximations for handling exp (-iW(j)). We shall address this point again in Section 4.) - Below, for sake of brevity, for the phonon 'displacements' $\lambda(q;j)$ the usual 'small polaron' (see, e.g.[10,11,18]) expression $\lambda(q;j) = \gamma(q;j)/\omega(q)$ is used. Compared with the variational result this approximation leads to minor errors of the order of $\Delta E/\omega_D$ (since the phonon density of states goes to zero sufficiently fast for ω going to zero).

3. Bare Hydrogen Transfer Integrals and their Renormalization by Phonons

For quantitative estimates the phonon expectation value in (5) may be written [12,18]

$$< \exp(iQ(r)) J(\underline{u}(m))\, \exp(-iQ(l)) >_{ph} = <J> \exp(-S_{ph}) \quad . \quad (6)$$

Here $\langle J\rangle$ is the bare tunnel integral suitably averaged over $\{\underline{u}(m)\}$. $\exp(-S_{ph})$ is the theoretically well-known reduction of $\langle J\rangle$ due to the necessity to shift the lattice deformations of the particle during a tunnel step of the latter. - If $\langle J\rangle$ depends on the displacements $\underline{u}(m)$ of only a finite number of surrounding atoms then from [12,18] we get for $\langle J\rangle$ in the harmonic approximation of the lattice

$$\langle J\rangle = \int ..d\underline{u}_m .. \, J(..,\underline{u}_m+\underline{\bar{u}}_m^o,..) \exp(-\tfrac{1}{2}\Sigma_{mm'} \, \underline{u}_m \underline{\underline{D}}_{mm'}^{-1} \, \underline{u}_{m'}) \, / \sqrt{\det(2\pi \underline{\underline{D}})} \quad , \tag{7}$$

$$\underline{\underline{D}}_{mm'} = \langle \underline{u}(m) \otimes \underline{u}(m')\rangle_{ph} \quad , \quad \underline{\bar{u}}_m^o = (\underline{\bar{u}}^l(m)+\underline{\bar{u}}^r(m))/2 \quad .$$

$\underline{\bar{u}}^j(m)$ denotes the displaced equilibrium position of atom m due to the averaged force exerted by the particle when residing on the interstitial site j. For H in Nb values of the $\bar{u}^j(m)$ have been estimated by a number of authors [3,5,6,12,20] from experimental data on the hydrogen force dipole tensor in Nb, by using suitable KANZAKI force models and the static lattice Green's function. Our further treatment is based on the values from [5,6,12], although there is some recent controversy about the appropriate force model (in particular for D in Nb [21]).

According to (7) evaluation of $\langle J\rangle$ needs knowledge of $J(\underline{u}(m))$ for $\{\underline{u}(m)\}$ near $\{\underline{u}_m^o\}$. For H (and D) in the bulk of Nb (Ta) estimates of J have been deduced by solving numerically the Schrödinger equation for H in the '2T-configuration' [4-6] of the lattice (which is near the $\{\underline{u}_m^o\}$-configuration), modelling the potential relief by a phenomenological hydrogen-lattice interaction. Our recent values for J(2T) [22] which agree fairly well with those from [5,6] are displayed in table 1 together with the earlier estimates [4]. The effect of fluctuation $\underline{u}_m$ on J has been studied in [5,7]. Evaluating $\langle J\rangle$ along the lines described there yields the values given in table 1, where the increase of $\langle J\rangle$ compared to J(2T) is due to the exponential dependence [5,6] of J on the fluctuations $\underline{u}_m$.

In the harmonic approximation the phononic renormalization parameter becomes [12]

$$S_{ph} = \frac{1}{2} \langle (Q(r)-Q(l))^2\rangle_{ph} = \Sigma_{mm'} \Delta\underline{\bar{u}}(m) \, \underline{\underline{S}}(m,m') \, \Delta\underline{\bar{u}}(m') \quad . \tag{8}$$

$$\underline{\underline{S}}(m,m') = M/4 \; N^{-1} \Sigma_q \underline{e}(q) \otimes \underline{e}(q) \; \omega(q) \; \exp(i\underline{q}\underline{R}_{mm'}) \tag{9}$$

is the momentum - momentum correlation function of the lattice ($\Delta\underline{\bar{u}}(m) = \underline{\bar{u}}^r(m)-\underline{\bar{u}}^l(m)$). For Nb (and Ta) estimates of the matrices $\underline{\underline{S}}(m,m')$ (and their temperature dependence) are given in [12], calculated from eight - (resp. seven -) neighbour force models for the phonons. These values together with the displacements $\bar{u}^j(m)$ [5,6,12] yield for H in Nb the renormalization parameter S_{ph} = 3.33 (3.58 for H in Ta) used in Table 1.

Table 1. Theoretical estimates of tunnel integrals, renormalization factors, and level splittings. (J(2T) in brackets from [4], ΔE_s(expt) from [2]; energies in meV)

	J(2T)	$\langle J\rangle$	$\exp(-S_{ph})$	$\exp(-S_{el}^s)$	ΔE_s	ΔE_s(expt)
H in Nb	0.33 (3.0)	0.56	0.036	0.68	0.027	-
D in Nb	0.01 (0.3)	0.034	0.052	0.68	0.0024	-
H at O in Nb	0.51	0.85	0.034	0.68	0.039	0.23
D at O in Nb	0.017	0.054	0.046	0.68	0.0034	0.02

4. Level Splitting Renormalization by Electrons

Estimation of the electronic renormalization factor demands evaluation of

$$\exp(-S_{el}) = \langle \exp(iW(r))\exp(-iW(1))\rangle_{el} \tag{10}$$

(and of similar quantities when evaluating the ground state energy to apply RITZ's principle). We have deduced approximate expression for S_{el} in a 'mean field' approach, where we substitute quadratic terms in the operators $W(j)$ by their expectation values. For higher powers of these operators this implies a contraction into pairs and for the commutator $[W(r), W(1)]$ the substitution $\langle [W(r), W(1)]\rangle$. Moreover, when calculating the ground state energy the electron part of V in (1) is treated like the W(j). This procedure yields

$$S_{el} = \frac{1}{2} \langle (W(r)-W(1))^2 \rangle_{el} , \tag{11}$$

$$\Delta w(\underline{k},\underline{k}') = -i\Delta V(\underline{k}-\underline{k}') \,/\, (\eta(\underline{k})-\eta(\underline{k}')+\Delta E \,\mathrm{sgn}(\eta(\underline{k})-\eta(\underline{k}'))), \tag{12}$$

(where $\Delta w = w(r)-w(1)$ and ΔV similar). Introducing, as proposed by KONDO [9,16], the Fermi surface average

$$K = N(o)^2 \;\overline{(V(\underline{k}_F-\underline{k}_F';r)-V(\underline{k}_F-\underline{k}_F';1))^2}^{\,FS} \tag{13}$$

(N(o) density of states per spin at the Fermi surface) and using a constant density of states for carrying out the energy integrations then leads to

$$S_{el} = K \{ \ln(E_F/\Delta E) + \ln(1-E_F/D) \} \tag{14}$$

(D effective bandwidth, E_f Fermi energy), as one would obtain from KONDO's [16] 'most divergent' terms.

We also have applied our variational approach to a BCS superconductor with energy gap Δ and excitation energies $E(\underline{k}) = \sqrt{(\eta(\underline{k})^2+\Delta^2)}$. Changing the electronic excitations via a Bogoljubow tranformation to those of the superconducting state results in

$$S_{el}^{s} = K \{\ln(E_F/(\Delta E+2\Delta)) + \ln(1-E_F/D) + I(\Delta E/(2\Delta))\}. \tag{15}$$

Here I(x) takes into account the $E/\sqrt{(E^2-\Delta^2)}$ singularity in the excitation spectrum of the superconductor and the coherence factors [23] which reflect the equivalence of electron and hole-type excitations in the paired state. Numerical integrations give $I(x) \approx 1.1\ (1-x)$ for $x \ll 1$.

Concerning the ratio of level splittings ΔE in the normal resp. superconducting state (below distinguished by indices n, s) the present results together with the assumption that K, $\langle J\rangle$, S_{ph} are independent of the electronic state lead to

$$\ln(\Delta E_n/\Delta E_s) = K \{ \ln(\Delta E_n/(\Delta E_s+2\Delta)) + I(\Delta E_s/(2\Delta)) \} . \tag{16}$$

From (16) K may be estimated once values for the gap and for the level splittings are known. Using in (16) for H at O in Nb the low-temperature value $\Delta E_s = 0.23$ meV [2], $(\Delta E_s-\Delta E_n)/\Delta E_s = 9\%$ (recent analysis of the experimental data favours a value less than 10% [24] rather than 15% preliminarily reported in [8]), and $\Delta = 1.53$ meV (as in [17]) yields $K \approx 0.054$. Thus the nonadiabatic coupling of the electrons to H in Nb seems to be much smaller than that to μ^+ in Cu, where K=0.3 has been estimated by KONDO [9]. Moreover, approximating D by the d-band width in Nb of about 0.6 Ry [25] (with $E_F \approx D/2$) predicts from (14) and (15) for H at O in Nb the renormalization factors $\exp(-S_{el}^{n}) \approx 0.61$, $\exp(-S_{el}^{s}) \approx 0.68$. -In addition, for deuterium at O in Nb equ.(16) with K=0.054 predicts $(\Delta E_s-\Delta E_n)/\Delta E_s \approx 20\%$, if for deuterium the experimental value $\Delta E_s \approx 0.02$ meV [1] is used.

5. Concluding Remarks

With the above expressions the tunnel level splitting becomes

$$\Delta E = 2 \langle J\rangle \exp(-S_{ph}) \exp(-S_{el}) \ . \tag{17}$$

For H(D) in superconducting Nb, estimates of the corresponding quantities are shown in table 1, where K=0.05 is used throughout, deduced here for H at O in Nb. (The results are restricted to the low-temperature regime since our variational approach for the electronic renormalization demands $kT \gg \Delta E$). The first two lines give the data for self-trapped H(D) in the bulk.

In order to see to which extent the presence of O modifies $\langle J\rangle$ and S_{ph} for H at O in Nb we have calculated these quantities in a microscopic model [12,22]. In this model the O impurity is simulated by a point-like perturbation as proposed by TEWARY [26] and the H-Nb interaction is modelled by a phenomenological potential similar to [5-7]. The resulting $\langle J\rangle$ and S_{ph} are given in table 1. For H at O in superconducting Nb they predict $\Delta E \approx 0.04$ meV which might be compared with the experimental low-concentration value of 0.23 meV [2]. (However, the effect of O should be considered with caution, since recent neutron scattering experiments [27] for O in Nb indicate that the here adopted force model [26] might be insufficient). - The remaining discrepancy between ΔE from theory and experiment may be due to an overestimation of S_{ph}. (Larger values of J would yield a drastic overestimation of the low-temperature diffusion coefficient for H in Nb [5,7]).

The author acknowledges stimulating dicussions with H. Wipf and he wishes to thank T.H. Metzger for making available Ref. [27].

References

1 H. Wipf and K. Neumaier: Phys. Rev. Letters 52, 1308 (1984)
2 A. Magerl, A.J. Dianoux, H. Wipf, K. Neumaier and I.S. Anderson: Phys. Rev. Letters 56, 159 (1986)
3 H. Sugimoto and Y. Fukai: Phys. Rev. B 22 , 670 (1980)
4 H. Sugimoto and Y. Fukai: Hyperfine Interactions 17-19, 139 (1984)
5 H. Teichler and A. Klamt: Phys. Letter 108A, 281 (1985)
6 A. Klamt and H. Teichler: phys. stat. sol. (b) 134, 103 (1986)
7 A. Klamt and H. Teichler: phys. stat. sol. (b) 134, 533 (1986)
8 H. Wipf, A. Magerl and K. Neumaier: In Atomic Transport and Defects in Metals by Neutron Scattering, ed. by C. Janot, B. Petry, D. Richter and T. Springer, Proc. in Phys. 10 (Springer, Berlin, Heidelberg 1986) p. 161
9 J. Kondo: Physica 126B, 377 (1984)
10 Y. Kagan and M.I. Klinger: J. Phys. C 7, 2791 (1974)
11 C.P. Flynn and A.M. Stoneham: Phys. Rev. B 1, 3966 (1970)
12 H. Teichler: Phys F, in press
13 C.C. Yu and P.W. Anderson: Phys. Rev. B 29, 6165 (1984)
14 L.-D. Chang and S. Chakravarty: Phys. Rev. B 31, 154 (1985)
15 H. Grabert, S. Linkwitz, S. Dattagupta and U. Weiss, Europhysics Letters, in press
16 J. Kondo, Physica 124B, 25 (1984)
17 C.C. Yu and A.V. Granato: Phys. Rev. B 32, 4793 (1985)
18 H. Teichler: phys. stat. sol (b) 104, 239 (1981)
19 D. Emin: Adv. Physics. 22, 57 (1973)
20 K. Kehr: Report JUL-211, Jülich (1975)
21 H. Dosch and J. Peisl: Phys. Rev. Letters 56, 1385 (1986)
22 H. Teichler: to be published
23 J.R. Schrieffer: Theory of Superconducting (Benjamin, New York 1964)
24 H. Wipf: personal communication
25 L. F. Mattheis: Phys. Rev. B 1. 373 (1970)
26 V.K. Tewary: J. Phys. F 3, 1515 (1973)
27 J. Peisl, H. Dosch and A. v. Schwerin: to be published

Debye-Waller Peculiarity in Phonon-Assisted Tunneling

M. Wagner

Drittes Institut für Theoretische Physik, Universität Stuttgart,
Pfaffenwaldring 57, D-7000 Stuttgart 80, Fed. Rep. of Germany

The Fulton-Gouterman transcription (=parity ordering) of the archetype Hamiltonian of phonon-assisted tunneling reveals a reflective property, by means of which the Debye-Waller peculiarity is established as a topological phenomenon. By means of a reflective unitary transformation it is shown that this phenomenon establishes a logarithmic suppression of localization for an "Ohmic dissipation" coupling law. Further consequences ("mode mixing") for quantum diffusion are also indicated.

1. Introduction

In our work we have followed up two lines of investigation, which mutually supplement each other and both of which emphasize *mode mixing* in the "phonon bath". Each of these two approaches is guided by a separate physical phenomenon: the Fano phenomenon on the one hand, and the "Debye-Waller peculiarity" on the other. In this presentation we will describe the latter.

The Debye-Waller phenomenon refers to the astonishing peculiarity of phonon-assisted transport setups that all results of first order in the tunneling parameter Δ display an overlap screening $\sim \exp[-A\alpha]$ ($\sqrt{\alpha}$=phonon coupling constant), whereas all 2nd order effects have no overlap screening but only one in proportion to $(1/\alpha)$. It is for this reason that in the strong coupling situation we may expect 2nd order influences onto the transmission dynamics to override those of 1st order. We consider the archetype model for phonon-assisted quantum diffusion, in which a tunneling mode (pseudo spin) is coupled to a quasi-continuous background of oscillatory modes, which are not mutually coupled. It is represented by the Hamiltonian

$$H = \Delta\sigma_z + \frac{1}{2}\sum_{\varkappa}\Omega(\varkappa)\left[P(\varkappa)^2 + Q(\varkappa)^2 + \sigma_x D(\varkappa)Q(\varkappa)\right] \tag{1}$$

which has been discussed by many workers in the fields of polaron motion, of particle diffusion, of tunneling centers and of exciton transfer. $Q(\varkappa)$, $P(\varkappa)$ are the oscillatory coordinates and momenta respectively, $[Q(\varkappa), P(\varkappa')] = i\delta(\varkappa,\varkappa')$ and a pseudospin description is used for the "left-right" tunneling Hamiltonian

$$\begin{aligned} \sigma_z &= -[|\ell\rangle\langle r| + |r\rangle\langle \ell|]/2, \\ \sigma_x &= [|\ell\rangle\langle \ell| - |r\rangle\langle r|]/2, \text{ etc.} \end{aligned} \tag{2}$$

2. Fulton-Gouterman Transformation

It has been first noted by Fulton and Gouterman /1/ that Hamiltonian (1) displays a reflection symmetry which establishes a *rigorous parity ordering* of the wavefunctions,

$$(3)\qquad |\Phi\rangle = |\ell\rangle\, \varphi(Q) + p|r\rangle\, G\varphi(Q), \qquad p=\pm 1$$

where $G=G(P, Q)$ is a reflection operator, $GQ(\varkappa)=-Q(\varkappa)G$, $G\varphi(Q)=\varphi(-Q)$ and it is Hermitian as well as unitary. More formally, the parity ordering can be introduced via the FG-transformation. For background information we refer to the original work /1/ and to a recent book /2/. The vibrational wavefunctions $\varphi(Q)$ satisfy the Fulton-Gouterman equation

$$(4)\quad H_{FG}\varphi(Q)\equiv\{\tfrac{1}{2}\sum_{\varkappa}\Omega(\varkappa)[P(\varkappa)^2+Q(\varkappa)^2+D(\varkappa)Q(\varkappa)]-\tfrac{1}{2}p\Delta G\}\varphi(Q)=E\varphi(Q).$$

We emphasize that this transcription is exact. It is the reflective term of this equation which most clearly elucidates the origin of the Debye-Waller peculiarity: Suppose we take a displaced oscillatory wavefunction $\chi_\mu(Q+D/2)$ as a starting choice for $\varphi(Q)$. Then G immediately creates a mirror image addition $\chi_\mu(-Q+D/2)$ with weight $(1/\alpha)$, which exhibits no exponential screening $\sim\exp[-A\alpha]$. This is illustrated in Fig. 1. $\sqrt{\alpha}$ has been used as a measure for the particle-phonon coupling strength (see below). This peculiar reflection also necessitates a persistent "nondiagonality" of oscillatory wavefunctions,

$$(5)\qquad \Gamma_o = \sum_{\mu(\neq 0)} |\langle\chi_o(Q+D/2)|H_{FG}|\chi_\mu(Q+D/2)\rangle|^2$$

$$= \frac{1}{4}\Delta^2\,[1-\exp[-\frac{1}{2}\sum_{\varkappa} D(\varkappa)^2]]\,.$$

Thus, with increasing displacements $D(\varkappa)$ the nondiagonality increases and is maximal in the case of "infrared divergency". In this sense eq. (5) can be considered as a kind of sum rule against localization.

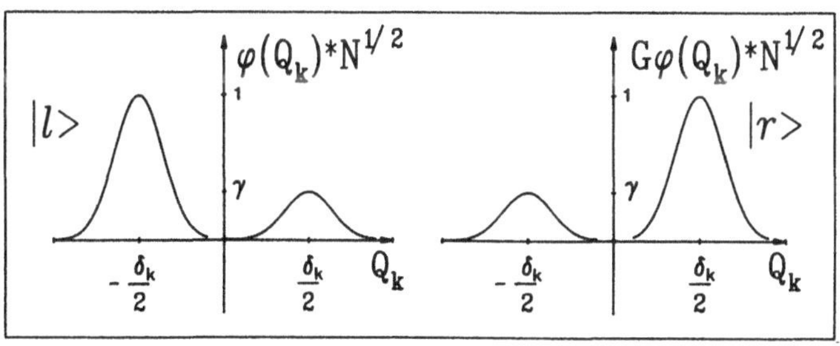

Fig. 1. Vibrational parts of the wavefunction (N=normalization constant). γ is of order Δ/α.

3. Reflective Projection

To avoid complications due to the infrared divergency of the quantities $D(\varkappa)$ we introduce the test-Hamiltonian

$$(6)\qquad \tilde{H} = \tilde{H}^{(0)} - \tfrac{1}{2}p\Delta G,\qquad \tilde{H}^{(0)} = \frac{1}{2}\sum_{\varkappa}\Omega(\varkappa)[P(\varkappa)^2 + Q(\varkappa)^2 + \delta(\varkappa)Q(\varkappa)]$$

where the quantities $\delta(\varkappa)$ are chosen to be

$$(7)\qquad \delta(\varkappa) = \frac{\Omega(\varkappa)D(\varkappa)}{\Omega(\varkappa)+\lambda\Omega_D}$$

and for simplicity we consider the coupling setup ("Ohmic dissipation")

(8) $D(\Omega)^2\rho(\Omega) = 4\alpha\Omega_D^{-1}\ (\Omega/\Omega_D)^{-1}, \quad \Omega < \Omega_D$.

We define a displaced oscillatory zero-order base

(9) $\tilde{H}^{(0)}\ \chi_\mu(q;\lambda) = \tilde{E}_\mu^{(0)}(\lambda)\chi_\mu(q;\lambda), \quad q(\varkappa) = Q(\varkappa) + \delta(\varkappa)/2$.

The most elegant way to diagonalize $\tilde{H}$ is by means of a unitary transformation /2/, which embodies the required reflective feature. We thus define an anti-Hermitian operator

(10) $S = G\ |u(q;\lambda)\rangle\langle\chi_o(q;\lambda)| - |\chi_o(q;\lambda)\rangle\langle u(q;\lambda)|G$

and require the groundstate wavefunction to be of the form

(11) $|\tilde{\varphi}_o\rangle = e^S|\chi_o(q;\lambda)\rangle$

$=[\cos a - \frac{1}{a}\langle u|G\chi_o\rangle \sin a]|\chi_o(q;\lambda)\rangle + \frac{1}{a}\sin a\ G|u(q;\lambda)\rangle$ where

(12) $a = [\langle u|u\rangle - \langle u|G|\chi_o\rangle^2]^{1/2}$.

Inserting expr. (11) in the Schrödinger equation we find

(13) $[\tilde{H}^{(0)} - \tilde{E}_0(\lambda) - \sum_\varkappa \Omega(\varkappa)\ \delta(\varkappa)\ q(\varkappa) + M(\lambda)]|u(q;\lambda)\rangle$

$= [a \operatorname{ctg} a - \langle u|G\chi_o\rangle]\ \{\frac{1}{2}p\Delta + (\tilde{E}_0 - \tilde{E}_0^{(0)})\ G\}|\chi_o\rangle + \frac{1}{2}p\Delta G|u\rangle$

where

(14) $M(\lambda) = \frac{1}{2}\sum_\varkappa \Omega(\varkappa)\ \delta(\varkappa)^2 = 2\alpha\Omega_D\ [1 - 2\lambda\ \ell n\frac{1+\lambda}{\lambda} + \frac{\lambda}{1+\lambda}]$.

We also note the overlap matrix element

(15) $L(\lambda) = \langle\chi_o|G|\chi_o\rangle = (\frac{\lambda}{1+\lambda})^\alpha \exp(\frac{1}{1+\lambda})$.

In the strong coupling case ($\alpha \gg 1$) the dominant term on the lhs of (13) will be the $M(\lambda)$-term, whereas on the rhs the first term will be the largest, whence in leading α-order we have

(16) $|u(q;\lambda)\rangle = \frac{1}{2}p\Delta|\chi_o(q;\lambda)\rangle/M(\lambda) + O(\frac{1}{\alpha^2})$.

This suggests the projection

(17) $|u(q,\lambda)\rangle = \gamma_o|\chi_o(q;\lambda)\rangle + \sum_\varkappa \gamma_\varkappa\ |\chi_\varkappa(q;\lambda)\rangle + \ldots$.

If this projection is inserted in eq. (13), one is left with a hierarchical set, which allows the calculation of the coefficients γ_o, $\gamma_\varkappa, \ldots$, yielding a value of order $1/\alpha$ for the zero-phonon coefficient γ_o, of order $1/\alpha^2$ for the one-phonon coefficients $\gamma_\varkappa$, etc. Suppose now that we have calculated $|u(q;\lambda=0)\rangle$ exactly. Then the leading deviation of eq. (13) from the $\lambda=0$ limit turns out to be of the form $\lambda\ \ell n\ \lambda$ for $\alpha>1$, which originates from the $M(\lambda)$-term, eq. (14),

(18) $[\tilde{H}^{(0)} - \tilde{E}_0 - \sum_\varkappa \Omega(\varkappa)\ \delta(\varkappa)\ q(\varkappa) + M]_{\lambda=0}\ \frac{\partial}{\partial\lambda}|u(q,\lambda)\rangle$

$= -4\alpha\Omega_D\ \ell n\ \lambda|u(q,0)\rangle + O(\lambda^0)$.

Thus the structural form of $u(q,\lambda)$ must be given as ($\alpha > 1$):

$$(19) \quad |u(q,\lambda)\rangle = |u(q,0)\rangle - \lambda\, \ell n\, \lambda |u^{(1)}(q,0)\rangle + 0(\lambda)$$

and we note that this result is of an *exact nature.* It is this result which necessitates a $\lambda\, \ell n\, \lambda$-term in the energy expectation value, and which eventually prevents localization.

4. Groundstate Expectation Value

Having found the exact limiting behaviour of $u(q;\lambda)$ for $\lambda \to 0$, it is straightforward to insert the projection (17) in eq. (13) and to evaluate the coefficients in an iterative manner. In this way the groundstate energy $\tilde{E}_o$ of the test-Hamiltonian may be found, which, however, still does not refer to the true FG-Hamiltonian (see expr. (4)). For the latter's groundstate we employ the Bogoliubov inequality

$$(20) \quad E_o < \tilde{E}_o + \langle \tilde{\varphi}_o | H_{FG} - \tilde{H} | \tilde{\varphi}_o \rangle$$

which yields

$$(21) \quad \begin{aligned} E_o < {} & \frac{1}{2} \sum_{\varkappa} \Omega(\varkappa) - \frac{1}{2}\alpha\Omega_D - \frac{1}{2} p \Delta \lambda^{\alpha} \\ & - \frac{\Delta^2}{8\alpha\Omega_D} [1 - \frac{1}{4\alpha} + 0(\frac{1}{\alpha^2})] - \frac{\Delta^2}{8\alpha\Omega_D} \lambda\, \ell n\, \frac{1}{\lambda} [1 - \frac{1}{2\alpha} + 0(\frac{1}{\alpha^2})] \\ & + 0(\lambda) + 0(\frac{\Delta^3}{\Omega_D^2}). \end{aligned}$$

The most conspicuous consequence of this result is the fact that $\lambda = 0$ does not yield the lowest energy value. The expression even displays a logarithmic descent near the limit $\lambda = 0$. This establishes a *suppression of localization* in the strict mathematical sense, since for $\lambda \neq 0$ the two groundstates of opposite parity are not degenerate. Hence a symmetry-broken ("localized") state is not stable at T=0. The behaviour of the groundstate in the neighbourhood of $\lambda = 0$ is drawn in Fig. 2.

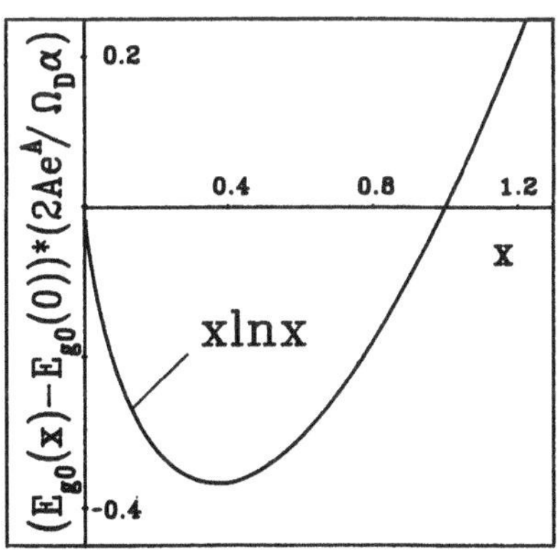

Fig. 2. Groundstate Dependence on the Localization Parameter λ. ($\lambda = 0$: "localization"). Coupling law: "Ohmic dissipation". Notation: $x = \lambda \exp A$, $A = 2(\alpha\Omega_D/\Delta)^2$, $(\Delta/\Omega_D) = 0.2$, $\alpha = 2.0$.

5. Perspectives

a) Although the present study shows that there cannot be localization in the strict mathematical sense, the tunneling experiences a modified exponential suppression by means of which in the strong coupling regime the particle mobility turns very small.

b) Therefore other consequences of the Debye-Waller peculiarity turn out to be of greater physical significance than the statement about localization. The most important of these is the non-exponentially screened nature of certain expectation values like $\langle\sigma_z\rangle$, etc., which are easily evaluated by use of the presented mirror-image projection device. It is the expectation value $\langle\sigma_z\rangle$ which also makes its appearance in a Green function approach /3, 4/ for phonon-assisted tunneling, if a Hartree-Fock factorization is introduced, and it may be evaluated there via HF-selfconsistency. Since both ways of evaluation yield the same result, the Debye-Waller peculiarity provides the justification for the HF-Green-function procedure /3, 4/.

c) This being achieved, the tunneling is traced back to the Ullersma decay integral, which completely pins down the time and temperature behaviour. One prominent result of this calculation is the general longtime behaviour $\simeq t^{-1}(\ln t)^{-2}\sin \Omega_D t$ of the correlation function $\langle\sigma_x(0)\sigma_x(t)\rangle$, which is valid for all power-law coupling forms and for all temperatures. Other results are given in refs. /3, 4/.

d) It is illuminating to have yet another view into the topological peculiarity of the FG-Hamiltonian (4). If a multi-oscillatory wavefunction is supplemented by a mirror-image contribution (see Fig. 1), this amounts to a *mode mixing* in the phonon bath. In this manner the Debye-Waller peculiarity may be viewed as a mode-mixing phenomenon. Correspondingly, any Fano-type of decay /3, 4/ involves a mode mixing ("antiresonances") in the bath. Thus, both our approaches emphasize the feedback influence of the disturbance created by the tunneling mode in the phonon bath onto the tunneling mode itself.

References

1. R. L. Fulton and Gouterman, J. Chem Phys. 35, 1059 (1961)
2. M. Wagner, Unitary Transformations in Solid State Physics, North-Holland, Amsterdam 1986
3. M. Wagner and J. Vâzquez-Mârquez, Solid State Comm., in print
4. M. Wagner and J. Vâzquez-Mârquez, J. Phys. C, in print

Quantum Diffusion Above $\Theta_{Debye}/2$

A. Klamt

Max-Planck-Institut für Metallforschung, Institut für Physik, Postfach 80 06 65, D-7000 Stuttgart 80, Fed. Rep. of Germany

Two different approaches to the quantum theory of diffusion, the perturbative approach of Flynn and Stoneham and the occurrence-probability ansatz of Holstein, are compared in their common range of validity $T \geq \Theta_{Debye}/2$. It will be shown that they give the same results for small transfer integrals J. For large J the perturbative approach breaks down while the occurrence-probability ansatz still holds and the corresponding jump rate obeys an Arrhenius law in agreement with experimental observations.

1. Introduction

During the last 20 years large efforts have been made in the field of the quantum theory of diffusion to describe the diffusive motion of a light interstitial particle in a lattice of heavier atom cores. Especially the temperature and mass dependence of the diffusivity of the light interstitial are of great interest. In the low-temperature regime the problem seems to be far from a final solution since several different temperature laws have been proposed. The experimental observations are not clear enough to decide unambiguously between the different theoretical results. On the other hand, for temperatures above half of the Debye-temperature Θ_{Debye} the situation cleared up during the last years. Here a consistent picture of the diffusion process can be drawn capable to explain the various experimental observations on the diffusivity of hydrogen and its isotopes in metals. This picture is based on Holstein's occurrence-probability ansatz [1], which was applied to hydrogen in metals for the first time by Emin, Baskes and Wilson [2]. They showed that the experimental data for the diffusion of H, D, and T in bcc metals can be understood if the various parameters in a generalized occurrence-probability ansatz are chosen properly. Recently their arguments have been confirmed by Klamt and Teichler [3], who calculated the diffusion coefficients of hydrogen and its isotopes in Nb and Ta, deriving the diffusion parameters in the occurrence-probability ansatz from a phenomenological model potential which had been adjusted to diffusion-independent experiments.

In this article a short review of the main concepts of the occurrence-probability ansatz will be given. Parallels and discrepancies with the well-known perturbative approach to the quantum theory of diffusion [1,4,5] are investigated and the adiabatic limit is worked out which is responsible for the Arrhenius-type diffusion law observed at high temperatures for all hydrogen-metal systems investigated so far.

2. Model

For the sake of lucidity and shortness we restrict ourselves to a simple model. Consider a linear chain of atoms with mass M bound harmonically to a nearly stiff background with a spring constant $M\omega^2$. The elastic coupling between the atoms is assumed to be small but non-vanishing to guarantee energy dissipation in the lattice. Between every two atoms i and i+1 there are localized states Φ_i of the interstitial particle. The energy of a state Φ_i depends linearly on the displacements Q_i and Q_{i+1} of the neighbouring atoms with the force constant A. Thus a particle on site i traps itself by displacing the neighbouring atoms by $\delta = A/M\omega^2$. The corresponding self-trapping energy is $E_{ST} = A^2/M\omega^2$. The minimum energy needed to produce a coincidence between a self-trapped particle state and its neighbouring state is $E_a = E_{ST}/4$.

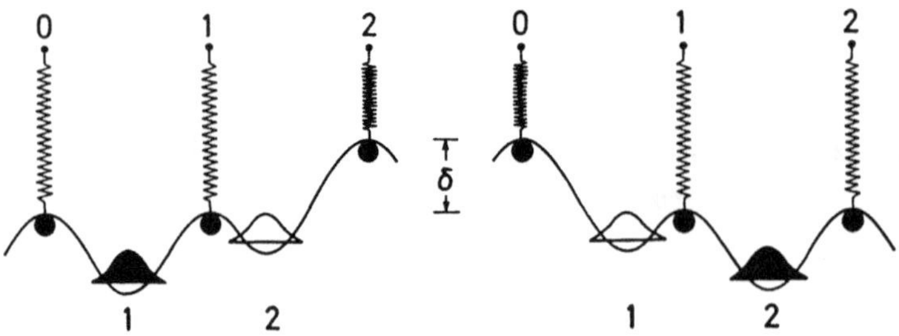

Fig. 1: Initial and final self-trapped states

The diffusion coefficient of the particle is the product of a well-known geometry factor and the jump rate ν from an initial site i to an adjacent site f; therefore the whole work is the calculation of this jump rate. Figure 1 shows the initial and the final self-trapped configurations of the system. These equilibrium configurations are connected by the transport mode $x = Q_2 - Q_0$ which is most important for the transition process. A second important mode is the symmetric mode $y = Q_1 - \delta$ which influences the barrier-height between the sites. Therefore the transfer integral J between the two states is a function of y. Although the Condon approximation of constant J has been used frequently in the quantum theory of diffusion, for realistic calculations a configuration-dependence J(y) has to be taken into account, which will be assumed to be of the form $J(y) = J_0 \exp[y/y_0]$ throughout this paper.

3. The Perturbative Approach

In this paragraph a short survey of the perturbative approach will be given. This has been used by Flynn and Stoneham [4] in their fundamental work of the QTD following the concepts of Holstein's small-polaron theory [1]. The total wavefunction of the system is written as a product of a localized particle wavefunction Φ_i and a lattice wavefunction $|[n_q]>_i$ where the n_q are the occupation numbers of the displaced phonon system. The displacement of the phonon system is given by the self-trapping distortion of the particle on site i. For small values of the transfer integral J the transition rate ν may be calculated with time-dependent perturbation theory yielding the golden-rule expression

$$\nu = \frac{2\pi}{\hbar} \left\{ \sum_{[n_q']} |_i <[n_q]| <\Phi_i| a_i^+ J a_f |\Phi_f> |[n_q']>_f |^2 \times \delta(\sum_q \hbar\omega_q (n_q - n_q')) \right\}_{Av([n_q])} \tag{1}$$

for constant J. For $T > \Theta_{Debye}/2$ this simplifies to

$$\nu \approx \frac{J^2}{2\hbar} \left[\frac{\pi}{E_a kT} \right]^{\frac{1}{2}} \exp\left\{ - \frac{E_a}{kT} \right\} . \tag{2}$$

This formula is frequently called Flynn-Stoneham formula. It deviates from an Arrhenius law by the factor $T^{-1/2}$ in the pre-exponential factor. The perturbative approach later has been extended beyond the Condon approximation [6,7,8]. For $J(y) = J_0 \exp[y/y_0]$ one finds the expression

$$\nu \approx \frac{J_0^2}{2\hbar} \left[\frac{\pi}{E_a kT} \right]^{\frac{1}{2}} \exp\left\{ - \frac{E_a}{kT} + \frac{kT}{E_b} \right\} \tag{3}$$

with $E_b = \frac{1}{2} M\omega^2 y_0^2$. Eq. (3) looks very similar to (2), but the temperature-dependence is changed due to the second term in the exponent which makes the jump rate diverging in the limit $T \to \infty$. This divergence is an artefact of the theory, since it is due to large transfer integrals which become increasingly important at high temperatures but which are out of the range of validity of the perturbative approach.

4. The Occurrence-Probability Ansatz

In his fundamental work on small polarons [1] Holstein proposed a second way to derive the Flynn-Stoneham formula, the occurrence-probability ansatz. In contrast to the perturbative approach here the lattice modes are treated by classical mechanics. Therefore the occurrence-probability ansatz is restricted to $T>\Theta_{Debye}/2$. The basic idea is that a jump of the particle from an initial state Φ_i to an adjacent state Φ_f can take place only when a coincidence of the two states is produced temporary by thermal fluctuations of the transport mode x. A coincidence crossing event is illustrated in fig. 2. According to Landau [12], Zener [11] and Morse and Stückelberg [10] the probability P for the particle to stay in the initial state during a coincidence crossing with velocity v is

$$P(H) = \exp\,[-H/2]\ , \tag{4}$$

where the transition parameter H is defined as

$$H = \frac{2\sqrt{2}\pi}{\hbar A}\ \frac{J^2}{v}\ . \tag{5}$$

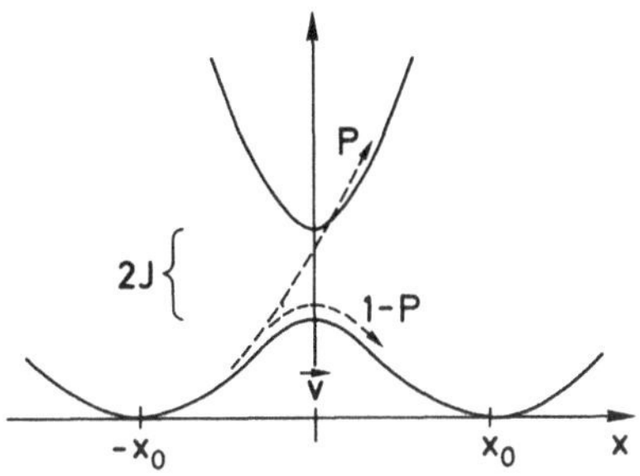

Fig. 2: Coincidence crossing event in a configuration potential diagram: Broken arrows indicate the alternative possibilities of staying on the initial site or passing over to the final site.

When the probabilities of all possible ways are summed the total probability W(H) to end at the final equilibrium state after a first coincidence crossing becomes

$$W(H) = (1 - P(H))\ /\ (1 - P(H)/2)\ . \tag{6}$$

For $H \ll 1$ W(H) is proportional to H, while for $H \gg 1$ it tends to 1. The region $H \gg 1$ is called the adiabatic region, since here the particle adjusts adiabatically to the lattice fluctuation throughout the coincidence crossing.

Since W(H) is a function of the crossing velocity v the jumprate ν has to be calculated as a thermal average:

$$\nu = \int_0^\infty R_c(v)\ W(H(v))\ dv\ . \tag{7}$$

The coincidence crossing rate $R_c(v)$ can easily be calculated using classical Boltzmann statistics:

$$R_c(v) = v\ \frac{M\omega}{2\pi kT} \exp\left\{-\left[\frac{M}{2}\,v^2 + E_a\right]/kT\right\}. \tag{8}$$

Using the exact expression for W(H), (7) has to be evaluated by numerical means. This yields approximately

$$\nu \approx \frac{\omega}{2\pi}\left[1 - \exp\left\{-\frac{J^2}{\hbar\omega}\left[\frac{\pi^3}{E_a kT}\right]^{\frac{1}{2}}\right\}\right] \exp\left\{-E_a/kT\right\}. \tag{9}$$

Eq. 9 becomes exact in the limits $J \ll J_c(T)$ and $J \gg J_c(T)$ with

$$J_c(T) = \left[\frac{\hbar^2\omega^2 E_a kT}{\pi^3}\right]^{\frac{1}{4}} = \frac{1}{2}\left[\hbar^3\omega^3 E_a\right]^{\frac{1}{4}} . \tag{10}$$

In the non-adiabatic limit $J << J_c(T)$ (9) is identical with the Flynn-Stoneham formula (2), while in the adiabatic limit $J >> J_c(T)$ it simplifies to

$$\nu = \frac{\omega}{2\pi} \exp[- E_a/kT] . \tag{11}$$

Since this just is the total coincidence crossing rate, (11) expresses the fact that in the adiabatic limit each coincidence event leads to a site jump.

The occurrence-probability ansatz can easily be generalized beyond the Condon approximation by performing an additional thermal average with respect to the symmetric mode y. In the non-adiabatic limit ($J << J_c(T)$, $kT << E_b$) the result again agrees with result (3) of the perturbative approach while in the adiabatic limit ($J >> J_c(T)$, $kT << E_b$) (11) still holds. For strong configuration dependence of the transfer integral or at high temperatures with $kT >> E_b$ the jumprate becomes half the coincidence crossing rate (11) since now the transition probability is nearly 1 for one half and 0 for the other half of the coincidence events.

With respect to their good agreement in their common range of validity, we may conclude that above $\Theta_{Debye}/2$ the occurrence-probability ansatz provides a more physical picture of the some transition mechanisms as described by the perturbative approach. Furthermore, the occurrence-probability ansatz allows a continuation of the results to large transfer integrals.

5. Comparison with Experimental Data

The experimental data on the diffusion of hydrogen and its isotopes in metals [9] show some striking common features:

(i) At high temperatures the diffusion coefficients obey Arrhenius laws.

(ii) Independent of the isotope mass the pre-exponential factors are in the range of 0.3 - 1.1 ν_{Debye} (see Table 1).

(iii) The activation energies show an isotope effect.

(iv) For H in the bcc metals Nb and Ta a clear bend in the Arrhenius plots of the diffusion coefficients can be observed at about room temperature.

Frequently the high-temperature Arrhenius-law is misinterpreted as a classical diffusion process. Apart from the fact that the hydrogen nuclei should not behave classically due to their large level spacings, a classical rate theory yields pre-exponential factors proportional to $m^{-1/2}$ and mass-independent activation energies, in clear disagreement with the experimental findings.

Not even within the perturbative approach can the experimental data be explained. The only Arrhenius law derived within the perturbative approach has been proposed by Flynn and Stoneham [4] assuming a configuration-dependent transfer integral of the form $J(y) = J_c\,\Theta(y-y_c)$, where $\Theta(x)$ is the unit step function. This configuration dependence seems to be quite artificial. Furthermore, their result

$$\nu = \frac{J_c^2}{8\hbar} (E_a E_c)^{-1/2} \exp[- (E_a + E_c)/kT] , \tag{12}$$

with $E_c = \frac{1}{2} M\omega^2 y_c^2$ cannot explain that the pre-exponential factor is about ν_{Debye} in all investigated metals.

In contrast, the experimental data can be understood consistently within the OPA as has been shown by Emin et al. [2]. Since with increasing temperature excited particle states with large

Table 1: Pre-exponential factors ν_0 for some hydrogen-metal systems. For bcc metals jumps between tetrahedral sites and in fcc metals jumps between octahedral sites are assumed. Values in brackets correspond to alternating jumps between octahedral and tetrahedral sites

metal	structure	isotope	D_0 [10^{-8} m^2/s]*	ν_0 [THz]	ν_{Debye} [THz]**
Nb	bcc	H/D/T	5.0 / 5.2 / 4.5	5.5 / 5.7 / 5.0	5.7
Ta	bcc	H/D	4.4 / 4.6	4.9 / 5.1	5.0
V	bcc	H/D	3.1 / 3.8	4.1 / 5.0	7.9
α-Fe	bcc	H	≈7.5	≈11.0	9.8
Pd	fcc	H	29.0	1.9 (3.8)	5.7
Ni	fcc	H	6.9	5.6 (11.2)	9.4

*) from [9]; **) from [13]

transfer integrals become more and more important, the diffusion process is dominated by adiabatic diffusion between these states at sufficiently high temperatures. Thus at high temperatures an Arrhenius law is expected where the pre-exponential factor is the frequency of the transport mode, which in metals is slightly less than ν_{Debye}. Since the excited states of different isotopes differ in energy, the activation energies depend on the isotope mass. Finally below the bend at room temperature the diffusion of H in Nb and Ta is dominated by non-adiabatic diffusion between ground states. This qualitative interpretations recently have been confirmed by detailed quantitiative calculations on H and D in Nb and Ta by Klamt and Teichler [3].

References

1. T. Holstein: Ann. Phys. (N.Y.) 8, 343 (1959)
2. D. Emin, M.J. Baskes, D. Wilson: Phys. Rev. Lett. 42, 791 (1979)
3. A. Klamt, H. Teichler: phys. stat. sol. (b) 134, 533 (1986)
4. C.P. Flynn, A.M. Stoneham: Phys. Rev. B 1, 3966 (1970)
5. Y. Kagan, M.I. Klinger: J. Phys. C 7, 2791 (1974)
6. Y. Kagan, M.I. Klinger: Sov. Phys. JETP 43, 132 (1976)
7. H. Teichler: phys. stat. sol. (b) 104, 239 (1981)
8. C. Schmidt: Thesis, Universität Stuttgart (1984)
9. J. Völkl, G. Alefeld: Hydrogen in Metals I, G. Alefeld and J. Völkl, eds. (Springer, Berlin 1978) p. 321
10. P. Morse, E.C.G. Stückelberg: Ann. der Physik 9, 579 (1931)
11. C. Zener: Proc. Roy. Soc. A137, 696 (1932)
12. L.R. Landau: Phys. Z. U.S.S.R. 1, 426 (1932)
13. C. Kittel: Einführung in die Festkörperphysik (Oldenbourg, München 1980) p. 159

Hydrogen Diffusion by Tunneling in Metals

I. Svare

Physics Department, University of Trondheim NTH,
N-7034 Trondheim, Norway

Most theories for hydrogen diffusion by tunnelling in metals are based upon the small polaron concept. It is assumed that the self-trapping energy of the hydrogens must be overcome with phonons before the tunnelling motion can occur [1 - 4]. However, these theories are complicated to use and they have not had striking success in fitting the diffusion data.

Recently [5] we have gone back to the old and simple formula for the phonon-assisted tunnelling rate via an excited hydrogen oscillation state E_n with tunnel splitting ΔE_{Tn}

$$\tau_{Tn}^{-1} = (\Delta E_{Tn}/\hbar)\exp(-E_n/k_BT) , \quad (1)$$

and with a one-dimensional sinusoidal potential applied it successfully to diffusion and internal friction (IF) in β-PdH_x with fcc structure. By choosing only one parameter, the height V_0 of the potential, we can easily calculate E_n, ΔE_{Tn} and τ_{Tn} in this model that does not explicitly consider the self-trapping, and compare with experimental data.

The potential is sketched in Fig.1. Some asymmetry A between the wells will combine with the tunnelling to split the states E_n by $\Delta E_n = [(\Delta E_{Tn})^2 + A^2]^{\frac{1}{2}}$. The ground states are more or less localised by A to left ψ_{0L} or right ψ_{0R}, while excited states with $\Delta E_{Tn} > A$ are delocalised as the ψ_{nS} and ψ_{nA} linear combinations of ψ_{nL} and ψ_{nR}. ψ_{nS} and ψ_{nA} interfere to give tunnelling motion as discussed in textbooks, and hence (1). We use the textbook interpretation to argue [6] that the self-trapping energy is constant during the interference since the self-trapped ψ_{nL} and ψ_{nR} have the same energy in a perfect lattice. Thus the self-trapping does not directly influence the tunnelling. A major theoretical complication in the treatment of tunnelling is removed if this interpretation is correct, and the proof will lie in the data fitting.

The asymmetry A is not caused by self-trapping but by lattice strain and binding effects near impurities and lattice defects. Increasing A permits tunnelling only in higher states E_n with larger ΔE_{Tn}, and for large A (1) changes into classical hopping over the barrier E_a with pre-exponential factor $\approx 10^{14}\ s^{-1}$.

The importance of the asymmetry is seen in β-PdH_x [5] where cold-work increases A and the magnitude of the H IF peak that lies on the E_a = 315 meV Arrhenius line through the high-temperature diffusion data. It corresponds to V_0 = 4000 K (in temperature equivalents) which gives E_1 = 56 meV and 4 ΔE_{T0} = 30 kHz, in excellent agreement with INS and NMR observations. In well-annealed PdH_x a lower temperature IF peak is most prominent and from (1) corresponds to tunnelling in E_2.

When we try to use (1) for hydrogen diffusion in the bcc metals V, Nb and Ta, we meet the difficulty that the apparent diffusion activation energy E_a^* is smaller than the known E_1 for H in V. For H in Nb and Ta we have this problem below room temperature. We solve this dilemma by assuming that in addition to the real excitations to E_n that give (1), we have virtual excitations to ψ_1 that give two-

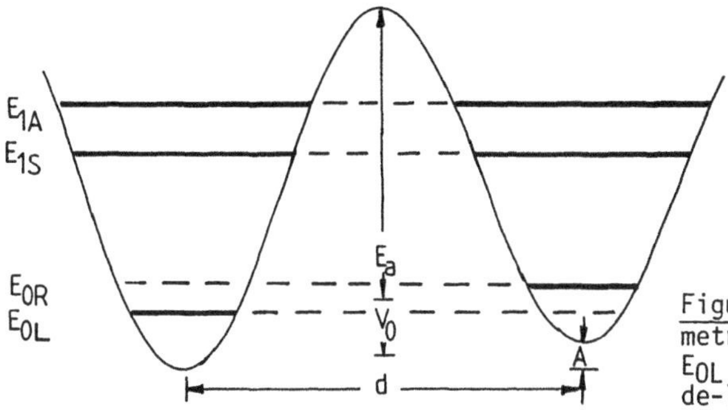

Figure 1. Sketch of potential. Asymmetry A localises ground states to E_{0L} and E_{0R}, while excited states are de-localised as E_{1S} and E_{1A}.

phonon tunnelling transitions with the approximate temperature dependence [7]

$$\tau_2^{-1}(T) \propto \int_0^{\omega_D} \frac{\omega^6 \exp(\hbar\omega/k_BT)\,d\omega}{[\exp(\hbar\omega/k_BT)-1]^2} \propto \begin{cases} T^2 & \text{for } T \gg \theta_D \\ T^7 & \text{for } T \ll \theta_D \end{cases} \tag{2}$$

Here $\omega_D = k_B\theta_D/\hbar$ is the Debye frequency of the lattice vibrations. It is easy to see by numerical integration that (2) will fit the diffusion data [8] for H in V with θ_D = 420 K just as well as an arbitrary Arrhenius line will. For D and T in V, and for H, D and T in Nb and Ta, we must also assume a contribution from (1) to fit the diffusion.

Adapting [7] we have calculated an improved version of (2) for transitions from the lowest of the asymmetry-split ground states by using the pair of excited states ψ_{1S} and ψ_{1A} as virtual intermediates, and find [6]

$$\tau_2^{-1} = \frac{3(\Delta E_{T1})^2 H_{01}^4}{32\pi^3\rho^2 v^{10}}$$

$$\times \int_{A/\hbar}^{\omega_D} \frac{\omega^3(\omega-A/\hbar)^3 \exp[(\hbar\omega-A)/k_BT]\,d\omega}{[E_1+A/2-\hbar\omega]^4[\exp(\hbar\omega/k_BT)-1][\exp((\hbar\omega-A)/k_BT)-1]} \tag{3}$$

Here ρ is the density of the metal, v is the velocity of sound, and H_{01} is the deformation potential for excitations to ψ_1. In fitting to diffusion data we take H_{01} as an adjustable parameter that turns out to have the same order of magnitude as V_0, as expected [6]. We see that the two-phonon rate goes to zero for $A \gtrsim \hbar\omega_D$. Eq.(3) is a crude approximation since dispersion is neglected and the phonon spectrum differs greatly from the Debye assumption.

In a sinusoidal potential of height V_0 the oscillation energies are $E_n = a_n - a_0$ where [9,5]

$$a_n = \frac{V_0}{k^2}\left\{(2n+1)k - \frac{1}{4}\left[n^2+(n+1)^2\right] - \frac{1}{16k}\left[n^3+(n+1)^3\right]\right\} + \dots \tag{4}$$

Here $k = (d/\pi\hbar)(2mV_0)^{\frac{1}{2}}$, d is the distance between the tetrahedral interstitial sites, and m is the hydrogen mass. The tunnel splitting due to interaction with one neighbouring well is

$$\Delta E_{Tn} = 8V_0 \frac{2^{3n} k^{(n-\frac{1}{2})} e^{-2k}}{\pi^{\frac{1}{2}} n!} \left(1 + \frac{b_n}{k} + \frac{c_n}{k^2}\right) \tag{5}$$

The correction coefficients b_n and c_n are tabulated in [9]. In the spirit of the one-dimensional model we consider only one excited state for each n .

To fit the free diffusion of hydrogen in small concentrations in pure V, Nb and Ta we multiply (1) and (3) with the four n.n. sites. Below 200 K the H will condense to ordered hydrides with little motion, unless each H is bound to an impurity atom O, N or other. In impure metals we thus expect two kind of H motion below 200 K : i) Tunnelling (3) permitted by small A between equivalent sites around the impurity, and there are probably two such n.n. sites. ii) Classical hopping over the barrier $E_a \approx V_0 - a_0$ away from the impurity since the large A ≈ Binding energy ≈ 1000 K [10] blocks the tunnelling in this direction. We use the same temperature-independent V_0 of order 3000 K and the same d for calculating all H motions in each metal, although A ≈ 1000 K occurring in one step would add ≈ 500 K to V_0.

In Fig.2 we show the fit for Nb where d = 1.166 Å and θ_D = 280 K. The diffusion data [8] can be fitted by adding a suitable amount of τ_2^{-1} from (3) with A ≈ 0 , to $4\tau_{T1}^{-1}$ calculated from (1) with V_0 = 3000 K. This V_0 gives E_1 = 108 meV which coincides with the lowest INS peak [11]. The coincidence is encouraging, but the real wavefunction in the tetrahedral site [12] is of course much more complicated than in our one-dimensional model. We calculate from this V_0 E_a = 2240 K which gives hopping rates close to the IF peaks for H at O and N impurities [13]. Some "leaking" through the barrier in spite of the large A give IF peak deviations from E_a at lower temperatures. We also find $2\Delta E_{T0}$ = 3.0 K, close to the 2.3 K splitting measured with INS [14] and specific heat [15]. These measurements also reflect the distribution in small A values. We are not aware of a better fit to so many experimental data for H in Nb from any other theory.

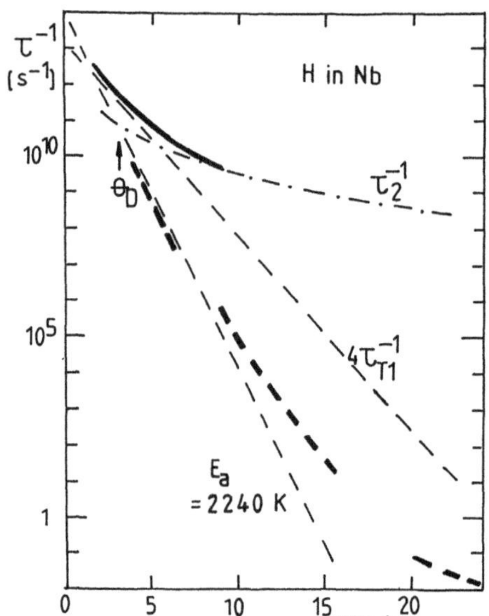

Figure 2 Motion of H in Nb. Heavy full line is diffusion [8]. Heavy dashed line is IF peaks [13]. Dash-dot line is two-phonon rate (3). Thin dashed lines are calculated tunnelling $4\tau_{T1}^{-1}$ and hopping over E_a.

In a similar fitting we find V_0 = 3700 K for H in Ta with d = 1.168 Å and θ_D = 240 K . That gives E_1 = 121 meV, E_a = 2940 K and $2\Delta E_{T0}$ = 1.4 K in good agreement with experimental data [8,16-18]. We note that the extrapolation of $\tau_2^{-1}/2$ from diffusion [8] comes very close to some T^7 data [19] around $\theta_D/10$.

From the more limited data for H in V with d = 1.069 Å, we fit V_0 = 2600 K which gives E_a = 1920 K, E_1 = 105 meV and $2\Delta E_{T0}$ = 9.1 K. This ground state splitting may be too large to be clearly seen in specific heat [20], but it might be observable with INS.

More detailed accounts of how the model fits the tunnelling motion of H, D and T in V, Nb, Ta and other metals will be published. We will also try our theory on muon μ^+ diffusion in metals, since μ^+ is often considered to behave like a light hydrogen isotope. From the good fits of so many different experimental results to our very simple theory of tunnelling of hydrogen in metals, we conclude that our method and the neglect of self-trapping in tunnelling must be basically correct.

References

1. C.P. Flynn, A.M. Stoneham: Phys. Rev. B1, 3966 (1970)
2. L.L. Dhawan, S. Prakash: Phys. Rev. B28, 7294 (1983), B29, 3661 (1984)
3. H.R. Schober, A.M. Stoneham: In Atomic Transport and Defects in Metals, ed. by C. Janot et al., Springer Proc. in Phys. Vol.10 (Springer, Berlin, Heidelberg 1986) p. 149
4. A. Klamt, H. Teichler: Phys. Stat. Sol. b134, 533 (1986)
5. I. Svare: Physica 141B, 271 (1986)
6. I. Svare: To be published
7. R. Orbach: Proc. Roy. Soc. A264, 458 (1961)
8. Z. Qi, J. Völkl, R. Lässer, H. Wenzl: J. Phys. F: Met. Phys. 13, 2053 (1983)
9. T.P. Das: J. Chem. Phys. 27, 763 (1957)
10. P.E. Zapp, H.K. Birnbaum: Acta Metallurgica 28, 1275 (1980)
11. A. Magerl, J.J. Rush, J.M. Rowe, D. Richter, H. Wipf: Phys. Rev. B27, 927 (1983)
12. A. Klamt, H. Teichler: Phys. Stat. Sol. b134, 103 (1986)
13. C.G. Chen, H.K. Birnbaum: Phys. Stat. Sol. a36, 687 (1976)
14. A. Magerl, A.J. Dianoux, H. Wipf, K. Neumaier, I.S. Anderson: Phys. Rev. Letters 56, 159 (1986)
15. H. Wipf, K. Neumaier: Phys. Rev. Letters 52 1308 (1984)
16. J.J. Rush, R.C. Livingston, L.A. de Graaf, H.E. Flotow, J.M. Rowe: J. Chem. Phys. 59, 6570 (1973)
17. U. Rodrian, H. Schultz: J. de Phys. 42, C5-79 (1981)
18. G.J. Sellers, A.C. Anderson, H.K. Birnbaum: Phys. Rev. B10, 2771 (1974)
19. A. Weidinger, R. Peichl: Phys. Rev. Letters 15, 1683 (1985)
20. G.J. Sellers, M. Paalanen, A.C. Anderson: Phys. Rev. B. 10, 1912 (1974)

Optical Studies of Proton Tunneling and Relaxation in Benzoic Acid

M. Pierre, H.P. Trommsdorff, and R.M. Hochstrasser

Laboratoire de Spectrométrie Physique associé au C.N.R.S., Université Scientifique, Technologique et Médicale de Grenoble, B.P. 87, F-38402 St. Martin d'Hères Cedex, France, and Department of Chemistry, University of Pennsylvania, Philadelphia, PA 19104, USA

Abstract. The double proton transfer, leading to the tautomerization of carboxylic acid dimers, is an example of a translational tunneling system. It provides a model for the interplay of tunneling and relaxation in a condensed phase environment. At low temperatures the study of such systems was made possible by the discovery that the optical spectra of dilute mixed crystals of indigo dyes in benzoic acid reflect the proton structure and dynamics of neighboring benzoic acid dimers. Fluorescence line narrowing and time-resolved spectral measurements lead to the determination of the tunneling matrix element (0.16 cm^{-1}) and the rate of tautomerization by tunneling (1.7 - 4.5 x 10^{+8} sec^{-1}).

1. INTRODUCTION

The study of hydrogen tunneling systems in condensed phases is of considerable interest regarding the transition from quantum motion at low temperature to classical transport at higher temperatures. Translational tunneling systems are of particular interest in relation to transport properties or chemical hydrogen transfer reactions for example ; in contrast to rotational systems, no spin selection rules are imposed on the population relaxation between tunneling levels. The study of such systems, however, is hampered by the fact that the different potential wells usually are inequivalent in a condensed phase environment, in contrast to rotational systems where the potential is symmetric even in an asymmetric environment. To the extent that the energy difference of the potential wells is larger than the tunneling matrix element. the wavefunctions are localized and direct transitions between them become too weak (because of the small overlap of the nuclear wavefunctions) to be spectrally observable. Thermally activated transitions between the wells can occur at higher temperatures, but at low temperature the system is in its most stable configuration and all dynamics is frozen out.

The tautomerization of carboxylic acid dimers at low temperatures is an example of a translational tunneling system. The degeneracy of the two tautomers is lifted by the crystal field and at low temperatures only the most stable form is populated. Previous studies by nuclear magnetic resonance (NMR) /1-3/ and inelastic neutron scattering (INS) /4,5/ techniques, therefore, were limited to temperatures where the tautomerization occurs as a thermally activated process, even though the NMR measurements suggested that at lower temperatures direct transitions become important. The rate of this process, as well as the magnitude of the tunneling matrix element, remained however inaccessible in these experiments.

A new approach of such systems was made possible by the discovery of dilute mixed crystal systems of indigo dyes dispersed in benzoic acid. It was found that the optical spectra of the dye molecules consist of triplets of lines which correspond to different proton configurations of benzoic acid dimers next to the dye. In proto benzoic acid the interconversion between these configurations or sites by tautomerization was found to occur on the timescale of the lifetime of the excited state of the dye. The rate of these processes could therefore be determined from time-resolved measurements of the emission corresponding to each of these proton configurations after selective excitation of one of them /6,7/.

In pure benzoic acid crystals the energy difference of the two tautomers (which are equivalent and therefore isoenergetic in an isolated free dimer) equals 35 cm^{-1} (4.4 meV) /1,2/. The doping by impurity molecules alters slightly the potential of neighboring benzoic acid dimers : the energy difference of the two tautomer forms of dimers next to the impurity may be larger or smaller, depending upon the relative position of the neighbor, the chemical nature and also the electronic state of the impurity molecule. For thioindigo there are two physically acceptable ways in which a guest molecule can replace substitutionally a benzoic acid dimer. For one of these sites the energy difference of the two tautomer forms of benzoic acid dimers next to the dye in its ground electronic state is very small, and both forms are populated even at liquid helium temperatures. When the dye is electronically excited, these energy differences are larger and in thermal equilibrium the more stable tautomer is populated only.

We present here new measurements for this site, which demonstrate that the asymmetry of the double well potential increases from beeing smaller to being much larger than the tunneling matrix element, such that the wavefunctions are delocalized and localized respectively. The overlap of these wavefunctions and their energy separations are determined using fluorescence line-narrowing techniques ; the value of the tunneling matrix element and of the asymmetry of the double well is thus determined. In addition we discuss measurements of the relaxation dynamics of these very close lying levels, which were performed using picosecond transient grating techniques /8/ : these complement previous fluorescence decay measurements on systems with large asymmetries /6,7/ and are of special interest regarding the transition from a symmetric to an asymmetric tunneling system /9/.

2. LEVEL STRUCTURE OF THIOINDIGO IN BENZOIC ACID

Each thioindigo molecule is sandwiched between a pair of translationally equivalent benzoic acid dimers separated by two lattice spacings. As illustrated in Fig. 1, such a structure gives rise to four different configurations of the acid protons. If the wavefunctions of the two tautomer forms of a benzoic acid dimer are labelled $|\alpha\rangle$ and $|\beta\rangle$, the four proton configurations of a pair of dimers are $|\alpha,\alpha\rangle$, $|\alpha,\beta\rangle$, $|\beta,\alpha\rangle$, and $|\beta,\beta\rangle$. In the centrosymmetric crystal field the states $|\alpha,\beta\rangle$ and $|\beta,\alpha\rangle$ are equivalent and the Hamiltonian of the system can be written as :

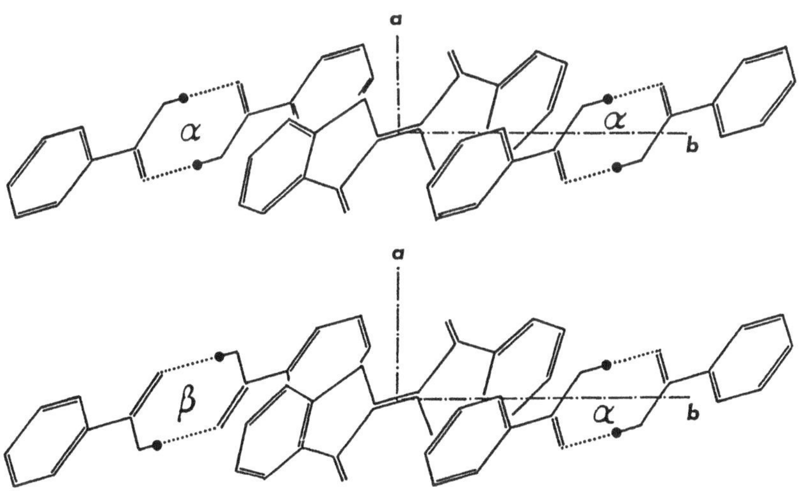

Figure 1 : Projection on the a,b crystallographic plane of a thioindigo molecule embedded substitutionally between two benzoic acid dimers separated by two lattice spacings. Due to tautomerization there exist four proton configurations, two of which are shown in the figure.

	$\vert\alpha,\alpha\rangle$	$\vert\alpha,\beta\rangle$	$\vert\beta,\alpha\rangle$	$\vert\beta,\beta\rangle$
$\langle\alpha,\alpha\vert$	-A	T	T	0
$\langle\alpha,\beta\vert$	T	A	0	T
$\langle\beta,\alpha\vert$	T	0	A	T
$\langle\beta,\beta\vert$	0	T	T	B

For thioindigo in the excited electronic state the values of A and B are much larger than the tunneling matrix element T, and the Hamiltonian is to a very good approximation diagonal in the above basis set of localized wavefunctions. The energy degeneracy of the $|\alpha,\beta\rangle$ and $|\beta,\alpha\rangle$ levels, for example, is lifted by approximately 4×10^{-4} cm^{-1} only ; this is less than the contribution to the optical linewidths from the finite lifetime of the excited state of the dye. For ground state thioindigo, temperature-dependent absorption measurements determined the value of B to be about 1.8 cm^{-1}, while A and T were found to be approximately 10 times smaller and of comparable size. The eigenfunctions of the system are now a superposition of the localized basis wavefunctions and are more or less delocalized over both wells of each dimer, depending upon the relative magnitude of the tunneling matrix element and the asymmetry.

3. FLUORESCENCE LINE NARROWING

In optical experiments, when the dye is electronically excited, the system is projected from a set of delocalized to a set of localized basisfunctions : the spectrum consists of several lines, the relative position of which is given by the variation upon excitation of the eigenvalues of the above Hamiltonian and with intensities proportional to the square of the overlap of the corresponding eigenfunctions. The inhomogeneous (strain) broadening of electronic transitions in mixed molecular crystals is typically of the order of 1 cm^{-1}. It is therefore not possible to resolve in conventional absorption or emission spectra the closely spaced levelstructure expected to arise for near-symmetric double wells. Fluorescence line-narrowing techniques overcome this limitation : to the extent that the inhomogeneous effects on the different spectral lines are correlated, frequency selective excitation of a subset of sites within the inhomogeneous distribution leads to line-narrowed emission.

Results of such experiments are shown in Fig. 2 : the splitting of the lines directly reflects the level structure corresponding to a ground state impurity, while the intensity distribution measuring the overlap of the wavefunctions depends also on the Boltzmann factor and, as illustrated by the difference of spectra a) and b), upon the excitation density, i.e. the position of the exciting laser frequency within the inhomogeneous absorption line. As both these quantities are measured independently, the simulation of the observed spectra involves only the two parameters to be measured, namely the tunneling matrix element T and the asymmetry A. The values determined are $T = 0.16 \pm 0.01$ cm^{-1} and $A = 0.09 \pm 0.01$ cm^{-1} : the double well potential is nearly symmetric and its eigenstates are significantly delocalized.

4. RELAXATION AND TAUTOMERIZATION KINETICS

The levels of an asymmetric double well potential represent different tautomer forms, and the relaxation between them the tautomerization reaction. The determination of the rate of tautomerization from time-resolved fluorescence measurements /6,7/ correspond to situations where the asymmetry of the double well i.e. the energy difference of the two tautomers lies between 15 and 40 cm^{-1} and is therefore very large as compared to the tunneling matrix element (0.16 cm^{-1}).

In the case of a near-symmetric potential, as discussed above, the wavefunctions are delocalized and the relaxation between the levels changes character. Such a situation has only been found to be induced by ground state thioindigo, and the same experimental technique cannot be used to measure the relaxation kinetics. An indirect evaluation of this rate has been obtained from measurements of the homogeneous optical linewidth using holeburning techniques and attributing the width to the

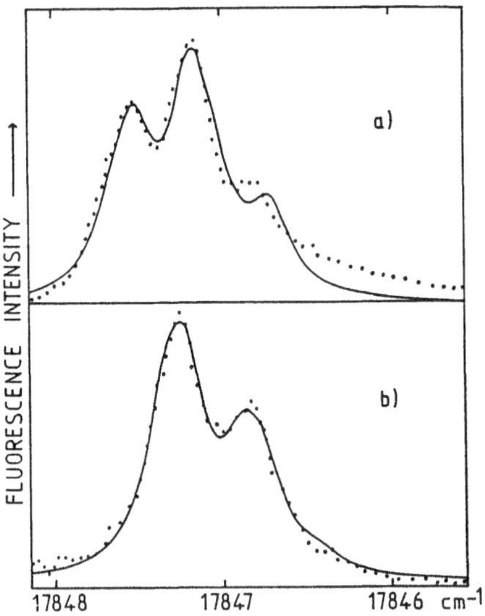

Figure 2 : Line-narrowed fluorescence spectra of thioindigo in benzoic acid at 1.4K. The narrowband laser excitation occurs from the closely spaced ground state levels to the near degenerate $|\alpha,\beta\rangle$, $|\beta,\alpha\rangle$ level pair. In a) and b) the excitation wavelength lies to the high and low energy side of the inhomogeneously broadened unresolved absorption line. After relaxation (by tautomerization) to the $|\alpha,\alpha\rangle$ level, the emission to the ground state levels is measured and can be spectrally resolved.

finite lifetime of both the excited and the ground state of the dye : the latter being due to the relaxation between the levels of the benzoic acid dimer neighbors.

In view of the importance of this value, an independent and more direct measure using socalled transient grating techniques was made. In these experiments two time coincident laser pulses of equal frequency cross in the sample and form a pattern of interference fringes. This leads to a spatially periodic excitation of the sample and produces variations of the index of refraction, which have contributions from both the excited and ground state populations of the dye molecules. This index grating is probed by a time-delayed third pulse, and the intensity of the diffracted signal, which is proportional to the square of the index variations, is measured as a function of the time delay. The signal therefore maps the time evolution of these populations.

Figure 3 shows the results of such measurements for proto and deutero (acid protons exchanged) benzoic acid. If the recovery of the ground state were solely determined by the lifetime of the excited state the signal should decay exponentially with half the excited state lifetime. This is indeed observed for the deuterated compound because the rate of deuterium transfers is orders of magnitude slower and does not occur on the timescale of the measurement. In proto benzoic acid the deviations from the exponential decay are obvious and can be explained by taking into account the relaxation between the levels of the benzoic acid dimers near the ground state thioindigo impurity. In the analysis of the data the description of the levels as obtained in the preceeding section was used : a single relaxation rate was introduced as the signal-to-noise ratio did not warrant a multiparameter fitting required if two relaxation rates are used for the three thermally populated levels. The value obtained for this rate ($3.5 \times 10^{+8}$ sec^{-1}) agrees well with the value obtained from the optical linewidth measurements (see above) /6/.

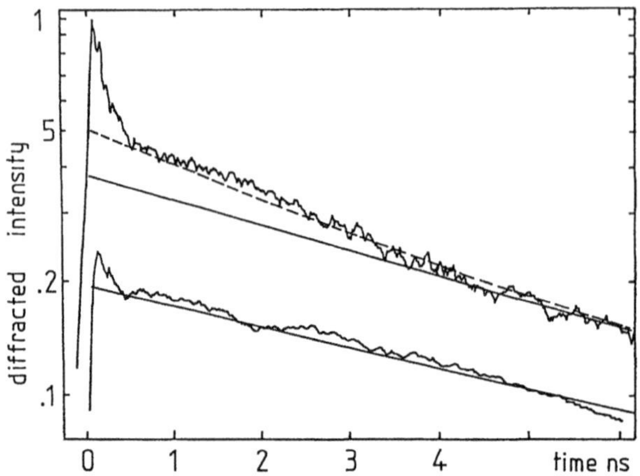

Figure 3 : Transient grating measurements in proto (top) and deutero (bottom) benzoic acid doped with thioindigo. The full and broken lines are calculated decays neglecting and including relaxation processes of benzoic acid dimers near ground state thioindigo molecules. The peak near t = 0 reflects a fast process, possibly related to multiphoton excitation of higher electronic states.

5. DISCUSSION

An important issue in the understanding of the tautomerization of carboxylic acids is the description of the reaction path. It is obvious that not only the hydrogen but also the carbon and oxygen atoms move in the reaction, as C-O double and single bonds are interchanged in the tautomerization. The structural relaxation of the skeleton during the hydrogen transfer must also be taken into account in order to explain the height to the potential barrier /10-12/. As a result the effective mass of the reaction coordinate is higher than for a pure hydrogen motion : the tunneling matrix element is lowered and the H/D isotope effect decreases. For the formic acid dimer, model calculations demonstrate the importance of these effects, but the tunneling splitting corresponding to barrier heights close to the measured values is 100 times smaller than determined here for benzoic acid /10/. The deuteration effect on this splitting also is calculated to be more than 10 times smaller than evaluated experimentally (see below). These large discrepancies may have several causes : one is the assumption that the barrier heights equal the activation energies for tautomerization at higher temperatures as determined by NMR and INS measurements, another is the crudeness of the energy level calculations which were performed in a one-dimensional straight cut through a potential energy surface modeled by ab initio calculations. Recent theoretical work on the hydrogen transfer in malonaldehyde stresses the importance of introducing more than one degree of freedom, especially when the minimum energy path on the potential energy surface is curved /13/.

Relaxation processes in double well potentials can be related to modulations by phonons of the asymmetry and the barrier i.e. the tunneling matrix element. Using Fermis golden rule and taking the phonon density of states to be proportional to the square of the energy, it is easy to see that rate of population relaxation is proportional to the square of the sum of two terms : one is the product of the tunneling matrix element and the modulations of the asymmetry, the second is the product of the asymmetry and the modulations of the tunneling matrix element /9/. Our results indicate that this rate is quite insensitive to the absolute value of the asymmetry and suggest therefore that the first term dominates : the measured relaxation rates are thus proportional to the square of the tunneling matrix element. The deuteration effect on these rates of about 10^{+3} indicates that the tunneling matrix element in deuterated benzoic acid is reduced by a factor of about 30 and is of the order of $5*10^{-3}$ cm^{-1} : in condensed phases it will be very difficult to measure such a small value as the asymmetry is likely to be always much greater and the system will be completely localized.

The temperature dependence of the rate of tautomerization has been evaluated at higher temperatures by NMR and INS measurements. The optical techniques explained here are limited in benzoic acid crystals to below about 10-15K, because of the large increase at higher temperatures of the linewidths (in part presumably due to the thermally induced proton disorder in the bulk of the crystal). In systems with large asymmetries the rate of relaxation was found to be temperature independent below 12K : it should be interesting to extend these measurements to the near symmetric case discussed here, as the thermal energy becomes equal or larger than the level spacing.

6. CONCLUSION

Optical techniques complement studies by NMR and INS of translational tunneling systems and, in the case of benzoic acid, extend these measurements to lower temperatures. The parameters determined here for benzoic acid should be typical for carboxylic acid dimers. The tunneling splitting in a symmetric dimer would thus approximately equal 0.3 cm^{-1} (40 ueV) : only if the asymmetry induced by the environment is smaller than this value (in matrix-isolated species for example) will it be possible to detect tunneling transitions in such systems by INS.

Acknowledgement

This research was supported by a grant from the Division of Materials Research of the NSF (DMR 85-07740).

References

1. S. Nagaoka, T. Terao, F. Imashiro, A. Saika, N. Hirota, and S. Hayashi : Chem. Phys. Letters 80 (1981) 580
2. S. Nagaoka, T. Terao, F. Imashiro, A. Saika, N. Hirota, and S. Hayashi : J. Chem. Phys. 79 (1983) 4694
3. B.H. Meier, F. Graf, and R.R. Ernst : J. Chem. Phys. 76 (1982) 767
4. B.H. Meier, R. Meyer, R.R. Ernst, A. Stöckli, A. Furrer, W. Hälg, and I. Anderson Chem. Phys. Letters 108 (1984) 522
5. B.H. Meier : Ph. D. Thesis, ETH Nr. 7620, Zürich (1984)
6. J.M. Clemens, R.M. Hochstrasser, and H.P. Trommsdorff : J.Chem.Phys. 80 (1984)1744
7. G.R. Holton, R.M. Hochstrasser, and H.P. Trommsdorff : Chem.Phys.Letters in press
8. M. Pierre and H.P. Trommsdorff : Int. Conf. on Dynamics of Molecular Crystals, Grenoble (1986), Proceedings (Elsevier, Amsterdam, in press)
9. H.P. Trommsdorff : 19th Jerusalem Symposium on Quantum Chemistry and Biochemistry, Proceedings (D. Reidel, Dordrecht, in press)
10. F. Graf, R. Meyer, T.-K. Ha, and R.R. Ernst : J. Chem. Phys. 75 (1981) 2914
11. S. Nagaoka, N. Hirota, T. Matsushita, and K. Nishimoto : Chem. Phys. Letters 92 (1982) 498
12. S. Hayashi, J. Umemura, S. Kato, and K. Morokuma : J. Phys. Chem. 88 (1984) 1330
13. T. Carrington, Jr. and W.H. Miller : J. Chem. Phys. 84 (1986) 4634.

Part V

Disordered Systems

How Universal Is the Spectral Density of Tunneling States in Glasses?

S. Hunklinger

Institut für Angewandte Physik II der Universität Heidelberg,
Albert-Überle-Str. 3–5, D-6900 Heidelberg, Fed. Rep. of Germany

1 Introduction

The low-temperature properties of amorphous solids have attracted the attention of physicists since the beginning of the century. More recently, in 1971, unambiguous evidence was given that below 1 K the thermal properties of amorphous insulating solids differ markedly from those of their crystalline counterparts /1/. In spite of intense experimental and theoretical efforts /2,3/ these "anomalies" can so far only be described on the basis of a phenomenological model /4/. In this so-called "Tunneling Model" (TM) it is assumed that in the amorphous network structural rearrangements are possible even at the lowest temperatures. More specifically: it is assumed that small groups of atoms can tunnel between two different but energetically nearly equivalent sites. Due to the randomness of the amorphous structure, such "Tunneling Systems" (TS) will exhibit a broad distribution of the intrinsic parameters. Thus a rather universal but unspecific interpretation of the low-temperature behaviour of thermal, elastic and dielectric properties of amorphous solids is possible. However, so far the nature of the moving entity is neither known in general nor in specific cases.

In the past it has often been formulated that the density of states of the TS is more or less a universal constant characterizing the amorphous state. In this article experimental results are reported which demonstrate, that this is by no means the case.

2 Tunneling Model

The Tunneling Model is based on two assumptions. Firstly it is assumed that certain particles, atoms or small groups of atoms, exist which are capable of occupying two different sites and are therefore moving in a double-well potential. The resulting splitting E of the ground state is thus given by $E^2 = \Delta^2+\Delta_o^2$. Δ is the asymmetry energy given by the difference in the depth of the wells. The splitting Δ_o is caused by the tunneling motion and is proportional to $\exp(-\lambda)$, where the tunneling parameter λ reflects the probability of tunneling and is determined by the mass of the tunneling particle and the shape of the potential. Appropriate distribution functions for the parameter of the TS are introduced by the second basic assumption of the TM. Δ and λ are assumed to be independent of each other and uniformly distributed: $P(\Delta, \lambda)\, d\Delta\, d\lambda = \bar{P}\, d\Delta\, d\lambda$, where $\bar{P}$ is a constant.

For our further discussion the coupling of TS to external strain fields is of particular interest. A sound wave travelling through an ensemble of TS will be attenuated by two different mechanisms.

Firstly it will be resonantly absorbed by those TS having an energy splitting corresponding to the phonon energy. Because of the wide distribution of energy splittings this process occurs at all frequencies. However, in this article the emphasis is lying on the second mechanism. It is the relaxation absorption /5/ which is due to the modulation of the asymmetry energy Δ by the sound wave. In this way the equilibrium of the TS is disturbed and a relaxation process restores it again. Because of the finite value of the relaxation time τ the sound wave is attenuated. The main contribution to this attenuation process is due to those TS which have an energy splitting $E \simeq kT$ and relax on a time scale comparable with the period of the sound wave, i.e. for which $\omega\tau \simeq 1$ holds. Because of $\tau \propto (E/\Delta_o)^2$ /2, 3/ a distribution of relaxation times exists even if E is kept constant, since Δ_o is already distributed. Therefore an integration has to be carried out with respect to both E and Δ_o. For the relaxation time τ two limiting cases can be distinguished: for $\Delta_o \rightarrow 0$, τ approaches infinity and for $\Delta_o = E$ a minimum value τ_m exists. Analytic solutions can be found for $\omega\tau_m >> 1$ and $\omega\tau_m << 1$, which I shall briefly discuss.

At low temperatures where $\omega\tau_m >> 1$ holds an absorption is expected, which is frequency independent but increases strongly with temperature. It should be given by Eq. 1, provided that the TS relax via absorption emission of a single thermal phonon (one-phonon process)

$$\alpha_{rel} = \frac{\pi^4 A C k^3 T^3}{12v} . \tag{1}$$

Here $C = \bar{P}\gamma^2/\rho v^2$ and $A = (\gamma_\ell^2/v_\ell^5 + 2\gamma_t^2/v_t^5)/2\pi\rho\hbar^4$, where ℓ and t refer to the longitudinal and transverse phonon branch. $\bar{P}$ represents the density of states introduced above, γ is the deformation potential of the TS, ρ the mass density and v the sound velocity.

At "high" temperatures, where $\omega\tau_m << 1$, the absorption should become temperature independent, i.e. a "plateau" is expected to be observed:

$$\alpha_{rel} = \frac{\pi\omega C}{2v} , \tag{2}$$

It should be pointed out that the occurrence of the plateau and also its height is independent of the way the TS relax. Therefore it allows direct conclusions on the validity of the distribution function $P(\Delta, \lambda)$. Supposition for a temperature-independent absorption is the energy independence of the density of states. In other words, with increasing temperature the number of TS contributing to the absorption process has to remain constant. Furthermore in the regime $\omega\tau_m << 1$ only TS with $\Delta_o << E$ are able to fulfil the condition $\omega\tau \simeq 1$. In this way information on the slowly relaxing TS is obtained.

In Fig. 1 measurements on PMMA and vitreous silica are shown /6, 7/. Instead of the attenuation α the internal friction $Q^{-1} = \alpha v/\omega$ is plotted as a function of temperature. Besides the low temperature rise, well-developed plateaus are observed at higher temperatures. They show that in both materials the distribution function $P(\Delta, \lambda)$ is a good approximation although the microscopic structures of these two amorphous solids are fundamentally different.

Intimately related with absorption processes is a change in the velocity of sound. Applying the Kramers-Kronig-relation to the resonant interaction, a logarithmic temperature variation of the velocity is found for $\hbar\omega << kT$

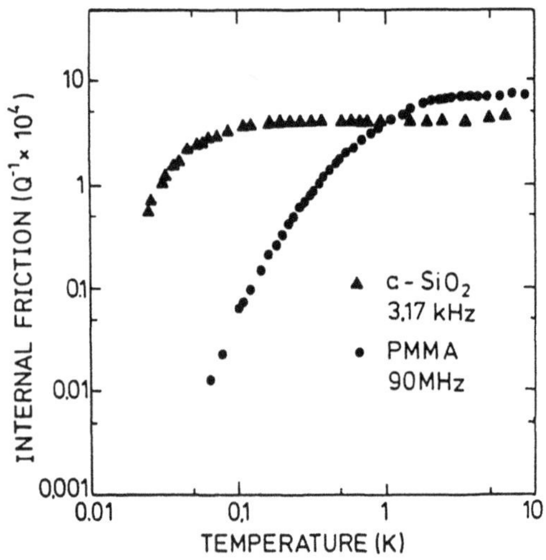

Fig. 1: Internal friction $Q^{-1} = \alpha\, v/\omega$ of polymethylmethacrylate and vitreous silica as a function of temperature (from /2/)

$$\left.\frac{\Delta v}{v}\right|_{res} = C \ln \frac{T}{T_o} \qquad (3)$$

where $\Delta v/v = (v(T) - v(T_o))/v(T_o)$ and T_o is an arbitrary reference temperature. The validity of Eq. 3 has been verified in a large number of insulating glasses. Clearly the logarithmical variation of the velocity gives an ideal measure of the coupling constant C /2/.

The contribution of the relaxation process is negligible compared to that of the resonant process as long as $\omega\tau_m > 1$. At high enough temperatures, however, when $\omega\tau_m < 1$, the relaxation process becomes dominant. Since its contribution is negative, the sound velocity starts to decrease in dielectric glasses. At low frequencies perfect agreement with the predictions of the TM has been found /2/, but we do not want to discuss this aspect any further.

3 Correlation Between Low-Temperature Anomalies and the Glass Transition Temperature

For a long time the density of states was thought to vary not too much when various glasses are considered. However, recent experiments have shown that in glasses with similar composition the specific heat changes when the glass transition temperature T_g is varied /8/. From the data the correlation $\bar{P} \propto T_g^{-1}$ was deduced which was also supported by theoretical arguments /9/. A similar conclusion had already been drawn from the analysis of the thermal conductivity of a variety of amorphous solids /10/.

Such a correlation between the density of states and the glass transition temperature is rather fascinating. A simple way of dealing with the problem is to consider T_g as the determining factor for the energy scale of the problem /4, 8, 9/. In this case, if a fixed number N of configurational states were distributed in some form over the energy range kT_g, the quantity $\bar{P}$ would be proportional to N/kT_g and the inverse scaling of $\bar{P}$ with T_g could be understood. But two important questions arise: are there TS with an energy splitting up to kT_g and is the total number of TS really independent of the chemical nature of the amorphous solid? Specific heat measurements are not suitable to answer the first question because at higher temperatures the contribution of the TS is covered by that of the phonons. Information on this question can be obtained by measurements of the acoustic attenuation. As mentioned in the previous section the temperature dependence in the plateau region reflects the energy dependence of the density of states. The expected plateau is in fact observed over a wide temperature range, for example in the case of the electrolyte glass $LiCl \cdot 7H_2O$ /11/ which exhibits a glass transition temperature of 140 K. Up to more than 50 K the absorption is constant, indicating that the density of states is constant at least up to energies corresponding to that temperature. At still higher temperatures absorption increases probably because of viscoelastic damping which is seen in all glasses in the vicinity of T_g. Unfortunately such an agreement with theoretical expectation is not found for vitreous silica /2/. Measurements at low frequencies show that the plateau is missing at temperatures above 100 K. Therefore it is likely that TS with the corresponding energy splitting do not exist in this material. Thus we may state that the first question raised above is still open. Also the second question cannot be answered unambiguously. In random structures with a local rigidity, i.e. in covalently bonded amorphous solids with higher coordination numbers, the density of states is smaller than in twofold coordinated silicate glasses or polymers /2/.

In general the density of states of excitations is determined by measurement of the specific heat. However, in such an experiment not only the TS contribute but also all other excitations of comparable energy like phonons or impurity modes. Another more suitable way to determine $\bar{P}$ are acoustic measurements /12/. Thereby different combinations of attenuation and velocity measurements can be used. Most of the values for $\bar{P}$ given below have been deduced in the following way. From measurements of the logarithmic variation of the sound velocity at low temperatures $\bar{P}\gamma_\ell^2$ or $\bar{P}\gamma_t^2$ are determined depending on the polarization of the sound wave (see Eq. 3). An additional measurement of the low-temperature rise of the attenuation yields $\bar{P}\gamma_\ell^2 \cdot A$ or $\bar{P}\gamma_t^2 \cdot A$ (see Eq. 1). A short inspection of the numerical values determining the quantity A shows that the predominant contribution to the relaxation rate is due to the transverse branch. Therefore we may approximate the constant A by $A \simeq \gamma_t^2/\pi\rho\hbar^4 v_t^5$. Thus $\bar{P}$, γ_ℓ and γ_t can be determined separately from such a set of acoustic data. An alternative way is to estimate τ_m and consequently γ_t from the position of the maxima in the velocity of sound so that velocity measurements alone are used for the determination of the relevant parameters.

The first result of such calculations is that the deformation potential γ increases linearly with T_g /12, 13/. Even more interesting is the result for the density of states $\bar{P}$. As indicated by the full line in Fig. 2 the relation

$$\bar{P} = \bar{P}_o \, e^{\varepsilon/kT_g} \qquad (4)$$

is obtained, where $\bar{P}_o = 5.3 \cdot 10^{44} J^{-1} m^{-3}$ and $\varepsilon/k = 630$ K. Obviously $\bar{P} \propto T_g^{-1}$ is a good approximation only for those glasses having a transition temperature of the order ε/k. The open circles representing polymers and metallic glasses do not fall onto the line. It is tempting to assume that Eq. 4 would also hold for these materials with their completely different microscopic structure, although the constants $\bar{P}_o$ and ε would be different in this case. So far not enough data are available for these groups, especially the glass transition temperature of polymers does not vary over a wide enough temperature range to confirm such an idea. Nevertheless such a relation is indicated in Fig. 2 by a dashed line.

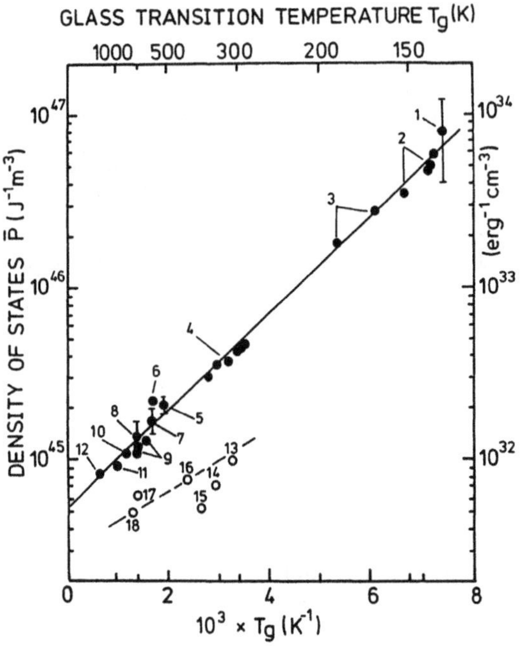

Fig. 2: Logarithm of the density of states $\bar{P}$ as a function of the inverse glass transition temperature. 1: H_2O, 2: LiCl-H_2O, 3: $ZnCl_2$-H_2O, 4: $Ca(NO_3)_2$-$K(NO_3)$, 5: BALNA, 6: Zn-Glass, 7: V 52, 8: LAT, 9: SF 4, 57, 59, 10: BK 7, 11: LaSF 7, 12: SiO_2, 13: Se, 14: PEMA, 15: PMMA, 16: PC, 17: PdSiCu, 18: CuZr (after /13/). For more details see /13/

Although the relation between $\bar{P}$ and T_g is rather spectacular, there is no real understanding of the underlying physics so far. One may replace T_g by the melting temperature T_m since the "two-thirds-rule" $T_g \simeq 2/3\ T_m$ is well satisfied by most glasses. Using Lindemann's formula for a rough estimate of T_m, the correlation $\varepsilon/kT_g = B/mv_D^2$ is found, where B = 3.2 eV, m is the mean atomic mass and v_D the Debye velocity. Since Lindemann's formula is only a rough approximation, the scattering of the values of B is not surprising. Of course, the question arises whether the glass transition temperature is an appropriate parameter. It may be that a more fundamental quantity determines both, $\bar{P}$ and T_g, but such a question cannot be answered on an experimental basis only.

4 Influence of Annealing on Low-Temperature Acoustic Properties

It has been proposed that annealing of metallic glasses near the crystallization temperature should have a strong influence on the low-temperature properties /14/. In particular it was thought that the density of states of the TS should be reduced by heat treatment. Experimental studies of the influence of annealing led to confusing results (for more references on this subject see /2/). In brief, reversible as well as irreversible effects have been observed. Reversible effects are much smaller and can only be seen after the sample has been annealed long enough so that no irreversible changes occur anymore. Here I shall report on two very recent studies of the irreversible changes caused by thermal treatment. In the metallic glass PdSiCu the influence of annealing as well as of a variation of the cooling rate were investigated /15/. In a further experiment annealing effects were studied on quench-condensed "amorphous water" a-H_2O /16/.

Let us first consider the results obtained for metallic glasses. The interaction of the TS with the free electrons of metallic glasses changes their behaviour in comparison with that in dielectric glasses /17/. Especially at low temperatures the relaxation times are much shorter in glassy metals, so that $\omega\tau_m << 1$ down to the Microkelvin range if the experiment is carried out at low frequencies. Therefore the plateau in the acoustic attenuation described by Eq. 2 should be observed in the whole attainable temperature range. The acoustic absorption should not depend on the fact whether relaxation of the TS via interaction with electrons or phonons is dominant.

In Fig. 3 the influence of annealing on the acoustic attenuation is shown. All samples were prepared with a cooling rate of $6 \cdot 10^5$ K/s. The untreated sample exhibits the highest attenuation. Annealing for 30 min at 450 K or 600 K reduces the attenuation considerably. At the same time the absorption peak at 13 K disappears.

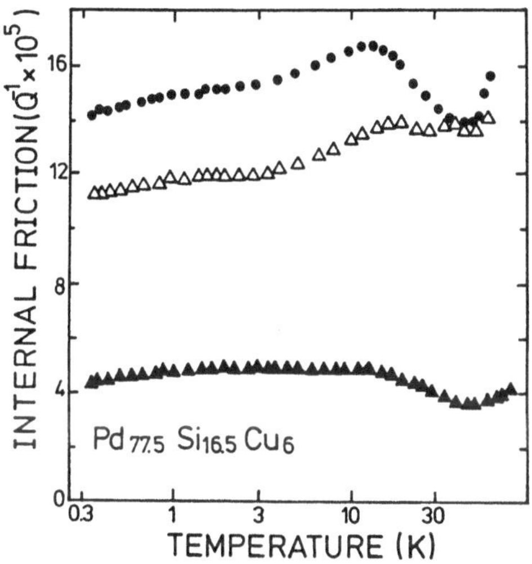

Fig. 3: Internal friction of PdSiCu splat-cooled with a rate of $6 \cdot 10^5$ K/s versus temperature: (●) untreated sample, (Δ) annealed at 450 K for 30 minutes, (▲) annealed at 600 K for 30 minutes (after /15/)

From Eq. 3 it is immediately clear that in both cases the constant C is decreased, but no distinction between the different parameters can be made. Measurements of the temperature variation of the velocity would provide us with additional information. In fact such measurements have been carried out /15/ and some of the results will be used in the discussion.

The mass density ρ remains roughly constant but the velocity increases by about 15%. While γ remains unchanged the density of states $\bar{P}$ decreases steadily, which can be understood qualitatively within the framework of the free volume theory /18/. According to this theory the total number of existing TS is related to the free volume. Since this quantity decreases steadily on annealing /19/, the observed variation of $\bar{P}$ is in qualitative agreement with this prediction. Similar measurements have also been carried out using samples prepared with different cooling rates. The overall behaviour is similar to the annealing experiment: a decrease of the attenuation is observed with decreasing cooling rate. In this case, however, both quantities $\bar{P}$ and γ are changed. Whereas γ decreases steadily with cooling rate, the density of states goes through a minimum value in contradiction to the prediction of the free volume theory. The reason for its failure could be that differences in sample preparation result in slightly different short-range orders.

Amorphous water (a-H_2O) is an extremely suitable candidate for studies of the influence of sample preparation and thermal treatment because preparation and crystallization temperature can differ by a great amount. Thin films of a-H_2O can be prepared at helium temperatures by quenched condensation and crystallization does not occur before 135 K. Measurement of both attenuation and velocity changes have been carried out using acoustic surface waves /16/.

The absorption data shown in Fig. 4 were obtained on a 0.6 μm thick film prepared by quenched condensation onto a $LiNbO_3$ substrate held at 4.4 K. Annealing at 5.6 K had no noticeable influence on the acoustic properties of the film. In all other cases attenuation was changed and reached a new constant value after annealing for two hours. Attenuation and velocity varied reversibly below, but irreversibly above the annealing temperature. At the lowest temperature absorption increases with temperature as known from conventional amorphous solids. Above 6 K a shoulder is seen which can be interpreted as the superposition of an absorption plateau and a steep rise of the absorption starting at about 10 K. Heating up to 20 K reduces the absorption below 5.6 K irreversibly by a small amount.

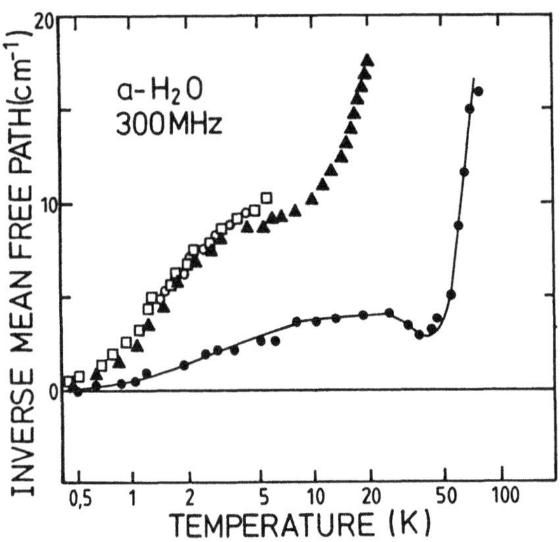

Fig. 4: Temperature dependence of the attenuation of 300 MHz acoustic surface waves of a 0.6 μm thick film of a-H_2O. The experimentally observed value of the attenuation has been scaled up to infinite film thickness. (o) Freshly condensed film, (□) after annealing at 5.6 K, (▲) at 20 K and (●) at 84 K (from /16/).

Annealing at 84 K led to a drastic change of the absorption behaviour. Both the plateau and the steep rise are shifted towards higher temperatures. In addition, the attenuation is reduced in the whole temperature range. Like in metallic glasses the coupling constant C is reduced by annealing. Simultaneously an increase of the sound velocity by about 20% is registered, indicating that the short-range order is changed. The striking difference is, however, the observation of the steep rise close to the annealing temperature. It cannot simply be attributed to the interaction of the sound wave with TS, but an additional absorption process must be present. A similar increase of the absorption is also found in conventional glasses, but only in the vicinity of the glass transition where viscoelastic damping is observed. It is tempting to explain the behaviour of a-H_2O in a similar way. We assume that close to the annealing temperature thermally activated motions of certain atoms or groups of atoms between different configurational states are possible, causing the observed high absorption. A further increase of the temperature will allow these structural units to surmount even higher potential barriers and to attain new sites with lower potential energy. Energetically unfavourable bonds can now be broken and rearranged, thus lowering the potential energy even further. The original structure will transform towards the crystalline state in these areas and will no longer contribute to the relaxation absorption. But during the annealing process always new units will become mobile and take over the part of those which have already undergone the transition to more symmetric structures. Therefore the steep rise in the absorption is found at all temperatures when the sample is warmed up.

Obviously the annealing behaviour of quench-condensed a-H_2O differs from that of "ordinary" glasses. The reason could be that its structure has never been in thermal equilibrium, neither during the formation process nor afterwards. Therefore one might argue that glasses cooled down from the melt exhibit a "unique" glass transition temperature but that in quench-condensed amorphous solids this quantity varies spatially. From the quenching temperature up to the temperature where the amorphous structure undergoes the transition to the crystalline state, a "local glass transition temperature" could exist.

5 Influence of the Superconducting Transition

As last example demonstrating that the density of states $\bar{P}$ of glasses can be varied, the attenuation of PdZr is shown in Fig. 5 /20/. In the superconducting state the attenuation at low temperatures rises because of the increase of the relaxation rate τ^{-1} due to the rising number of phonons and quasiparticles. Between 1.5 K and 3 K a plateau is observed. Above this temperature an additional relaxation process sets in which I am not discussing here. A completely different behaviour is observed when superconductivity is suppressed, because relaxation by free electrons dominates in the whole temperature range. Surprisingly, the attenuation in the normal state is always smaller than in the superconducting state although $\omega\tau_m << 1$ holds in the whole temperature range, and the nature of relaxation of the TS should be without influence.

According to Eq. 2 which should hold in the plateau region C is reduced by the presence of quasiparticles. Additional velocity measurements support the interpretation that this effect is mainly due

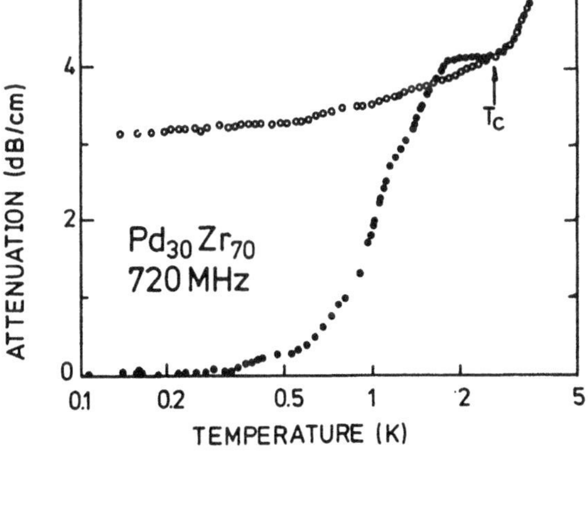

Fig. 5: Ultrasonic absorption in PdZr at 720 MHz. Data indicated by open circles were taken after applying a magnetic field of 5.6 T to suppress superconductivity (after /20/).

to a variation of $\bar{P}$. Recently the "standard" description of the TS-electron interaction /21/ has been extended /22/ assuming that "bound" states are formed below a certain temperature if the coupling between TS and conduction electrons is strong enough. Although a reduction of $\bar{P}$ can be explained qualitatively by this theory, quantitative agreement cannot be obtained /19/.

6 Conclusion

The experimental studies done so far have shown that the dynamics of the TS are understood in principle. The TM provides a good phenomenological basis on which various observations can be explained. This is definitely the main achievement of the past decade of intensive investigation.

In the first studies of the anomalous properties of amorphous solids the emphasis was lying on the universality of the observed behaviour. More recently the influence of sample preparation, thermal treatment and correlation to other physical parameters like the glass transition temperature has gained rising interest. Such experiments allow to establish rules associating the low-temperature behaviour with other physical properties, so that more informational data is provided for facilitation of the theoretical development. In short, what is still missing is the theoretical description of the dynamics of structural units in an amorphous matrix, which contain more atoms than just the next nearest neighbours.

References

1. R.C. Zeller, R.O. Pohl: Phys. Rev. B 4, 2029 (1971)
2. see for example: S. Hunklinger, A.K. Raychaudhuri: In Progress in Low-Temperature Physics, de. by D.F. Brewer, Vol. IX, p. 265 (Amsterdam, Oxford, New York, Tokyo: North-Holland 1986)
3. see for example: Amorphous Solids - Low-Temperature Properties, Topics in Current Physics, ed. by W.A. Phillips, Vol. 24, p. 81, (Berlin, Heidelberg, New York: Springer 1981)
4. W.A. Phillips: J. Low Temp. Phys. 7, 351 (1972);
   P.W. Anderson, B.I. Halperin, C. Varma: Phil. Mag. 25, 1 (1972)
   V.G. Karpov, M.I. Klinger, F.N. Ignatev: Sov. Phys. JETP 57, 439 (1983);
5. J. Jäckle: Z. Phys. 257, 212 (1972)
6. G. Federle, S. Hunklinger: J. Phys. (Paris) 43, C9-505 (1982)
7. A.K. Raychaudhuri, S. Hunklinger: J. Phys. (Paris) 43, C9-483 (1982)
8. A.K. Raychaudhuri, R.O. Pohl: Solid State Commun. 37, 105 (1980)
9. M.H. Cohen, G.S. Grest: Solid State Commun. 39, 143 (1981)
10. C.L. Reynolds Jr.: J. Noncryst. Solids 30, 371 (1978)
11. G. Kasper, V. Röhring: In Proc. 17th Int. Conf. on Low-Temperature Physics, LT 17, ed. by U. Eckern, A. Schmid, W. Weber, H. Wühl, p. 385, (Amsterdam, Oxford, New York, Tokyo: North-Holland 1984)
12. P. Doussineau, M. Matecki, W. Schön: J. Phys. (Paris) 44, 101 (1983)
13. U. Reichert, M. Schmidt, S. Hunklinger: Solid State Commun. 57, 315 (1986);
    U. Reichert, Ph.D. Thesis, University of Heidelberg, 1986
14. M. Banville, R. Harris: Phys. Rev. Lett. 44, 1136 (1980)
15. T. Rabenau: Ph.D. Thesis, University of Heidelberg, 1986;
    T. Rabenau, G. Weiss, D. Herlach: to be published
16. H. Härdle, G. Weiss, S. Hunklinger, F. Baumann: Z. Physik B (1986) to be published

17. B. Golding, J.E. Graebner, A.B. Kane, J.L. Black: Phys. Rev. Lett. 41, 1817 (1978)
18. M.H. Cohen, G.S. Grest: Phys. Rev. Lett. 45, 1271 (1980)
19. D.M. Herlach, H.W. Gronert, E.F. Wassermann: Europhys. Lett. 1, 23 (1986)
20. P. Esquinazi, H.M. Ritter, H. Neckel, G. Weiss, S. Hunklinger: Z. Phys. B, 64, 81 (1986)
21. J.L. Black, P. Fulde: Phys. Rev. Lett. 43, 453 (1979)
22. K. Vladar, A. Zawadowsky: Phys. Rev. B28, 1564; B28, 1582; B28, 1596 (1983)

Very Low Frequency Elastic Loss of PMMA Below 5 K

M. v. Schickfus, H. Tietje[1,2], *and E. Gmelin*[1]

Institut für Angewandte Physik II der Universität Heidelberg, Albert-Überle-Str. 3–5, D-6900 Heidelberg, Fed. Rep. of Germany
[1]Max-Planck-Institut für Festkörperphysik, Heisenbergstr. 1, D-7000 Stuttgart 80, Fed. Rep. of Germany
[2]now at: GKSS Forschungszentrum, Max-Planck-Straße, D-2054 Geesthacht, Fed. Rep. of Germany

1. Introduction

In the past decade the universal low-temperature behaviour of amorphous materials has received increasing attention. If there is no contribution from conduction electrons, e.g. in dielectrics, the thermal and acoustic properties of pure crystalline materials in this temperature range can be very well predicted from macroscopic quantities like density and elastic constants. In amorphous materials, however, a broad spectrum of excitations in addition to the phonons causes drastic deviations from the low-temperature properties of crystals. Typical features of a "glassy" low-temperature behaviour /1/ are a linear temperature dependence of the specific heat, a quadratic temperature dependence of the thermal conductivity, and a high ultrasonic attenuation which is saturable in certain temperature and frequency ranges.

A widely accepted explanation of these anomalies is given in the Tunneling Model /2,3/ which ascribes the low-energy excitations to the tunneling of yet unspecified entities on an atomic scale. One prediction of this model is a distribution of relaxation times of the tunneling states /4/ which is unlimited towards high values. Of course a major research effort is to identify the tunneling entities at least in selected cases. Therefore a number of experiments has been performed on amorphous polymers, where the low-temperature anomalies can be found as well, in order to find out whether tunneling can be traced to structural entities of the polymer (e.g. sidegroup or backbone motions).

Our work was focused on the question whether there are differences in the low-temperature elastic anomalies between the organic polymer glass Polymethylmetacrylate (PMMA) and "classical" anorganic glasses. Therefore we have analyzed the long-time relaxation behaviour of the tunneling states by measuring the relaxation absorption in the range of a few mHz to 10 Hz.

In this paper we will first shortly introduce the Tunneling Model and discuss its predictions for elastic absorption. Then, after having shown our experimental method, we present and discuss the results and their implications onto the Tunneling Model.

2. The Tunneling Model

The Tunneling Model /2,3/ assumes that in an amorphous network atoms or groups of atoms can occupy two energetically nearly equivalent sites. These sites which are assumed to be harmonic potential wells are characterized by their ground state energy $\hbar\Omega$, a potential wall of height V and a distance d between them. The tunneling entities have a mass m. The energy of this double-well potential is determined by an energy difference Δ in the minima of the potential wells and a tunnel splitting $\Delta_o = \hbar\Omega \exp(-\lambda)$, where $\lambda = d(2mV/\hbar^2)^{1/2}$. An ad hoc assumption of the model, which nevertheless has proven very successful, is a constant distribution $p(\Delta,\lambda) = \bar{P}$ of the parameters Δ and λ.

The interaction of the tunneling states with strain fields is characterized by a deformation potential γ. This coupling has three effects: It causes a resonant interaction with ultrasonic strain fields which we shall not consider here, the energy $E = (\Delta^2 + \Delta_o^2)$ of the tunneling states will be modulated by a strain field, and finally the coupling to phonons provides the main relaxation channel for the tunneling states in dielectric solids. The relaxation time is given by /4/

$$\tau^{-1} = \left(\frac{\gamma_l^2}{v_l^5} + 2\frac{\gamma_t^2}{v_t^5} \right) \frac{\Delta_o^2 E}{2\pi\rho h^4} \coth(E/2kT) \qquad (1)$$

where ρ is the mass density of the sample, c the sound velocity, and the indices l and t stand for the phonon polarization. Using (1) the distribution $p(\Delta,\lambda)$ can be converted into

$$p(E, \tau^{-1}) = \bar{p}\tau\, (4(1-\tau/\tau_m))^{-1/2} \qquad (2)$$

with τ_m^{-1} the rate of the fastest relaxing states with $\Delta_o/E=1$. This distribution diverges for $\tau^{-1} \to 0$ leading to an infinite number of tunneling states wich is clearly impossible. One aim of our experiment was therefore to find out whether a lower bound for τ^{-1} can be found.

The relaxation of the tunneling states leads to an elastic loss Q^{-1} /4/

$$Q^{-1} = \pi \bar{P} \gamma^2 / (2\rho v^2) \qquad (3)$$

for the limit $\omega\tau_m^{-1} \ll 1$ which is always fulfilled under our experimental conditions. The temperature independence of Q^{-1} over a wide range of temperatures and frequencies is therefore an important parameter to test the validity of the Tunneling Model.

3. Experimental Technique

The absorption at very low frequencies was measured using the elastocaloric effect /5/. This means that we have used the heating of the sample associated with a periodic deformation to determine the absorption coefficient. If the sample is subject to a sinusoidal stress $\sigma = \sigma_o \cos(\omega t)$, the absorption process generates a heat

$$q_{abs} = Q^{-1} (1/2Y)\ \omega\sigma_o^2 (1+\cos(2\omega t)) \qquad (4)$$

where Y is the Young's modulus. In our case of a rod-shaped sample held at a constant temperature at both ends the temperature increase due to the time-independent term in (4) is

$$T_{abs} = \dot{q}_{abs} L^2 / (\kappa\pi^2) \qquad (5)$$

in the steady-state limit. Here L is the length and κ the thermal conductivity of the sample. A typical example for the temperature variation with an applied stress of $\sigma_o = 2.5$ MPa is shown in Fig.1 together with the stress measured by a load cell.

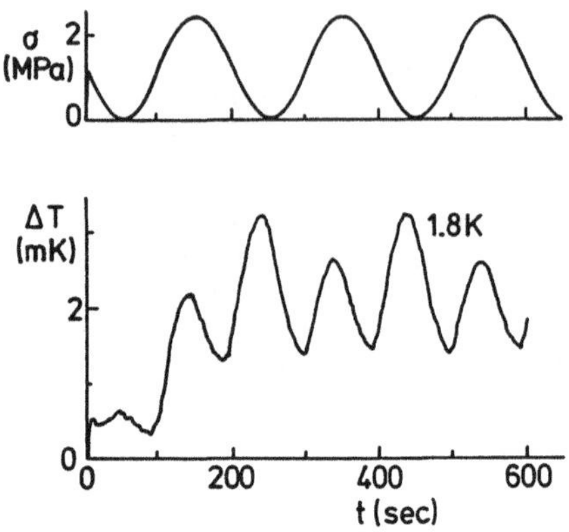

Fig.1: Thermal response of a PMMA sample at 1.8 K. The upper part of the figure shows the applied stress with a frequency of 5 mHz and an amplitude of 2.5 MPa. Besides the temperature variation at twice the stress frequency (4) a component at the fundamental frequency caused by elastic nonlinearity of the sample /5/ is observed.

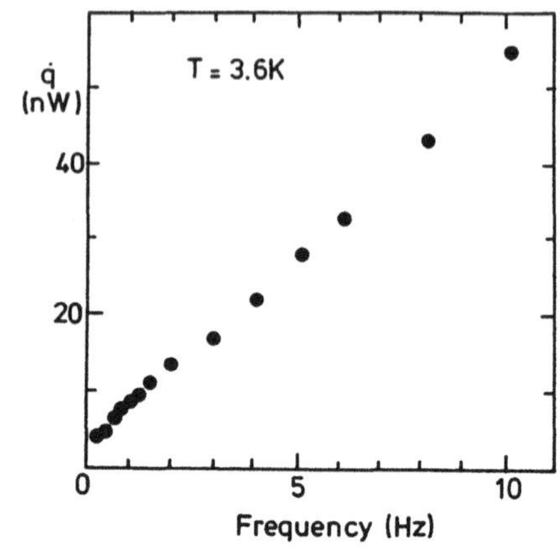

Fig.2: Frequency dependence of the elastic loss in PMMA at T=3.6 K between $5 \cdot 10^{-3}$ and 10 Hz.

In our experiment the rod-shaped sample had a length of 120 mm and a diameter of 4 mm and was placed in the vacuum chamber of a ^{3}He cryostat. Stress could be applied through a bellows-sealed wire by an electrodynamic vibration exciter situated outside the cryostat. The ends of the sample were cemented into Invar cylinders soldered to the pulling wire. These mountings were thermally anchored to the ^{3}He chamber. With this system stress levels of 10 MPa could be reached. Temperature variation (typically a few mK) was detected with the bare sensor of a germanium thermometer attached to the center of the sample. For calibration purposes the sample could be uniformly heated with a resistive wire.

4. Experimental Results and Discussion

In Fig.2 we show the frequency dependence of the absorption at a temperature of 3.6 K. Obviously the prediction of (3) is very well obeyed for a frequency range from $5 \cdot 10^{-3}$ Hz to 10 Hz. There is, however, an unexpected feature in the frequency dependence of Q^{-1}: Absorption does not extrapolate to zero with frequency. This problem shall be discussed a little later.

The temperature dependence of Q^{-1} is shown in Fig.3. The data points have been taken from the slope of the curves for which Fig.2 is an example. In quite good agreement with (3) the absorption is only very weakly temperature dependent. Almost the same temperature dependence, but a higher absolute value of Q^{-1} has been found in a vibrating reed experiment at 1 KHz /6/ and in ultrasonic measurements between 15 and 100 MHz /7/.

Our results show that even at very low frequencies and at temperatures below 1 K an amorphous polymer like PMMA exhibits a remar-

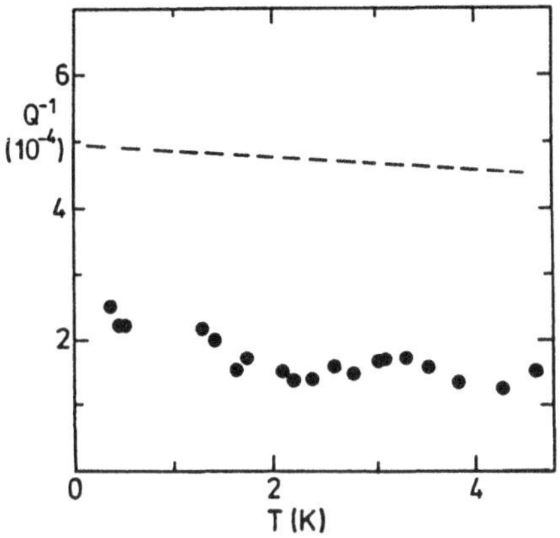

Fig.3: Temperature dependence of the elastic absorption of PMMA at frequencies between $5\cdot10^{-3}$ Hz and 10 Hz. Data from a vibrating reed experiment /6/ are included for comparison.

kable mechanical loss. Similar results have been obtained for anorganic and metallic glasses /5/. Although this result agrees qualitatively with the Tunneling Model, there remain a few puzzles: Why does the sample produce heat at zero frequency and why is our absorption lower than that at higher frequencies? A cutoff of the distribution $p(E,\tau^{-1})$ for small values of τ^{-1} could explain both phenomena: If an applied stress can both increase or decrease the relaxation rate of the tunneling states, a cutoff of $p(E,\tau^{-1})$ will cause an imbalance of both contributions and thus an apparent "creep" of the sample. This creep has been observed not only at the lowest frequencies in our experiment, but also if constant strain was applied to the sample /5/. Also the fact that our absorption is lower than that at higher frequencies can be understood if the minimum relaxation rate τ_{min}^{-1} ($\approx 0.02\ s^{-1}$ /5/ around T=1 K) is of the order of the experimental frequency /8/.

In conclusion we want to state that our experiment on PMMA confirms the mechanical relaxation on very long time scales predicted by the Tunneling Model for amorphous materials. In addition, our results suggest a cutoff in the distribution of relaxation times which has not been observed so clearly in other amorphous materials. On the other hand the question remains open whether this cutoff is connected with peculiarities of the amorphous polymer structure or whether it is shifted to higher values of τ^{-1} with regard to anorganic glasses because of the much lower sound velocity.

References

1. For a review see: Amorphous Solids. Low-Temperature Properties ed. by W.A. Phillips, Springer Ser. Topics in Current Phys., Vol.24 (Springer, Berlin, Heidelberg, New York 1981)
2. W.A. Phillips: J. Low Temp. Phys. 7, 351 (1972)
3. P.W. Anderson, B.I. Halperin, C.M. Varma: Philos. Mag. 25, 1 (1972)
4. J. Jäckle: Z. Phys. 257, 212 (1972)
5. H. Tietje, M. v.Schickfus, E. Gmelin: Z.Phys.B 64, 95 (1986)
6. N. Geis, G. Kasper, S. Hunklinger: In Nonmetallic materials and composites at low temperatures, ed. by G. Hartwig and D. Evans, Vol.3 (Plenum Press, New York 1986)
7. G. Federle, S. Hunklinger: J. Phys. (Paris) 43, C9-505 (1982)
8. The slight temperature dependence of the absorption can also be explained with a cutoff in $p(E,{}^{-1})$. (N. Geis, private communication)

Ultrasonic Attenuation Study of Two-Level Systems in Neutron-Irradiated Quartz upon Heat Treatment

A. Vanelstraete and C. Laermans

K.U. Leuven, Dept. Physics, V.S.H.D., Celestijnenlaan 200 D,
B-3030 Leuven, Belgium

1. Introduction

At low temperatures amorphous solids exhibit dynamical properties which are quite different from those of their crystalline counterparts[1]. They can be attributed to the existence of two-level systems (TLS) [2], and can formally be described by the tunnelling model [3]. Similar anomalies as in amorphous solids were observed in neutron-irradiated quartz [4-8]. They were explained by the presence of TLS similar to those in vitreous silica, but with a density of states which is smaller than in vitreous silica and which increases with neutron dose. The microscopic origin of the TLS is however not known yet [1].

About the origin of the TLS in neutron-irradiated quartz two different opinions have been put forward. Diffuse X-ray scattering studies performed by Grasse et al. [9] showed the presence of 20 Å-defective clusters, which were induced in the crystalline structure during fast neutron-irradiation, and they found evidence that these regions were amorphous. From a comparison of the amorphous fraction with the fraction of TLS, relative to vitreous silica, they concluded that the TLS are contained in these amorphous clusters. This opinion was contradicted by Anderson et al [10] on account of thermal conductivity measurements [6] performed before and after a heat treatment at 840° C. These studies gave evidence that the amorphous clusters disappear upon such a heat treatment, but not the TLS. Whether or not the two-level systems are contained in amorphous clusters is an important question. In order to contribute to this discussion, we have performed ultrasonic attenuation measurements on a comparable neutron-irradiated quartz sample as the one used in [6]. The attenuation was studied before and after a heat treatment at 840° C, which was also similar to the one of [6], in a temperature range 1.4 to 10 K, for two different annealing times. The results are discussed in [11]. They indicate that the TLS are still present after this annealing. However, the density of states has decreased drastically. The observations were discussed in view of the results of both studies cited above. We found evidence that the TLS in neutron-irradiated quartz cannot be explained only in terms of the amorphous clusters, as was originally put forward by Grasse et al. Smaller defects than the 20 Å -amorphous clusters should also give rise to TLS.

Here we report on a more detailed study of the effect of heat treatment on the TLS in neutron-irradiated quartz. Ultrasonic attenuation measurements were performed after an annealing at 840° C for different frequencies: 640 and 420 MHz. It is shown that the low temperature T^3-attenuation is frequency independent. This behaviour was also seen in amorphous solids and in neutron-irradiated quartz prior to heat treatment [12] and is typical for the existence of TLS. This observation therefore provides additional evidence for the presence of TLS after this annealing.

2. Theoretical Considerations

The tunnelling model [3] assumes the existence of a broad spectrum of low-energy excitations associated with two-level tunnelling systems. Each two-level-system can be described by an asymmetric double-well potential of energy splitting Δ and tunnelling parameter λ. One of the basic assumptions concerns the distribution of the parameters. According to the tunnelling model the parameter Δ and λ are independent of each other and uniformly distributed:

$$p(\Delta , \lambda)\, d\Delta d\lambda = \bar{P}\, d\Delta\, d\lambda \tag{1}$$

with $\bar{P}$ a constant, called the density of states of TLS.
For rather high intensities, as is the case in our experiment, the ultrasonic wave changes the energy splittings of the TLS. Their occupation numbers are brought out of thermal equilibrium. A return to the equilibrium state is accomplished by energy transfer to the thermal phonons. This effect gives rise to the so-called relaxation attenuation α and is for low temperatures (regime $\omega\tau_m >> 1$) given by [13] :

$$\alpha = \frac{\Pi^3 k^3\, \bar{P}\, \gamma_\ell^2}{24\rho^2\hbar^4 v_\ell^3} \left(\frac{\gamma_\ell^2}{v_\ell^5} + \frac{2\gamma_t^2}{v_t^5} \right) T^3 ; \tag{2}$$

τ_m is the smallest relaxation time of the TLS. The parameters γ_ℓ and γ_t, respectively, represent the coupling of the longitudinal, transverse thermal phonons with the TLS. Expression (2) indicates that the relaxational attenuation at the lowest temperatures varies as T^3 and is expected to be independent of frequency and directly proportional to the density of states of TLS,$\bar{P}$.

3. Experimental results and discussion

A single crystal of synthetic quartz, labelled N_4, x-cut and of high purity was neutron irradiated up to a dose of 20 X 10^{18}n/ cm² (E $\gtrsim$ 0.1 MeV) or 12 X 10^{18}n/cm² (E $\gtrsim$ 0.3 MeV). Neutron activation analysis yielded as impurity content in µg/g:

Na	2.3	;	Fe	0.7	;	Ga	0.06	;	Ge	0.0016 ;
Cℓ	9.8	;	Zn	0.4	;	Sb	0.06	;	Co	0.0022.

The Al-impurity-content was below detectable limit which is 10 µg/g. The mass density of the sample was measured by a hydrostatic method: ρ = 2.638 ± 0.002 g/cm³.

The longitudinal ultrasonic attenuation of this sample has been measured before and after heat treatment in a temperature range 1.4 to 10 K, using the pulse echo technique. The annealings were performed at 840° C in an argon atmosphere. The sample was brought to temperature in 22 min, kept at 840° C for about 15 min, then allowed to cool, reaching 500° C in 35 min and 200°C in about 2 hours. After heat treatment, the attenuation has been measured for two frequencies: 640 and 420 MHz. The low-temperature results are plotted in fig. 1 on a log-log-scale. A temperature-independent residual attenuation α_0, attributed to geometrical factors,is subtracted from the data. Before heat treatment, the attenuation follows a T^3-law at the lowest temperatures. This behaviour, which is similar to that found in many glasses, was already observed before in neutron-irradiated quartz [12,14] and very recently also in electron-irradiated quartz [15]. It is typical for the relaxation attenuation due to TLS, given by expression (2).

The effect of heat treatment on the low-temperature attenuation in neutron-irradiated quartz has already been discussed in detail in [11] for one frequency and for different annealing times. After heat treatment, a reduction of the ultrasonic attenuation was observed and the decrease was found to be dependent on annealing time. As can be seen in fig. 1, the present attenuation data performed on neutron-irradiated quartz after an annealing at 840° C for two different frequencies show a T^3-behaviour at the lowest temperatures, as was also observed in [11]. This behaviour is also observed before heat treatment and indicates the presence of two-level systems. Moreover, the low-temperature T^3-behaviour is frequency independent (see fig.1). This observation is in agreement with the predictions of the tunnelling model for the low-temperature relaxational attenuation (see (2)) and therefore gives additional evidence for the presence of TLS after this heat treatment. The drastic decrease of the attenuation after annealing,

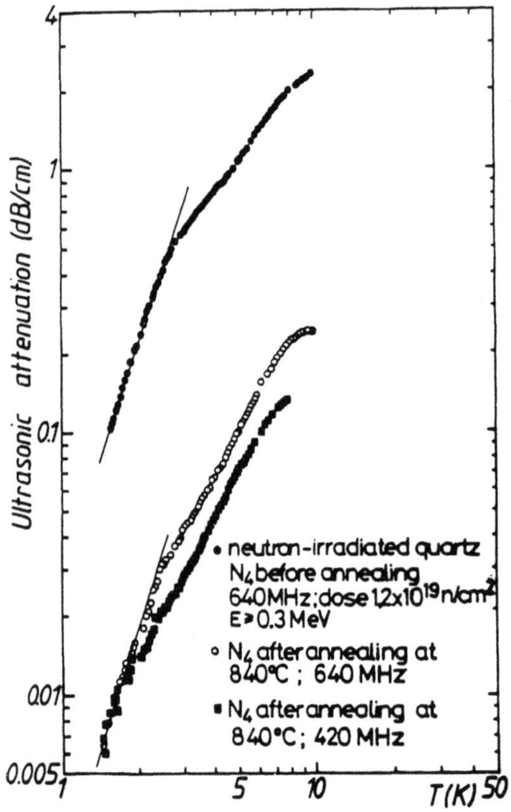

Fig. 1.

Ultrasonic attenuation as a function of temperature for neutron-irradiated quartz N_4, dose: 1.2×10^{19} n/cm² (E > 0.3 MeV)

● before annealing, f = 640 MHz

○ after annealing at 840°C, f = 640 MHz

■ after annealing at 840°C, f = 420 MHz.

however, indicates a decrease of the density of states of TLS. Indeed, from (2) it can be deduced that the density of states of TLS, $\bar{P}$, has decreased with a factor of about twelve for this annealing. This reduction might be related to structural changes known to be the consequence of this annealing [9]. More annealing studies are in progress to investigate the effect of annealing temperature on the two-level systems in neutron-irradiated quartz.

4. Conclusion

The present ultrasonic attenuation measurements performed on neutron-irradiated quartz after a heat treatment at 840° C show additional evidence for the presence of TLS after this annealing. Indeed the ultrasonic attenuation data performed for two different frequencies show the frequency independence of the T^3-behaviour, which is predicted by the tunnelling model for the low-temperature relaxation attenuation. Although the TLS are present, the density of states of TLS has decreased with a factor of twelve for this annealing.

Acknowledgements

The authors thank the Studiecentrum voor Kernenergie, Mol (Belgium) for performing the irradiations and the Oak Ridge National Laboratory (Tennessee, U.S.A.) and especially R. Weeks for neutron activation analysis. They also want to thank the Belgian Interuniversitair Instituut voor Kernwetenschappen for financial support.

References

1. For a review of the low-temperature anomalies in amorphous solids see: W.A. Phillips (ed.) Amorphous Solids: Low-Temperature Properties, Springer-Verlag (1981).
2. S. Hunklinger and W. Arnold, in Physical Acoustics, Vol.12, W.P. Mason and R.N. Thurston, eds., Academic Press, Oxford (1976).
3. P.W.Anderson, B.I. Halperin and C.M. Varma, Philos. Mag.,25, 1 (1972); W.A. Phillips, J. Low Temp. Phys., 7, 351 (1972).
4. C. Laermans, Phys. Rev. Lett., 42, 250 (1979).
5. B. Golding, J.E. Graebner, W. H. Haemmerle and C. Laermans, Bull. Am. Phys. Soc. 24, 495 (1979); B. Golding and J.E.Graebner, in Phonon Scattering in Condensed Matter, H.J. Maris, ed., Plenum Press, 11(1980).
6. J.W. Gardner and A.C. Anderson, Phys. Rev. B, 23, 474 (1981).
7. A.M. de Goer, M. Locatelli and C. Laermans, J. Physique,42, C6-78 (1981).
8. For a review see: C. Laermans in Structure and Bonding in Non-Crystalline Solids eds. G. Walrafen and Revesz, (Plenum Press, 1986) p. 329.
9. D. Grasse, O. Kocar, H. Peisl, S.C. Moss and B. Golding, Phys. Rev. Lett., 46, 261 (1981).
10. A.C. Anderson, J.A. McMillan and F.C.Walker, Phys. Rev. B, 24, 1124 (1981).
11. C. Laermans and A. Vanelstraete, Phys. Rev. (to be published).
12. C. Laermans, V. Esteves and A. Vanelstraete, Rad. Effects, 97, 175 (1986).
13. A. Raychaudhuri and S. Hunklinger, Z. Phys. B, 57, 113 (1984).
14. C. Laermans, V. Esteves, in Phonon Scattering in Condensed Matter, W. Eisenmenger, K. Lassman, S. Döttinger, eds., 407 (1983).
15. C. Laermans and A. Vanelstraete, Phys. Rev. B, 34, 1405 (1986).

The One-Dimensional Quantum Rotator H_2S^- in Several Alkali Halides

D. Schoemaker and E. Goovaerts

Physics Department, University of Antwerp, B-2610 Wilrijk, Belgium

I. Introduction

Reorientation motions of molecules in solids can vary from classical thermally activated jumping motions between equivalent orientations to quantum mechanical free rotation. In between stand the tunneling- and phonon activated tunneling-motions[1]. In the classical regime the reorientation rates are strongly temperature dependent and the motion of the molecule can vary from standing still to average reorientation rates approaching the quantum rotation frequency of the free molecule. A glance at the literature reveals that thermally activated motion is a common occurence.

Genuine free quantum rotation of molecules in solids is much less common. A well-known example is H_2 in solid molecular hydrogen which even at very low temperatures exhibits unhindered rotation[2]. This is proven by the observation in solid hydrogen of sharp rotational Raman lines at the free H_2 rotation energy. Furthermore, the free rotation leads to the existence of the para-and ortho-modifications of molecular hydrogen,the origin of which is the interdependence of the nuclear spin- and the molecular rotational-states imposed by the nuclear spin statistics[2]. However, observing para-and ortho-states in molecules does not necessarily imply free rotation: Nuclear permutation symmetry is equally important when molecules exhibit hindered quantum rotation.

The present paper deals with an electron-spin-resonance (ESR) study of the quantum rotation of a H_2S^- molecule in several alkali halides. The ESR spectra, which had already been observed earlier[3], have only recently been interpreted correctly and extensively discussed by us[4], and we will present here some additional data.

The alkali halide samples used in these experiments contained SH^- impurities introduced by growing the crystals in a $5 torr$ atmosphere of H_2S. The SH^- concentrations were estimated to be about $0.1 mol\%$.

II. Analysis of the H_2S^- ESR spectra.

The substitutional H_2S^- molecule is produced as follows. First interstitial hydrogen atoms, H_i^0, and substitutional S^- ions are produced by photodecomposition of SH^-. This is achieved by irradiating for about an hour a, e.g., $KCl : SH^-$ sample at $20K$ with the UV light of a deuterium lamp, filtered by a $206nm$ broad band interference filter. Subsequently the sample is warmed to $120K$ for a few minutes. At this temperature the H_i^0 atoms have become mobile and some of them are trapped by substitutional SH^- to form substitutional H_2S^- molecules. In ESR one then observes two simple spectra which are intimately related. One is a single

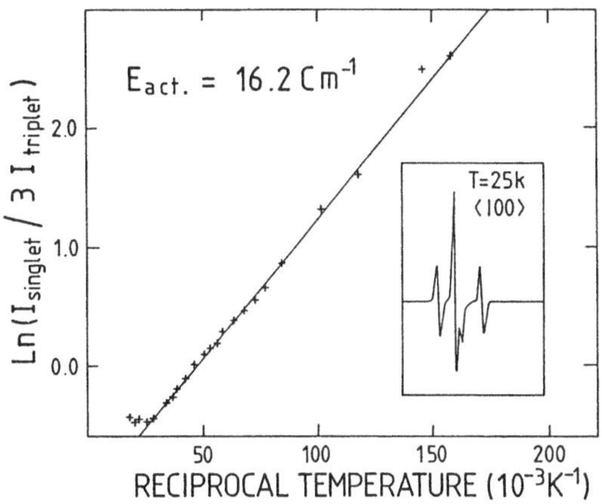

Fig. 1 : Arrhenius plot of the relative intensity $I(singlet)/3I(triplet)$ of the H_2S^- ESR spectrum in KCl (given in the insert), showing that with increasing temperature the triplet lines are enhanced at the expense of the singlet line, with an activation energy of $16.2cm^{-1}$.

anisotropic line (the singlet) and the other consists of three equidistant hyperfine (hf) lines of equal intensities (the triplet) and both spectra possess axial symmetry around $< 111 >$. An example of an ESR spectrum is shown in the insert of Fig. 1. Table I presents the g-and proton hyperfine (hf) parameters resulting from a fit to a standard spin-Hamiltonian. It is seen that the g-parameters of the singlet and triplet spectra in KCl are very similar. The hf parameters corresponding to ^{33}S (nuclear spin 3/2 and 0.78% natural abundance) are not presented in Table I and were determined in Ref. 3 for KCl samples containing 11.56% enriched $^{33}SH^-$ and the results are given in Ref. 4.

A second correspondence between the two spectra concerns their thermal formation and decay properties: A pulse anneal experiment reveals that the formation and decay curves of both spectra coincide perfectly. This already strongly points to the fact that the singlet and triplet spectra originate from the same center. The

Table I : The spin-Hamiltonian parameters of the H_2S^- and H_2Se^- centers in several alkali halides and the rotational constant B_z (in cm^{-1}). The hydrogen hyperfine parameters and the linewidth ΔH are given in mT. The ^{33}S hf parameters for KCl can be found in Ref. 4. Note that the quoted proton hf values are not the principal values (see Ref. 4).

Crystal	Defect		$g^a_{\parallel}$ $< 111 >$	$g^a_{\perp}$	$A^b_{\parallel}$ $< 111 >$	$A^b_{\perp}$	ΔH	B_z
KCl	H_2S^-	Singlet	1.985	2.211			0.40 ± 0.05	16.2 ± 0.7
		Triplet	1.986	2.206	3.29^b	2.17^b		
KBr	H_2S^-		1.987	2.221	2.58^c	2.11^c	1.7 ± 0.1	10.1 ± 1.0
KI	H_2S^-		1.987	2.246	–	–	2.6 ± 0.2	–
RbCl	H_2S^-		1.966	2.283	–	–	2.3 ± 0.2	–
KCl	H_2Se^-		1.767	2.548	–	–	0.9 ± 0.2	–

[a] Precision : ± 0.001 for H_2S^- in KCl and ± 0.002 for the others.
[b] Precision : $\pm 0.02mT$.
[c] Precision : $\pm 0.05mT$.

third and most significant fact confirms this. The relative intensities of the singlet and the triplet change with the ESR observation temperature, the singlet being strongest at lower temperatures. If one plots the logarithm of the singlet/triplet intensity ratio versus $1/T$, one obtains a straight line over more than a decade as shown in Fig. 1 for KCl. Clearly, the triplet grows at the expense of the singlet with an activation energy of $(16.2 \pm 0.7) cm^{-1}$. The conclusion is inescapable: The two ESR spectra originate from the same molecule, the singlet from the ground state and the triplet from an excited state at $16.2 cm^{-1}$.

III. The one-dimensional quantum rotator model.

The above observations are straightforwardly explained by assuming that one is dealing with a H_2S^- molecule rotating around C_{2v} axis which is oriented along a $< 111 >$ direction. As a result the $H-H$ axis is perpendicular to the rotation axis and only the two protons determine the moment of inertia of the molecule. A schematic model of the center is presented in Fig. 2. Such a one-dimensional quantum rotator is described by

$$H_{rot} = B_z J_z^2$$

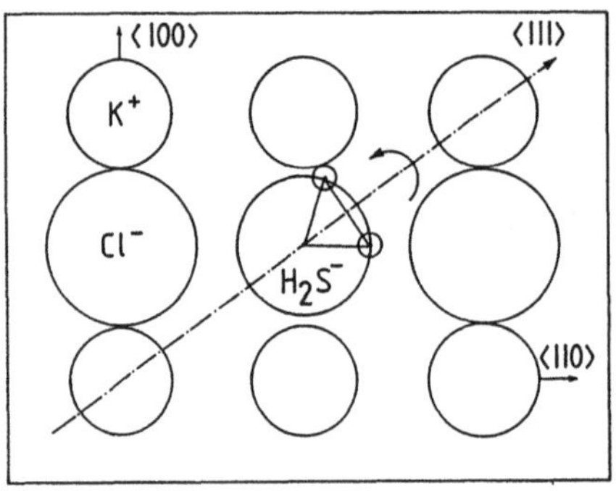

Fig. 2: Schematic model in a $\{110\}$ plane of the subsitutional H_2S^- center in KCl also showing the $z \parallel < 111 >$ axis around which the molecule is rotating.

Here J_z is the molecular angular momentum around the rotation axis, $B_z = \hbar^2/2I$ is the rotational constant, in which $I = 1/2md^2$ is the moment of inertia with m the proton mass and d the distance between the two protons. The energy levels are given by

$$E(M_J) = B_z M_J^2$$

Because of the fermion character of the two H_2S^- protons the product of the molecular rotation wavefunction and the nuclear spin wavefunctions, $\Omega(M_J)\Psi(I, M_I)$, must be antisymmetric under exchange of the two protons whose nuclear spins $1/2$ combine to $I = 0, 1$. The ground state is the nuclear singlet para-state $\Omega(M_J = 0)\Psi(I = 0)$, and the first excited state at B_z is the nuclear triplet ortho-state $\Omega(M_J = \pm 1)\Psi(I = 1, M_I = 0, \pm 1)$. This corresponds exactly to the experimentally observed behavior of the singlet and triplet ESR lines (Fig. 1) and there is no doubt that we are indeed observing a H_2S^- molecule which is rotating around a $< 111 >$ direction.

The electronic configuration of H_2S^- is[4] $...(1b_2)^2(3a_1)^2(1b_1)^2(4a_1)^1$. The ground orbital $4a_1$ is composed out of an anti-bonding combination of a $s-p_z$ hybrid on the sulfur and two $s-p$ hybrids on the hydrogens. We refer to Ref. 4 for an in-depth analysis of the g-components and the ^{33}S hf interaction based on this configuration scheme. The analysis establishes that the rotation axis coincides with the $< 111 >$ oriented C_{2v} axis of the molecule as depicted in Fig. 2.

Because it is taking place around the C_{2v} axis along which the sulfur $3p_z$ lobe is oriented, the rotation does not affect the measured ^{33}S hf interaction. However, the rotation does influence the measured proton hf interaction. As a result, the proton hf values quoted in Table I do not constitute the principal values of the hf tensor. They merely represent the proton hf separations measured parallel and perpendicular to the C_{2v} axis. A proper analysis has to take the rotational motion into account. Such an analysis would yield interesting structural information about the H_2S^- molecule such as, e.g., the $H-S-H$ angle. Unfortunately, our attained accuracy did not warrant such an analysis and none was attempted[4].

The $B_z = 16.2cm^{-1}$ result allows one to calculate the $H-H$ distance in H_2S^- provided the rotation is unhindered. One finds $d = 1.4Å$. This may be compared to[5] $d = 1.87Å$ for the neutral H_2S molecule and $0.74Å$ for H_2. Placing, as an approximation to the molecular structure, the two hydrogens on the surface of a S^- sphere (radius $\approx 1.35Å$) yields an $H-S-H$ angle of roughly 55°. In the neutral H_2S molecule this angle is[5] 92.2^o. The fact that the singlet and the triplet have slightly different $g_\perp$ values illustrates the effect that the spin-rotation interaction has on the $g_\perp$ factor of the triplet state.

In a KCl crystal containing a mixture of SH^- and SD^- impurities the ESR spectra of D_2S^- and HDS^- molecules have been observed[4] whose properties are consistent with those observed for H_2S^-. For HDS^-, the nuclear spin statistics do not play a role.

IV. Hindered versus unhindered rotation.

The foregoing analysis shows that the molecule is rotating around the $< 111 >$ oriented C_{2v} axis. The question arises whether it is freely rotating or to what extent it is experiencing a hindering potential of C_{3v} symmetry. Turning on a static hindering potential will progressively reduce the rotation frequency until it is more appropriate to speak of a tunneling reorientation motion. The transition is a gradual one, but the importance of the nuclear spin statistics in determining the symmetry of the molecular states remains in the tunneling case. Qualitatively the two regimes cannot be distinguished.

If an independent determination of the $H-H$ distance in H_2S^- were available either from spectroscopic measurements or from molecular structure calculations, one would be able to compare the calculated moment of inertia with the one derived from the experimental rotational constant $B_z = 16.2cm^{-1}$. However, such information is not available. As a result little can be said about the absence or presence of a hindering potential on H_2S^- in KCl from the KCl ESR data alone. The value $d = 1.4Å$ derived above for the $H-H$ distance is an upper limit and corresponds to the case of genuine free one-dimensional rotation of H_2S^-. A hindering potential will lead to a smaller value. A lower limit is undoubtedly set by the proton internuclear distance $d = 0.78Å$ of molecular H_2 because it corresponds to an exclusively covalent bond between two hydrogen atoms. The rotational constant for the latter molecule is[2] $B = 59.1cm^{-1}$ and B_z of H_2S^- is unlikely to exceed this value. In any case, the proton distance should be smaller than $d = 1.87Å$ in neutral H_2S and this is indeed the case.

In order to get an idea of the presence of a hindering potential, two type of experiments were performed. First we attempted to measure the Raman scattering spectrum of the H_2S^- in KCl. The $M_J = 0$ to $M_J = 2$ transition is Raman-active and should be at $4B_z = 64.8cm^{-1}$ if the H_2S^- rotator is indeed completely unhindered, and deviations from this value would give an idea about how much the rotator is actually hindered. However, several careful searches failed to yield a Raman line in a broad region around this value. Undoubtedly, both the concentration of the H_2S^- centers in the crystals available to us and the scattering cross section were too low.

Secondly, it was tried to observe the H_2S^- quantum rotator in other alkali halides. Performing the same production procedure as for KCl we were able to observe in SH^- doped KBr, KI and $RbCl$ crystals simple ESR spectra exhibiting axial symmetry around $< 111 >$ and possessing very similar g-factors as in KCl. On the basis of the observed symmetry all these spectra are attributed to rotating H_2S^- molecules rather than substitutional S^- ions. The latter possess orthorhombic symmetry with $z \parallel < 110 >$[6]. In KBr the triplet was positively identified and although its central line was not resolved from the singlet one, it was possible to perform an Arrhenius analysis. The result, $B_z(KBr) = (10.1 \pm 1.0)cm^{-1}$, is included in Table I. In KI the triplet was seen as unresolved structure on the flanks of the singlet line but an analysis was not readily feasable. In $RbCl$ no triplet was observed either because it is not resolved from under the singlet line or because the B_z value is so high that the triplet state is not sufficiently populated. For completeness, Table I includes the parameters of a $< 111 >$ symmetry center observed in a $KCl : SeH^-$ crystal and which might possible originate from a H_2Se^-. The much lower B_z value in KBr compared to KCl and the possibility that B_z in $RbCl$ may even be larger than in KCl strongly indicate that even in KCl there is no reason to think that H_2S^- rotates unhindered. It would be highly desirable to have either independent spectroscopic data or reliable molecular structure calculations on the H_2S^- radical. A more careful analysis of the ESR spectra in KI, but especially in $RbCl$, may throw further light on this problem.

The authors want to thank L. O. Schwan for supplying some of the crystals used in these experiments, and A. Bouwen for experimental support. This work was supported by IIKW, the NFWO and the Geconcerteerde Acties to which the authors are greatly indebted.

1. V. Narayanamurti and R. O. Pohl, *Revs. Mod. Phys.* 42, 201 (1970)
2. I.F. Silvera, *Revs. Mod. Phys.* 52, 393 (1980)
3. A. Hausmann, *Z. Phys.* 192, 313 (1966)
4. D. Schoemaker, J.F. Zhou and E. Goovaerts, *Phys. Rev.* B33, 25 (1986)
5. G. Herzberg, *Infrared and Raman Spectra of Polyatomic Molecules*, (Van Nostrand Reinhold, New York, 1945)
6. L.E. Vannotti and J.R. Norton, *Phys. Rev.* 174, 448 (1968)

Relaxation Times of k≈ 0 Rotons in Parahydrogen Crystals at Low Concentrations of Orthohydrogen Impurities

E. Goovaerts, X.Y. Chen[1], C. Sierens, A. Bouwen, and D. Schoemaker

Physics Department, University of Antwerp (U.I.A.),
Universiteitsplein 1, B-2610 Wilrijk, Belgium

1. Introduction

The simplest of molecular crystals, solid hydrogen, has been very intensively investigated and was found to possess interesting and peculiar properties, among which is the quantum nature of both the rotational and translational motions of the molecules, which are highly decoupled from each other [2,3]. The two species of the H_2 molecule, para-H_2 and ortho-H_2 , with total nuclear spin $I = 0$ and $I = 1$, respectively, can only occur in states with given rotational quantum numbers J, which must be even and odd, respectively. Even in the solid, the molecules behave as nearly free quantum rotators only perturbed by the anisotropic intermolecular interactions which are weak ($\approx 1cm^{-1}$) compared to the rotational constant of H_2 ($B = 59cm^{-1}$). Very recently, a calculation was performed of the dephasing time (T_2) of the $J = 2$ Raman-active rotational excitations, called rotons, in para-H_2 resulting from the intermolecular interactions in para-H_2 [4].

In this paper we present an experimental investigation of the roton relaxation in very pure para-H_2 crystals (ortho-H_2 concentration $c_{ortho} \approx 0.05\%$) and of the influence of ortho-H_2 impurities at low concentrations on this process. Frequency domain measurements were achieved by high-resolution Raman experiments with a tandem Fabry-Perot–monochromator set-up [5], and direct time domain information was obtained from picosecond stimulated Raman experiments [6]. A comparison is made of the theoretical and experimental values of the relaxation times in pure para-H_2 . Also, a brief discussion is given of the mechanism of line broadening and corresponding shortening of the lifetime of the $k \approx 0$ rotons by ortho-H_2 impurities.

2. Measurements in the frequency domain

In Fig. 1 unpolarized high-resolution Raman spectra ($\lambda = 514.5nm, T = 9K$) of the $J = 2$ rotons in para-H_2 with different concentrations of ortho-H_2 impurities are shown. The three Raman lines correspond to rotons with different values of $|M| = 0,1$ and 2, with M the rotational quantum number quantized along the hexagonal axis of the hcp crystal structure. The lines are split by the anisotropic molecular interactions [3]. The experimental lineshapes are accurately fitted by a convolution of Lorentz profiles with a Gaussian instrumental transmission function (FWHM≈ $0.12cm^{-1}$). These measurements clearly demonstrate the strong dependence of the linewidth on ortho-H_2 concentration. Therefore measurements were performed in extremely pure para-H_2 crystals [$c_{ortho} = (0.05 \pm 0.03)\%$] (see Fig. 1a). The relaxational lifetimes of the rotons corresponding to the deconvoluted Raman linewidths are given in Table I and compared to their calculated values. The agreement in magnitude between the calculated and measured lifetimes is as good as can be expected for such a complex system. It is however interesting to note that, while the calculated lifetimes for the three $|M|$-values are nearly equal, a difference of up to 40% is found between the measured values.

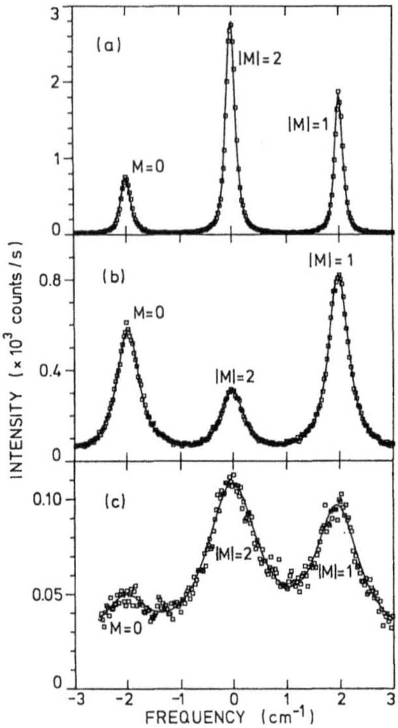

Fig. 1 : Experimental Raman spectra (open squares) and fitting result (solid line) of the $J = 0$ to $J = 2$ rotational excitations in para-H_2 (100mW Ar^+ laser excitation at $514.5nm$) measured at $T = 9K$ in single crystals with different concentrations of ortho-H_2 : (a) $c_{ortho} = 0.05 \pm 0.03\%$, (b) $c_{ortho} = 2.6 \pm 0.3\%$ and (c) $c_{ortho} = 6.3 \pm 0.6\%$. The frequency scale is relative to the position of the central peak at $355cm^{-1}$. The relative intensities of the peaks vary from one spectrum to another due to different orientations of the crystal with respect to the laser polarization.

Table I : Dephasing times, T_2 (in ps), of the $J = 2$ Raman-active rotons in pure para-H_2 determined, (a) from high resolution spectral measurements [5], (b) from time-resolved experiments, and (c) from theory [4].

	$M = 0$	$\|M\| = 1$	$\|M\| = 2$
a	97 ± 9	169 ± 16	123 ± 9
b	71 ± 5	102 ± 7	81 ± 9
c	88	93	92

The linewidths from a series of measurements at different ortho-H_2 concentrations are plotted in Fig. 2 as a function of c_{ortho}. First, this demonstrates that the results in our purest samples should be very close to those in crystals without ortho-H_2 . Second, one observes a near-linear behavior of the linewidths as a function of c_{ortho}. Such a behavior is obtained when plane-wave-like excitations, the rotons, interact with randomly distributed non-interacting scattering centers, in this case the ortho-H_2 impurities.

3. Picosecond stimulated Raman experiments

We have very recently performed time-resolved measurements (resolution $\approx 3ps$) of the phase relaxation of the $J = 2$ rotons in para-H_2 employing a pulse-and-probe technique of stimulated Raman scattering [6]. The excitation and the probing pulses each consist of a pair of picosecond pulses from two synchronuously pumped mode-locked dye lasers. The lasers are tuned such that their frequency

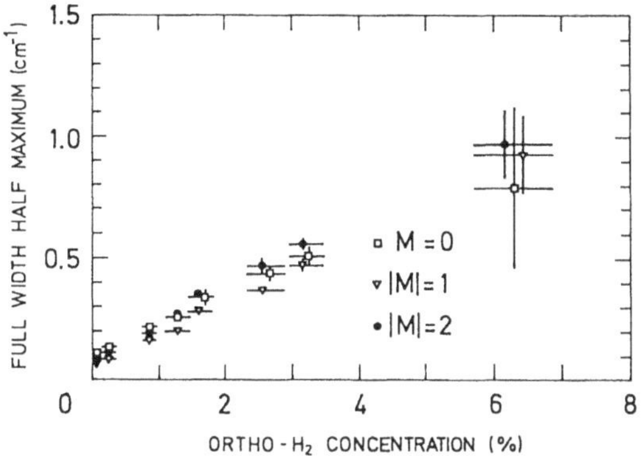

Fig. 2 : Experimental linewidth (FWHM) of the roton Raman scattering in para-H_2 as a function of concentration of ortho-H_2 impurities ($T = 9K$).

difference matches the Raman transition frequency. The stimulated gain signal is then measured as a function of the time delay between pump and probe pulses. These measurements yield relaxation times in generally good agreement with those obtained in the frequency domain experiments. As an example, the decay of the $M = 0$ stimulated Raman signal in very pure para-H_2 is shown in Fig. 3 in a semi-logarithmic plot.

The recent time-resolved experiments were found to be more convenient than the previously described spectral measurements, especially for the accurate determination of relatively long relaxation times ($\tau \geq 50ps$). Moreover, it was possible to select out a specific $|M|$ Raman transition by the choice of polarization of the laser pulses, without much loss of signal intensity. The fit of the decay curve in Fig. 3 very convincingly demonstrates the exponential behavior of the roton

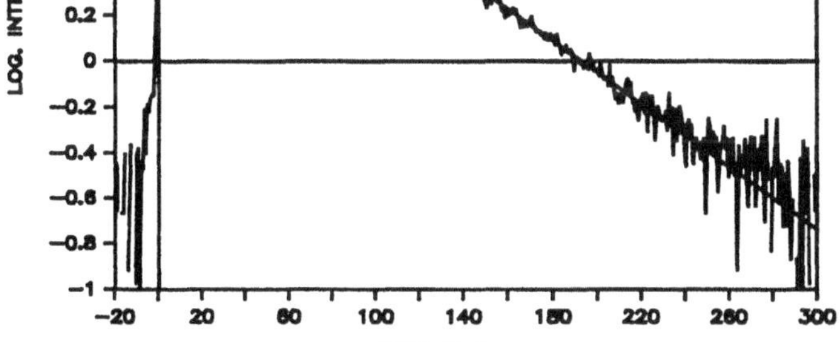

Fig. 3 : Dephasing of the $M = 0$ roton in pure para-H_2 [$c_{ortho} = (0.14 \pm 0.02)\%$] as measured by time-resolved stimulated Raman scattering ($T = 9K$). The plot is semi-logarithmic and the decay was fitted to an exponential decay with relaxation time : $T_2 = (63 \pm 2)ps$.

relaxation. The corresponding spectral linewidths are so small that the Lorentz lineshape could hardly be distinguished from, e.g., a Gaussian. Furthermore, at low c_{ortho}, and correspondingly long relaxation times a systematic discrepancy was found between the frequency- and time-domain results. This discrepancy is ascribed to a systematic error in the deconvolution of the measured lineshapes for very narrow lines. The dephasing times measured in the time-resolved experiments and extrapolated to zero ortho-H_2 concentration are also given in Table I and are seen to be in even closer agreement with the theoretical values. In spite of this better agreement, which however could be fortuitous, it is seen that the large differences in relaxation times for different $|M|$ are still maintained, in contrast with the theoretical results.

4. Discussion

Measurements of both the relaxational linewidth and the time-dependent relaxation of the $J = 2$ rotons in para-H_2 have demonstrated a strong influence of ortho-H_2 impurities on this process. The roton dephasing is found to be of exponential nature to a high degree, which essentially excludes the influence of inhomogeneous broadening effects. Extrapolation of the relaxation times for different c_{ortho} yields for pure para-H_2 crystals magnitudes of the dephasing times in good agreement with the theoretical values. A much larger difference is however found between the T_2-times for different $|M|$ rotons than expected from the calculation of Vanhimbeeck *et al.* [4]. This is possibly due to the approximations, both in the Hamiltonian of the system and in the methods of calculation. It is not expected, but not yet excluded, that in the experiments residual impurities still determine the relaxation times, e.g., the isotopic impurity HD ($c_{HD} = 0.03\%$). Experiments are being undertaken to eliminate this possibility.

The ortho-H_2 impurities can be considered as quasi-isolated scattering centers which scatter the Raman-active rotons at $k \approx 0$ into rotons with nearly equal energy at k-vectors throughout the Brillouin zone. Dispersion relations of the rotons can be calculated taking into account the electric quadrupole-quadrupole (EQQ) interaction between the molecules. A calculation of the transition probabilities between roton excitations is underway which treats the presence of one ortho-H_2 impurity as equivalent to the absence of one para-H_2 molecule.

The authors would like to thank M. Vanhimbeeck, H. De Raedt and M. De Mazière for many valuable discussions, and P. Casteels for excellent experimental support. One of us (X.Y.C.) is grateful to the Chongqing University (China) for a scholarship. Support from the I.I.K.W. and the Geconcerteerde Acties is gratefully acknowledged.

1. Permanent address: Physics Department, Chongqing University, Chongqing, Sichuan, China.
2. I.F. Silvera, *Rev. Mod. Phys.* 52, 393 (1980).
3. J. Van Kranendonk in *Solid Hydrogen* (Plenum Press, New York, 1983).
4. M. Vanhimbeeck, H. De Raedt, A. Lagendijk and D. Schoemaker, *Phys. Rev.* *B*33, 4264 (1986). A unit conversion error arising from confusion of the spectroscopic $\tilde{\lambda} = 1/\lambda$ and the wavevector $k = 2\pi/\lambda$ has resulted in this paper in T_2-values overestimated by a factor 2π.
5. E. Goovaerts, X.Y. Chen, A. Bouwen and D. Schoemaker, *Phys. Rev. Lett.*57, 479 (1986).
6. M. De Mazière and D. Schoemaker, *J. Appl. Phys.*58, 1439 (1985).

Index of Contributors

Abed, K.J. 65
Adel, H. den 113
Agulló-Rueda, F. 38
Alefeld, B. 113
Alsanoosi, A.M. 28

Baetz, L. 2
Barker, G.J. 65
Bartolomé, J. 33,38
Blank, H. 58,62
Bouwen, A. 217
Brom, H.B. 113
Buekenhoudt, A. 72
Burriel, R. 33

Calleja, J.M. 38
Cardona, M. 38
Cavagnat, D. 24
Chen, X.Y. 217
Chihara, H. 76
Clough, S. 24,65
Cockcroft, J.K. 58,62
Collinson, S.J.M. 86
Crits, E. 54
Cvikl, B. 81

Damay, P. 103
Dianoux, A.J. 103,153

Gilchrist, J. 118
Gmelin, E. 204
González, D. 33
Goovaerts, E. 212,217
Grabert, H. 130
Green, R.M. 28
Grondey, S. 98
Gutsmiedl, P. 153,158

Häusler, W. 107
Heger, G. 19
Heidemann, A. 44,98
Hochstrasser, R.M. 186
Horsewill, A.J. 28,65
Hoser, A. 19
Hüller, A. 2
Hunklinger, S. 194
Hunt, G.M. 86

Ingman, L.P. 122

Jencic, I. 81

Kearley, G.J. 58,62
Klamt, A. 177
Knözinger, E. 93
Koivula, E. 122

Laermans, C. 208
Lalowicz, Z.T. 122
Langel, W. 93
Leclercq, F. 103

Magerl, A. 24,153
Morr, W. 163

Navarro, R. 33
Neumaier, K. 153,158

Odin, J. 118

Palacios, E. 33,38
Peternelj, J. 81
Peyrard, J. 118
Pierre, M. 186
Pintar, M.M. 81

Prager, M. 93,98,107
Prandl, W. 19
Press, W. 98,103
Punkkinen, M. 122

Richter, D. 140

Schickfus, M. v. 204
Schiekhofer, M. 158
Schoemaker, D. 212,217
Schrems, O. 93
Sierens, C. 217
Smalley, M.V. 86
Stanislawski, J. 107
Statham, M.L. 86
Steinbinder, D. 153
Strößner, K. 38
Svare, I. 182
Syassen, K. 38

Takeda, S. 76
Teichler, H. 167
Thomas, R.K. 86
Tietje, H. 204
Trevino, S.F. 9
Trommsdorff, H.P. 186

Van Gerven, L. 54,72
Vandemaele, G. 72
Vanelstraete, A. 208
Vettier, C. 24

Wagner, M. 172
Weiss, G. 163
Wipf, H. 153,158,163

Ylinen, E.E. 122

Springer Proceedings in Physics

Volume 1

Fluctuations and Sensitivity in Nonequilibirum Systems

Proceedings of an International Conference, University of Texas, Austin, Texas, March 12-16, 1984
Editors: **W. Horsthemke, D. K. Kondepudi**
1984. 108 figures. IX, 273 pages. ISBN 3-540-13736-X

Volume 2

EXAFS and Near Edge Structure III

Proceedings of an International Conference, Stanford, CA, July 16-20, 1984
Editors: **K. O. Hodgson, B. Hedman, J. E. Penner-Hahn**
1984. 392 figures. XV, 533 pages. ISBN 3-540-15013-7

Volume 3

Nonlinear Phenomena in Physics

Proceedings of the 1984 Latin American School of Physics, Santiago, Chile, July 16-August 3, 1984
Editor: **F. Claro**
1985. 110 figures. IX, 441 pages. ISBN 3-540-15273-3

Volume 4

Time-Resolved Vibrational Spectroscopy

Proceedings of the 2nd International Conference Emil-Warburg Symposium, Bayreuth-Bischofsgrün, Federal Republic of Germany, June 3-7, 1985
Editors: **A. Laubereau, M. Stockburger**
1985. 187 figures. XI, 301 pages. ISBN 3-540-16175-9

Volume 5

Physics of Finely Divided Matter

Proceedings of the Winter School Les Houches, France, March 25-April 5, 1985
Editors: **N. Boccara, M. Daoud**
1985. 168 figures. IX, 365 pages. ISBN 3-540-15885-5

Volume 6

Aerogels

Proceedings of the First International Symposium, Würzburg, Federal Republic of Germany, September 23-25, 1985
Editor: **J. Fricke**
1986. 168 figures. VIII, 205 pages. ISBN 3-540-16256-9

Volume 7

Nonlinear Optics: Materials and Devices

Proceedings of the International School of Materials Science and Technology, Erice, Sicily. July 1-14, 1985
Editors: **C. Flytzanis, J. L. Oudar**
1986. 185 figures. VIII, 249 pages. ISBN 3-540-16260-7

www.ingramcontent.com/pod-product-compliance
Lightning Source LLC
Chambersburg PA
CBHW061208240926
51846CB00006B/1144

* 9 7 8 3 6 4 2 7 1 9 1 6 5 *